FROM THE PRESHOWER TO THE NEW TECHNOLOGIES FOR SUPERCOLLIDERS

World Scientific Series in 20th Century Physics

Published

Vol. 1 Gauge Theories — Past and Future
edited by R. Akhoury, B. de Wit, and P. van Nieuwenhuizen

Vol. 2 Scientific Highlights in Memory of Léon Van Hove
edited by F. Nicodemi

Vol. 3 Selected Papers, with Commentary, of T. H. R. Skyrme
edited by G. E. Brown

Vol. 4 Salamfestschrift
edited by A. Ali, J. Ellis and S. Randjbar-Daemi

Vol. 5 Selected Papers of Abdus Salam (with Commentary)
edited by A. Ali, C. Isham, T. Kibble and Riazuddin

Vol. 6 Research on Particle Imaging Detectors
edited by G. Charpak

Vol. 7 A Career in Theoretical Physics
edited by P. W. Anderson

Vol. 8 Lepton Physics at CERN and Frascati
edited by N. Cabibbo

Vol. 9 Quantum Mechanics, High Energy Physics and Accelerators — Selected Papers of J. S. Bell (with Commentary)
edited by M. Bell, K. Gottfried and M. Veltman

Vol. 10 How We Learn; How We Remember: Toward an Understanding of Brain and Neural Systems — Selected Papers of Leon N. Cooper
edited by L. N. Cooper

Vol. 11 30 Years of the Landau Institute — Selected Papers
edited by I. M. Khalatnikov and V. P. Mineev

Vol. 12 Sir Nevill Mott — 65 Years in Physics
edited by N. Mott and A. S. Alexandrov

Vol. 13 Broken Symmetry — Selected Papers of Y. Nambu
edited by T. Eguchi and K. Nishijima

Vol. 14 Reflections on Experimental Science
edited by M. L. Perl

Vol. 15 Encounters in Magnetic Resonances — Selected Papers of Nicolaas Bloembergen (with Commentary)
edited by N. Bloembergen

Vol. 16 Encounters in Nonlinear Optics — Selected Papers of Nicolaas Bloembergen (with Commentary)
edited by N. Bloembergen

Vol. 17 The Collected Works of Lars Onsager (with Commentary)
edited by P. C. Hemmer, H. Holden and S. K. Ratkje

Vol. 18 Selected Works of Hans A. Bethe (with Commentary)
edited by Hans A. Bethe

Vol. 19 Selected Scientific Papers of Sir Rudolf Peierls (with Commentary)
edited by R. H. Dalitz and R. Peierls

Vol. 20 The Origin of the Third Family
edited by O. Barnabei, L. Maiani, R. A. Ricci and F. R. Monaco

Vol. 21 Spectroscopy with Coherent Radiation — Selected Papers of Norman F. Ramsey (with Commentary)
edited by N. F. Ramsey

Vol. 22 A Quest for Symmetry — Selected Works of Bunji Sakita
edited by K. Kikkawa, M. Virasoro and S. R. Wadia

Vol. 23 Selected Papers of Kun Huang (with Commentary)
edited by B.-F. Zhu

Vol. 24 Subnuclear Physics — The First 50 Years: Highlights from Erice to ELN
A. Zichichi; edited by O. Barnabei, P. Pupillo and F. Roversi Monaco

Vol. 25 The Creation of Quantum Chromodynamics and the Effective Energy
V. N. Gribov, G. 't Hooft, G. Veneziano and V. F. Weisskopf; edited by L. N. Lipatov

Vol. 26 A Quantum Legacy — Seminal Papers of Julian Schwinger
edited by K. A. Milton

Vol. 27 Selected Papers of Richard Feynman (with Commentary)
edited by L. M. Brown

Vol. 28 The Legacy of Léon Van Hove
edited by A. Giovannini

Vol. 29 Selected Works of Emil Wolf (with Commentary)
edited by E. Wolf

Forthcoming

Selected Papers of J. Robert Schrieffer — In Celebration of His 70th Birthday
edited by N. E. Bonesteel and L. P. Gor'kov

In Conclusion — A Collection of Summary Talks in High Energy Physics
edited by J. D. Bjorken

Formation and Evolution of Black Holes in the Galaxy
edited by H. A. Bethe, G. E. Brown and C.-H. Lee

Selected Papers of Kenneth G. Wilson
edited by M. E. Peskin and B. Baaquie

World Scientific Series in 20th Century Physics Vol. 31

FROM THE PRESHOWER TO THE NEW TECHNOLOGIES FOR SUPERCOLLIDERS

In Honour of Antonino Zichichi

Björn H. Wiik
Albrecht Wagner
Horst Wenninger
Editors

World Scientific
New Jersey • London • Singapore • Hong Kong

Published by

World Scientific Publishing Co. Pte. Ltd.
5 Toh Tuck Link, Singapore 596224
USA office: Suite 202, 1060 Main Street, River Edge, NJ 07661
UK office: 57 Shelton Street, Covent Garden, London WC2H 9HE

Library of Congress Cataloging-in-Publication Data
From the preshower to the new technologies for supercolliders : in honour of Antonino Zichichi / Björn H. Wiik, Albrecht Wagner, Horst Wenninger, editors.
p. cm. -- (World Scientific series in 20th century physics ; v. 31)
Includes bibliographical references.
ISBN 9812381996
1. Supercolliders--Technological innovations. I. Zichichi, Antonino. II. Wiik, Björn H., 1937-1999. III. Wagner, A. (Albrecht) IV. Wenninger, Horst. V. Series.

QC787.S83 F76 2003
539.7'36--dc21 2002191088

British Library Cataloguing-in-Publication Data
A catalogue record for this book is available from the British Library.

The editors and the publisher would like to thank the following publishers for their assistance and their permission to reproduce the articles found in this volume:

Elsevier Science Publishers B. V. (*Nuclear Instruments and Methods*), Plenum Press, Società Italiania di Fisica (*Nuovo Cimento, Il Nuovo Cimento, Rivista del Nuovo Cimento*)

First edition jointly published in 2000 by DESY, CERN, and Academy of Sciences and University of Bologna

This book is printed on acid-free paper.

Printed in Singapore by Mainland Press

CONTENTS

FOREWORD

During the year 2000, the City of Bologna will be the European Capital for Culture. For this reason, the University of Bologna and its Academy of Sciences, following the Gugliemo Marconi Centenary and the Luigi Galvani Bicentenary Celebrations, have decided to call attention to the major achievements of their most distinguished members, in the field of science and technology.

This volume is devoted to a series of inventions and technological developments, some thought of and directly implemented by Professor Antonino Zichichi, others suggested and developed under his leadership, all of them having contributed to the discovery of new particles and new phenomena in the field of Subnuclear Physics. The volume was conceived by an eminent scientist, Professor Dr. Björn H. Wiik, Director of Germany's most prestigious physics laboratory, DESY in Hamburg. It would not have been published were it not for Professor Dr. Albrecht Wagner, Chairman of the DESY Board of Directors and Dr. Horst Wenninger from CERN in Geneva, the greatest European physics laboratory. To all of them we express our gratitude.

Ovidio Capitani
President of the Bologna Academy of Sciences

Fabio Roversi Monaco
Rector of the University of Bologna

As stated by V.F. Weisskopf — CERN DG in the early sixties — new physics needs new technological developments.

INTRODUCTION

Albrecht Wagner
Chairman of the DESY Board of Directors

The future plans for DESY, initiated by Prof. Dr. Björn Wiik, have been left unfinished with his untimely death in early 1999 and, in particular, the ongoing R&D work for the TESLA project. This project was conceived by him as an international, interdisciplinary centre for research at DESY, based on a superconducting linear collider with an integrated free electron laser. Following my appointment as Björn Wiik's successor and chairman of the DESY Board of Directors, DESY will continue this work in the framework of the international TESLA collaboration.

Moving into Björn Wiik's office, I found on his desk the manuscript of the present book, prepared by him in honour of his colleague and friend, Prof. Dr. Antonino Zichichi. Björn Wiik was aware of the importance of pioneering ideas and developments for advanced detector and accelerator technologies, which have to precede and prepare for large-scale scientific projects in the field of subnuclear physics.

Stimulated by the LAA project, started by Prof. A. Zichichi in the mid-1980s, which paved the way for further detector R&D programmes at CERN which became of great value for the LHC experimental set-ups, Björn Wiik had prepared the manuscript for this book. It is intended to document the early basic work by A. Zichichi and his ongoing contributions to the development of new methods in experimental particle physics and to launch new detector R&D programmes for future subnuclear physics experiments in Europe.

I share Björn Wiik's view on the important contribution and initiatives by A. Zichichi for advanced technological developments for our field. I have asked

Horst Wenninger, a close collaborator of A. Zichichi on the LAA project at CERN, to help continue and complete the volume prepared by Björn Wiik.

Björn Wiik was a great believer in new technologies based on superconductivity. Not only did he develop and co-ordinate the construction of the superconducting proton ring for HERA – a flagship electron-proton accelerator and the base of the ongoing forefront physics programme at DESY – he also launched the development of superconducting RF cavity technology, a promising technique for future high-energy linear electron-positron colliders and free electron lasers.

The initiative, taken by A. Zichichi and Björn Wiik, to promote the development and production of high field superconducting magnets in Europe in collaboration with Italian Industries for the HERA superconducting proton accelerator ring now benefits the ongoing construction of the superconducting large hadron collider in Europe, the CERN LHC project.

In the early 1990s, based on the success of superconducting RF cavity development work and superconducting cavity operations in the large electron-positron collider LEP at CERN, Björn Wiik launched the development of high gradient superconducting cavities for the next generation electron-positron collider, in close collaboration with INFN in Italy, CEA in France and other institutes.

Björn Wiik had prepared the first part of the present book "From the Preshower to the Roots of the LAA Project" and the documentation of A. Zichichi's pioneering work as an experimental physicist; Horst Wenninger describes the importance and impact of the LAA project on the detector R&D for LHC. The book concludes with an "Outlook" on DESY's plans for a new detector R&D programme in the spirit of Björn Wiik's vision for the future, a vision based on the success of past experience.

Björn H. Wiik
DESY, Deutsches Elektron-Synchrotron, Hamburg, Germany

FROM THE PRESHOWER TO THE ROOTS OF THE LAA PROJECT

FROM THE PRESHOWER TO THE ROOTS OF THE LAA PROJECT

Björn H. Wiik
DESY, Deutsches Elektron-Synchrotron, Hamburg, Germany

During many years, the dominant detector technology used in experimental subnuclear physics has been the bubble-chamber, which contributed to the discovery of many dozens of mesons and baryons in strong interaction physics and in later years to the first evidence of neutral currents in a neutrino experiment with the heavy liquid bubble chamber GARGAMELLE. The high rate of discovery of new particles focused the attention of a large fraction of the physics community on the physics of hadrons and neutrinos and on the bubble chamber detector technology.

It is remarkable that while all this was going on, Antonino Zichichi's view was that lepton physics had to be actively studied and experimentally investigated and he started to study lepton pairs produced in hadronic interactions at CERN in the early sixties [1]. There were two reasons why this field was not generally studied in those times. One was of technical origin. It was difficult to disentangle the production of electrons and muons in a process where very many hadrons (essentially pions) were produced. The other was of theoretical nature. Why start a new technology to discriminate the direct production of electrons and muons if no theoretical guide-lines were available? For example, the production of new vector mesons was expected to be characterized by a large width and the search for narrow resonances decaying into (e^+e^-) and ($\mu^+\mu^-$) pairs was not given any emphasis: nor were the ($e\mu$) pairs [2].

The break-through came in 1974 with the so called "November-revolution", when the J-particle was discovered at Brookhaven, immediately confirmed at SLAC and followed by the discovery of many ψ's. For the J-particle discovery the "preshower" technology was essential and this is the beginning of the "post-bubble-chamber" era.

The present experimental technology in subnuclear physics is based on fast electronic detectors with imaging wire chambers and silicon tracking devices, replacing the slow bubble chambers, complemented by preshower, muon punch-through, calorimetry and high precision time measurements. These latter experimental physics technologies were developed in the sixties and many of them originated from CERN and represent the roots of the LAA project.

Let us briefly review:

1) The pioneering work on the new method to detect electrons in a high background level of hadrons. The method, originally called the "early shower development" method [3[a, b]], is now universally adopted in any experiment where electron detection with high rejection power is needed, and is called the "preshower" method.
2) The study of high energy muon range curves, in order to understand the discrimination power against pion penetration of a thick telescope [4[a, b]]. This is now called "muon punch-through" technology.
3) The studies of a telescope made with alternate layers of lead and scintillator — the so-called "Lead-Scintillator Sandwich Telescope" — in order to study the signal produced by hadronic (π, K, p) and by electromagnetic (e, γ) particles. This kind of telescope [5] [6] was the predecessor of what are now called calorimeters.
4) The study of TOF devices based on plastic scintillator counters, for high accuracy Time-Of-Flight (TOF) measurements. This technology was used in two instances:
    i) To measure the time-of-flight of a particle in a beam in order to determine its velocity, coupled with other measurements in order to establish some property of the same particle. This is how it became possible to discriminate antideuterons, from antiprotons, negative kaons and pions [7[a, b]] with a rejection power one order of magnitude higher [8] than previous measurements.

ii) To measure the light transition time in a scintillator in order to determine the position of an incident particle, be it charged [9] or neutral (neutron) [10a, b]. This technology was used for neutron detection, coupled with time and position determinations, thus allowing the build-up of a powerful "neutron missing-mass spectrometer" [10a, b].

The accuracy achieved with this TOF system, ± 70 psec, is still competitive.

All these developments took place at CERN in the years when the dominating field of research was the physics of hadrons and the dominating technology was the bubble-chamber.

The study of new methods in order to improve electron and muon detection in hadronic machines opened a new field [11]. The best sentence is probably that of Weisskopf – CERN DG at the time – who said: «As emphasized by Nino, the key to unblock the ever-increasing number of mesons and baryons was in the study of lepton pairs produced in hadronic processes». The "*November revolution*" was in fact nothing else but the discovery of a new particle produced in hadronic interactions and decaying into (e^+e^-) pairs with a very narrow width. Furthermore, the first example of quark in the 3rd family was, again, a new particle produced in hadronic interactions and decaying into $(\mu^+\mu^-)$ pairs with a narrow width.

The experimental proof that lepton pair production in hadronic processes could indeed be observed in a very high background environment was obtained, thanks to the invention of the "preshower" method [3a, b] and to the detailed studies of the "muon punch-through" [4a, b]. The new technological invention and development allowed a powerful and efficient electron detection, together with muon identification. Also the hadronic component of an interaction could be studied in its "global" behaviour rather than in a detailed analysis, as provided by the bubble-chamber technology. This was achieved with the first full exploitation of the Lead-Scintillator Sandwich Telescope technology [5] [6], now called "calorimetry", much used in all laboratories the world over. Another key element was the Neutron Telescope [$10^{a, b}$] which was coupled with the "preshower" [$3^{a, b}$] and the electromagnetic calorimeter [5] [6].

In this volume one can see how the "preshower" and the "muon punch-through" technologies started and how they were coupled to the Lead-Scintillator Sandwich Telescope and to the neutron missing mass spectometer in order to study some of the problems which later became very relevant. To achieve these goals in physics, the new technology needed was called by A. Zichichi the NBC (non bubble-chamber) technology. This is how it was proved that the X^0-meson (now η') could not be a tensor but a spin-zero particle; in fact the $X^0 \to \gamma\gamma$ decay was finally discovered thanks to NBC. And it is the NBC technology, invented and put to work by Zichichi and his group, which produced a set of interesting results [11], including the discovery – before the existence of (e^+e^-) colliders – of the (e^+e^-) decay of the vector mesons

$$\omega \to e^+e^-$$
$$\phi \to e^+e^-$$

thus giving the first direct measurement of the (ω–ϕ) mixing, confirming the breaking of the SU(3)-flavour global symmetry in a direct way, not using the masses. These problems could not be studied with the (dominating) bubble-chamber technology.

To sum up, the invention of the "preshower" method represents one of the most significant contributions to the progress of subnuclear physics. In fact, this technology, after having been applied for the discovery of the J-particle and more recently of the weak bosons, is now used worldwide wherever electrons need to be detected against the background of ordinary hadrons.

The first section of this volume contains the reproduction of the major contributions by A. Zichichi and his team concerning the "preshower", from its invention to its most recent applications, followed by the reproduction of the first original studies on the "muon punch-through" and on the Lead-Scintillator Sandwich Telescope technology (now called calorimetry).

The second section contains the original papers by A. Zichichi and his team on R&D for high-precision TOF systems, followed by the one on NBC neutron missing-mass spectrometer technology.

The third section contains the original papers on the origin of the LAA project built up and directed by A. Zichichi, followed by the report on its achievements.

The fourth section refers to the most recent work on R&D for detectors to be used in the next generation of supercolliders, the first one being LHC, which have many of their roots in the achievements of the LAA project.

THE ROOTS OF LAA

SUMMARY

① • *High Precision Magnetic Fields* $\frac{\Delta B}{B} = \pm 10^{-4}$.

② • *Preshower* => $\pi/e \leq 5 \times 10^{-4}$.

③ • *Muon Punch-Through.*

④ • *Calorimetry.*

⑤ • *TOF* => ± *75 psec.*

⑥ • *Neutron Missing-Mass Spectrometer (± 100 psec).*

⑦ • *High Precision* $\frac{dE}{dx}$ *with* λ_{abs} = *30 metres.*

Synthetic and basic information on each of these technical achievements is reported — in the following pages — together with the physics motivations which prompted each of these R&D works.

① *High Precision Magnetic Fields* $\frac{\Delta B}{B} = \pm 10^{-4}$

Physics Motivation	$(g-2)_\mu \rightarrow 5 \times 10^{-3}$

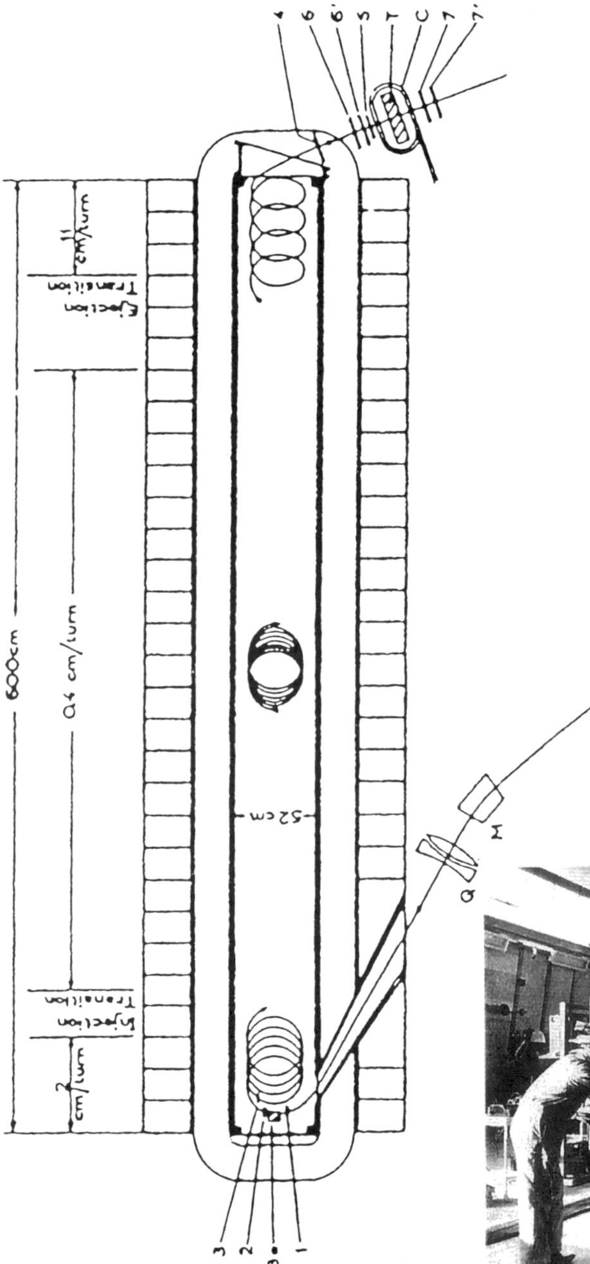

General plan of the six-metre "flat-magnet". *M*: bending magnet; *Q*: pair of quadrupoles; 1, Be, 2, 3: injection assembly consisting of Be-moderator and counters 1, 2, 3; *T*: methylene-iodide target; counters 66', 77': "backward" and "forward" electron telescopes. A stored and ejected muon is registered as a coincidence 4, 5, 66' $\overline{7}$, gated by a 1, 2, 3 and by either a forward or backward electron signal. (Figure from *Measurement of the Anomalous Magnetic Moment of the Muon*, G. Charpak, ... and A. Zichichi, *Phys. Rev. Lett.* 6, 128 (1961)).

A photo of the six-metre "flat-magnet" where a sequence of high precision magnetic fields has been implemented using the "shimming technology".

②	**PRESHOWER => $\pi/e \leq 5 \times 10^{-4}$**
③	**MUON PUNCH-THROUGH**

To simultaneously detect ($e^{\pm}\mu^{\mp}$) final states in ($\bar{p}p$) annihilation ≡ the 1st experiment to search for the THIRD LEPTON (HL ≡ τ).

Physics Motivation	$(HL^{\pm}) \equiv \tau^{\pm}$	&	$F_p^{em}(q^2)_{time\text{-}like}$

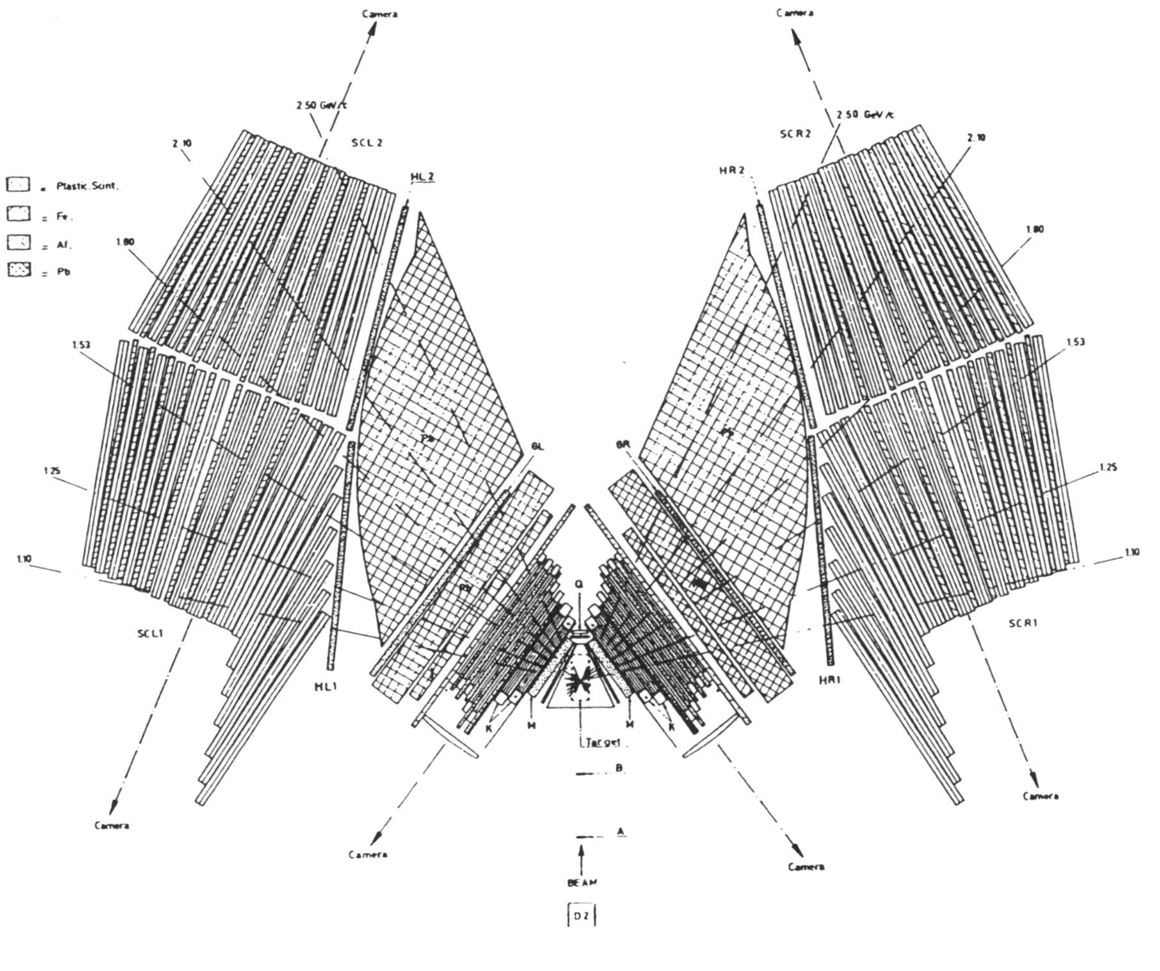

The experimental set-up implemented to search for the Third Lepton (HL ≡ τ) and for a time-like electromagnetic structure of the proton ($F_p^{em}(q^2)_{time\text{-}like}$). The set-up was able to detect simultaneously the (e^+e^-), the ($\mu^+\mu^-$) and the ($\mu^{\pm}e^{\mp}$) final states of the ($\bar{p}p$) annihilation.

④[a] CALORIMETRY

Physics Motivation	see ⑥

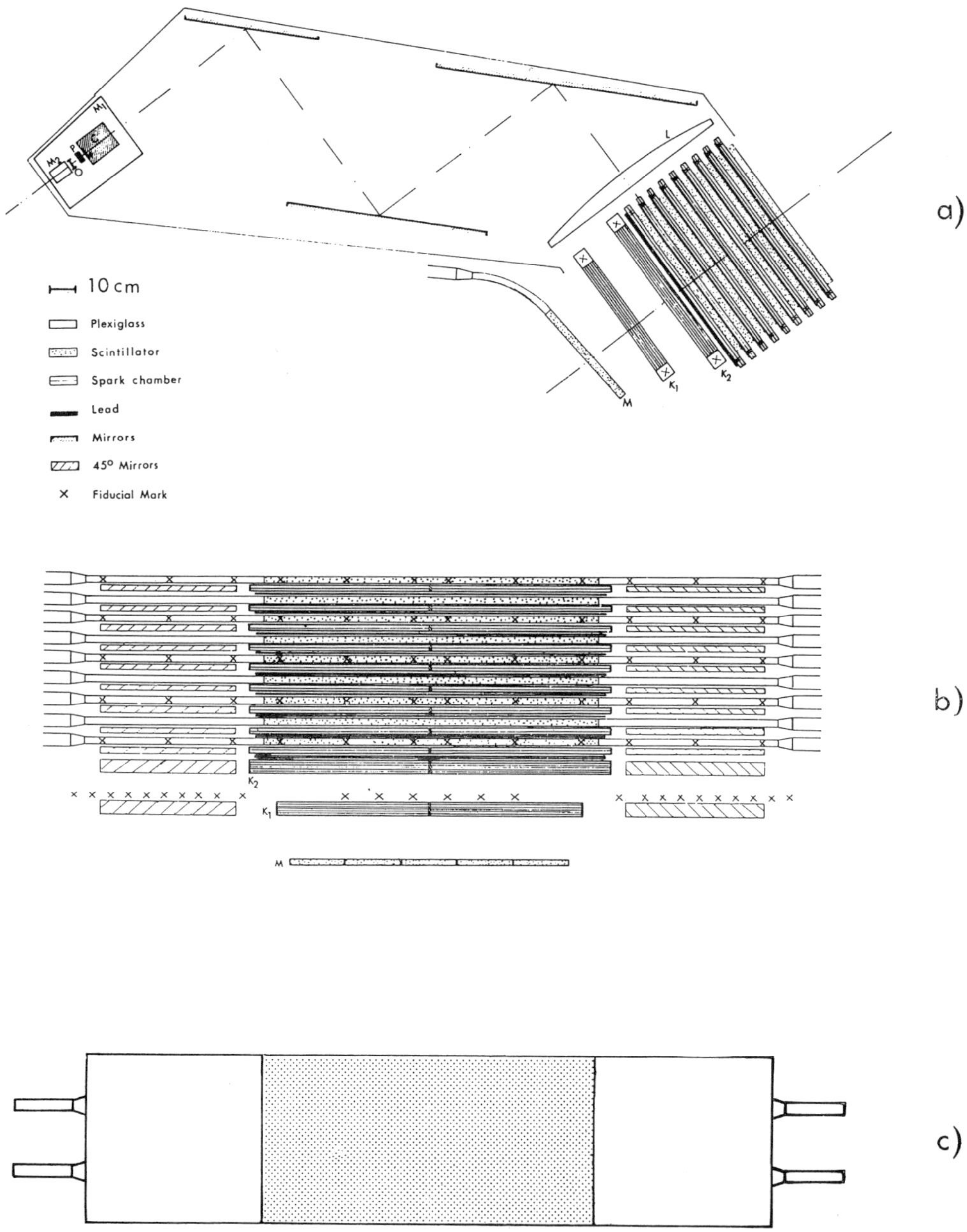

The basic design of the Lead-Scintillator-Sparkchamber “Sandwich-Calorimeter”.

④[b]	CALORIMETRY

Physics Motivation	see ⑥

A picture of the effective implementation of the "Sandwich-Calorimeter".

⑤	**TOF => ± 75 psec**

Physics Motivation	$\dfrac{N(\bar{D})}{N(\pi^{\pm})} = 10^{-8}$

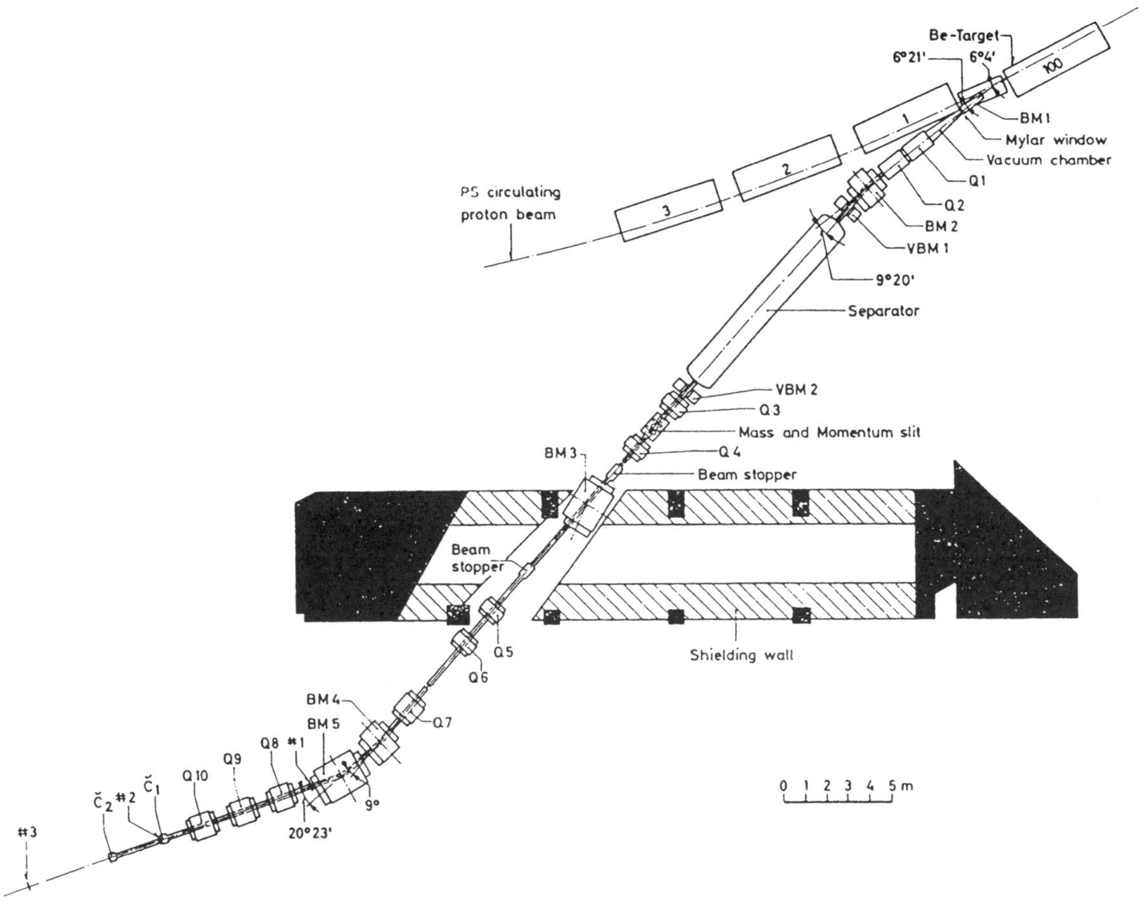

Zichichi's spectrometer to search for antideuterons. Figure from S.C.C. Ting's Lecture *The Discovery of Nuclear Antimatter and the Origin of the AMS Experiment*, presented at the *Symposium in honour of Antonino Zichichi to celebrate the 30th anniversary of the Discovery of Nuclear Antimatter*, Conference Proceedings 53, 21, Italian Physical Society, Bologna, Italy (1995). L. Maiani and R.A. Ricci (eds).

⑥	NEUTRON MISSING-MASS SPECTROMETER

TOF => ± 100 psec

PRESHOWER => $\pi/e \le 5 \times 10^{-4}$

CALORIMETRY

Physics Motivation	$\theta_{PS} \neq \theta_V$

The "**preshower**" technology implemented in the CERN experimental set-up for the study of the rare decay modes of the vector mesons (1963) in order to measure in a direct way — without the use of the mass formula — the pseudoscalar θ_{PS} and vector θ_V mesons mixing angles. This is the first example of what is now "standard" in experimental subnuclear physics: very large acceptance detectors.

⑦	*High Precision* $\frac{dE}{dx}$ *with* λ_{abs} = *30 metres*

TOF => ± 100 psec

Physics Motivation	DIS in (pp): no quarks

The Split-Field-Magnet interaction region at the CERN ISR. Clearly visible in the foreground are the large area scintillation counters for TOF. The dE/dx counters for quarks are inside the Split-Field-Magnet.

References.

[1] *Lepton Physics at CERN and Frascati*
N. Cabibbo (ed), 20th Century Physics Series, Vol 8, World Scientific (1994).

[2] C.S. Wu, T.D. Lee, N. Cabibbo, V.F. Weisskopf, S.C.C. Ting, C. Villi, M. Conversi, A. Petermann, B.H. Wiik and G. Wolf
The Origin of the Third Family, O. Barnabei, L. Maiani, R.A. Ricci and F. Roversi Monaco (eds), Academy of Sciences and University of Bologna - INFN - SIF (1997); and World Scientific (1998).

[3[a]] *A Telescope to Identify Electrons in the Presence of Pion Background*
T. Massam, Th. Muller and A. Zichichi
CERN Report 63-25, 27 June 1963.

[3[b]] *A New Electron Detector with High Rejection Power Against Pions*
T. Massam, Th. Muller, M. Schneegans and A. Zichichi
Nuovo Cimento 39, 464 (1965).

[4[a]] *Range Measurements for Muons in the GeV Region*
A. Buhler, T. Massam, Th. Muller and A. Zichichi
CERN Report 64-31, 24 June 1964.

[4[b]] *Range Measurements for Muons in the GeV Region*
A. Buhler, T. Massam, Th. Muller and A. Zichichi
Nuovo Cimento 35, 759 (1965).

[5] *Un Grand Détecteur E.M. à Haute Réjection des Pions*
D. Bollini, A. Buhler-Broglin, P. Dalpiaz, T. Massam, F. Navach, F.L. Navarria, M.A. Schneegans et A. Zichichi
Revue de Physique Appliquée 4, 108 (1969).

[6] *A Large Electromagnetic Shower Detector with High Rejection Power Against Pions*
M. Basile, J. Berbiers, D. Bollini, A. Buhler-Broglin, P. Dalpiaz, P.L. Frabetti, T. Massam, F. Navach, F.L. Navarria, M.A. Schneegans and A. Zichichi
Nuclear Instruments and Methods 101, 433 (1972).

[7[a]] *A High-Intensity Enriched Beam of Kaons and Antiprotons*
G. Brautti, G. Fidecaro, T. Massam, M. Morpurgo, Th. Muller, G. Petrucci, E. Rocco, P. Schiavon, M. Schneegans and A. Zichichi
Proceedings of the International Conference on "High-Energy Physics", Dubna, USSR, 5-15 August 1964 (Atomizdat, Moscow, 1966), Vol. II, 556.

[7[b]] *A High-Intensity, Partially Separated, Beam of Antiprotons and K-Mesons*
G. Brautti, G. Fidecaro, T. Massam, M. Morpurgo, Th. Muller, G. Petrucci, E. Rocco, P. Schiavon, M. Schneegans and A. Zichichi
Nuovo Cimento 38, 1861 (1965).

[8] *Experimental Observation of Antideuteron Production*
T. Massam, Th. Muller, B. Righini, M. Schneegans and A. Zichichi
Nuovo Cimento 39, 10 (1965).

[9] *Accurate Position Determination of Particle Trajectories Using Plastic Scintillator Counters*
D. Bollini, P. Dalpiaz, P.L. Frabetti, T. Massam, F. Navach, F.L. Navarria, M.A. Schneegans and A. Zichichi
Nuclear Instruments and Methods 81, 56 (1970).

[10[a]] *A New Large-Acceptance and High-Efficiency Neutron Detector for Missing-Mass Studies*
D. Bollini, A. Buhler-Broglin, P. Dalpiaz, T. Massam, F. Navach, F.L. Navarria, M.A. Schneegans, F. Zetti and A. Zichichi
Nuovo Cimento 61A, 125 (1969).

[10[b]] *An Apparatus of the NBC Type and the Physics Results Obtained*
A. Zichichi
Annals of Physics 66, 405 (1971);

[11] *The Basic SU(3) Mixing:* $\omega_8 \leftrightarrows \omega_1$
A. Zichichi
in "Evolution of Particle Physics" (Academic Press, New York-London, 1970), 299.

From
The "Early Shower",
The Muon Telescope,
The Lead-Scintillator Telescope,
The High Precision TOF
The Missing Mass Neutron Spectrometer
To
LAA
And
The New Technologies For Super Colliders

Section 1

THE PRESHOWER METHOD,
THE MUON PUNCH-THROUGH,
THE LEAD-SCINTILLATOR TELESCOPE TECHNOLOGY
(NOW CALLED CALORIMETRY)

The invention of the "preshower" method represents one of the most significant contributions to the progress of subnuclear physics. In fact, this technology has been applied to the search for new phenomena and new particles have been discovered, such as the J and the weak bosons.

This section contains the major contributions from its invention to its applications, together with the original papers on the "muon punch-through" studies and on the Lead-Scintillator Telescope technology.

T. Massam, Th. Muller and A. Zichichi

A TELESCOPE TO IDENTIFY ELECTRONS IN THE PRESENCE OF PION BACKGROUND

From
CERN Report 63-25, 27 June 1963

1963

CERN 63-25
Nuclear Physics Division
27th June 1963

ORGANISATION EUROPÉENNE POUR LA RECHERCHE NUCLÉAIRE
CERN EUROPEAN ORGANIZATION FOR NUCLEAR RESEARCH

A Telescope to Identify Electrons in the Presence of Pion Background

T. Massam, Th. Muller and A. Zichichi

G E N E V A

A Telescope to Identify Electrons in the Presence of Pion Background

T. Massam, Th. Muller and A. Zichichi

I. Introduction

Two instruments have recently been developed which can be used to distinguish electrons and pions of the same momenta in the GeV region : the lead glass total absorption Čerenkov counter [1] and the total absorption scintillation detector [2]. For both instruments the probability of mistaking a pion for an electron of the same momentum is of a few per cent. In this article we describe a simple method by which such a probability is decreased by nearly a factor of 10. The improvement is achieved by also looking at the first stage of the shower development, as it occurs in a lead slice of about two radiation lengths. This first stage of the shower development is observed in a scintillation counter following the Pb slice. The further development of the cascade showers occurs in a lead glass Čerenkov counter [1] which is an essential part of the telescope described here.

The telescope has been used in an actual experiment [3]. Electrons of momenta in the range 1 to 2.5 GeV/c were detected with an average efficiency of $\sim$80%. The corresponding pion contamination (i.e. the probability for a π^- in the same momentum range to appear as an electron) was 0.3%.

II. Description and Principle of Operation

The telescope is shown to scale in Fig. 1. It consists of a 1 cm thick slice of lead, followed by a 28 cm diameter, 1 cm thick scintillation counter (N), and finally by the 35 cm diameter, 30 cm long lead glass Čerenkov counter (Č) of Gatti et al. [1)].

If the pions lost energy only by ionization in the Čerenkov counter, then using the Čerenkov pulse height alone there would be a negligible chance of mistaking a pion for an electron of the same momentum. However, the pions can undergo nuclear interactions and convert into electromagnetic energy via charge exchange. Consequently the pion spectrum has a tail, in which about 2% of the pions extend under the electron peak.

The purpose of the lead sheet and counter N is to make a preliminary rejection of the pions. A high-energy electron will, in general, start a cascade shower in the ~2 radiation lengths of the Pb sheet. The electron pulse-height distribution in counter N will therefore have a much higher peak than the pulse-height distribution of a pion of the same momentum. The higher pulse heights produced by pions in counter N are caused by :

i) statistical fluctuations;
ii) nuclear stars;
iii) charge exchange followed by shower production.

In the Cerenkov counter

i) will give the normal pion spectrum and tail;
ii) will give small pulse heights;
iii) will give only the tail of the pion spectrum. This tail will appear with reduced intensity because the lead presents less material in which the pion may charge exchange than does the first part of the Čerenkov counter.

The thickness of lead is chosen so that there is a high probability of greater than unit multiplication of electrons. With this system, by discriminating to accept only large pulse height in N and in the Čerenkov counter, we may expect to obtain high electron efficiency and low pion contamination.

III. Calibration of the Telescope

The properties of the telescope were examined using the simple set-up shown in Fig. 1. The beam telescope DAB was used to separate pions or electrons from a 1.15 GeV/c pion beam containing ~5% electrons, of the CERN Proton Synchrotron. Counter D was a gas Čerenkov counter filled with hydrogen at three atmospheres and sensitive only to electrons. A and B were two small beam-defining counters.

The following measurements were made :

i) The beam was defined as described above and the pulse-height spectra of pions and electrons in counter N were measured using a TMC pulse-height analyser. The results are shown in Fig. 2. The true width of the pion curve was 37%, but there was some broadening caused by amplified noise before the pulse-height analyser input. The dotted curve shows the electron spectrum after a correction for the pion contamination of the selected electrons.

ii) Counter N was included in the beam telescope coincidence to gate the Čerenkov spectrum (Fig. 3). The discriminator level of N was set at minimum to accept all particles crossing N. Note that the electron curve shows quite clearly a pion contribution. This was estimated to be $(10.5 \pm 1.5)\%$ of the electron spectrum by comparing the integral spectra of pions and electrons. This contamination, we believe, was caused by low-energy knock-on electrons produced in 3 cm of scintillator placed upstream in the beam by another group. These electrons were coincident with the pions; they produced a count in the hydrogen Čerenkov counter, but were not energetic enough to produce large pulses in N or in the Čerenkov counter.

iii) The discrimination level of N was increased so as to accept only 10% of the pions, and the pion and electron spectra in the Čerenkov counter were remeasured. These results are also shown in Fig. 3. It may be seen that a large fraction of the pions which give large pulse heights in N give very small pulse height in the Čerenkov counters.

These results are summarized in Fig. 4 which shows the normalized integral spectra in the Čerenkov counter for measurements ii) and iii).

To use the telescope, in practice one would impose a discrimination level in the Čerenkov counter. It may be seen that for useful discrimination levels (channels 30 to 80), the normalized spectra are effectively the same for high or low pulse heights in counter N.

Use may be made of this independence to optimize the trigger levels to obtain minimum pion contamination for a fixed electron efficiency. Figure 5 shows the expected results of this optimization. As we see, we can get 80% electron efficiency with 0,3% π- contamination.

7043/p

References :

1) G. Gatti, G. Giacomelli, W.A. Love, W.C. Middelkoop and T. Yamagata.
Rev. Sc. Instr. 32, 949 (1961)

2) G. Backenstoss, B.D. Hyams, G. Knop and U. Stierlin.
Nuclear instruments and methods, 20, p. 294-296 (1963)

3) M. Conversi, T. Massam, Th. Muller and A. Zichichi.
Physics Letters 5, 195 (1963)

7043/p

Figure Captions :

Fig. 1. Counter arrangement and electronics

A,B,N = plastic scintillator counters

D = hydrogne Čerenkov counter

Č = lead glass Čerenkov counter

Pb = lead foil

d = delay

a = amplifier

t = trigger

ds = discriminator

c = coincidence input

v = anti-coincidence input

c.u. = coincidence unit

sc = scaler

Fig. 2. Pulse height spectra in counter N

Fig. 3. π and e spectra in the Čerenkov counter

Fig. 4. Integral spectra in the Čerenkov counter

Fig. 5. Showing the minimum π efficiency as a function of electron efficiency

7043/p

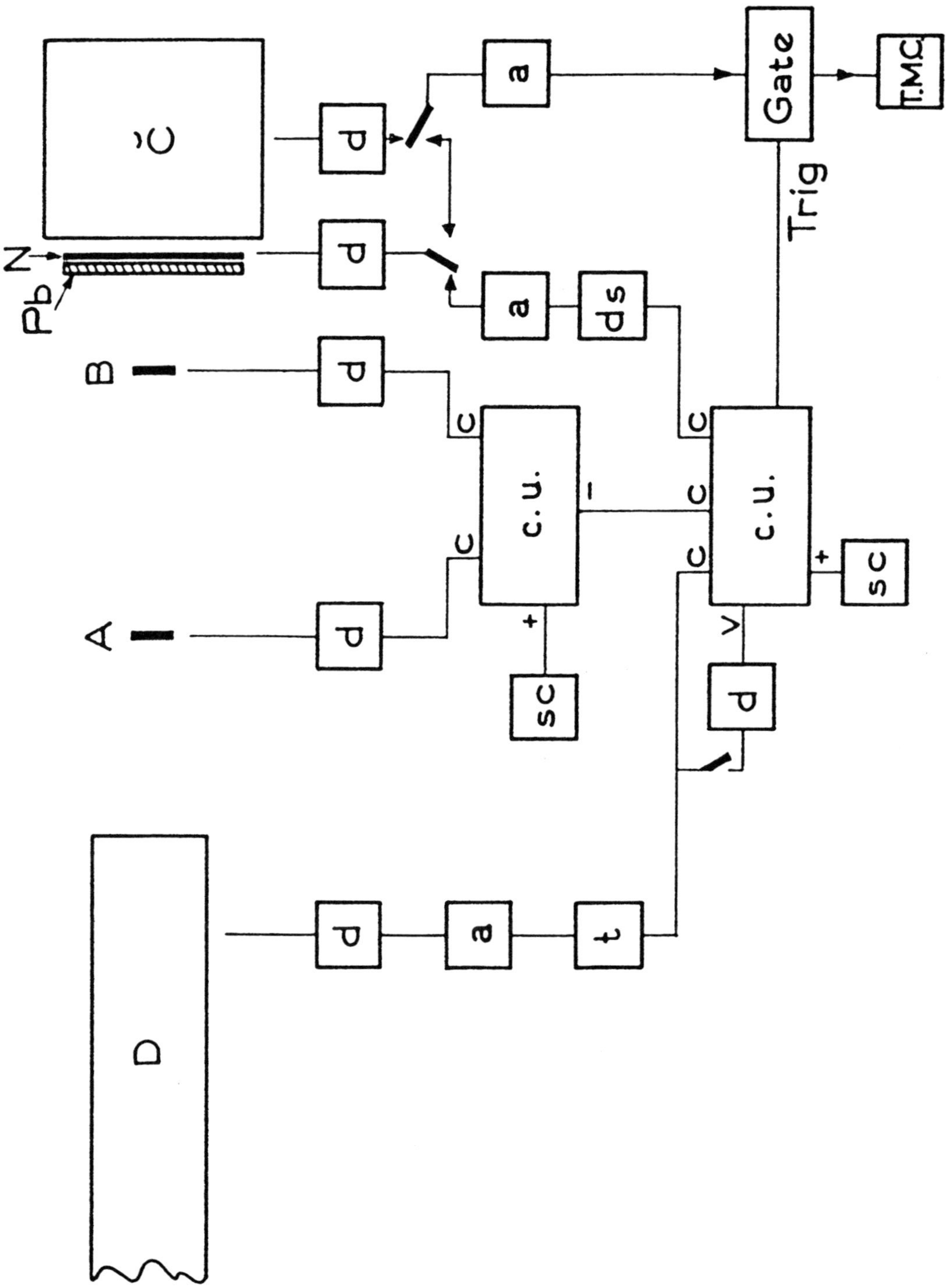

FIG. 1

FIG. 2 Pulse height spectra in counter N

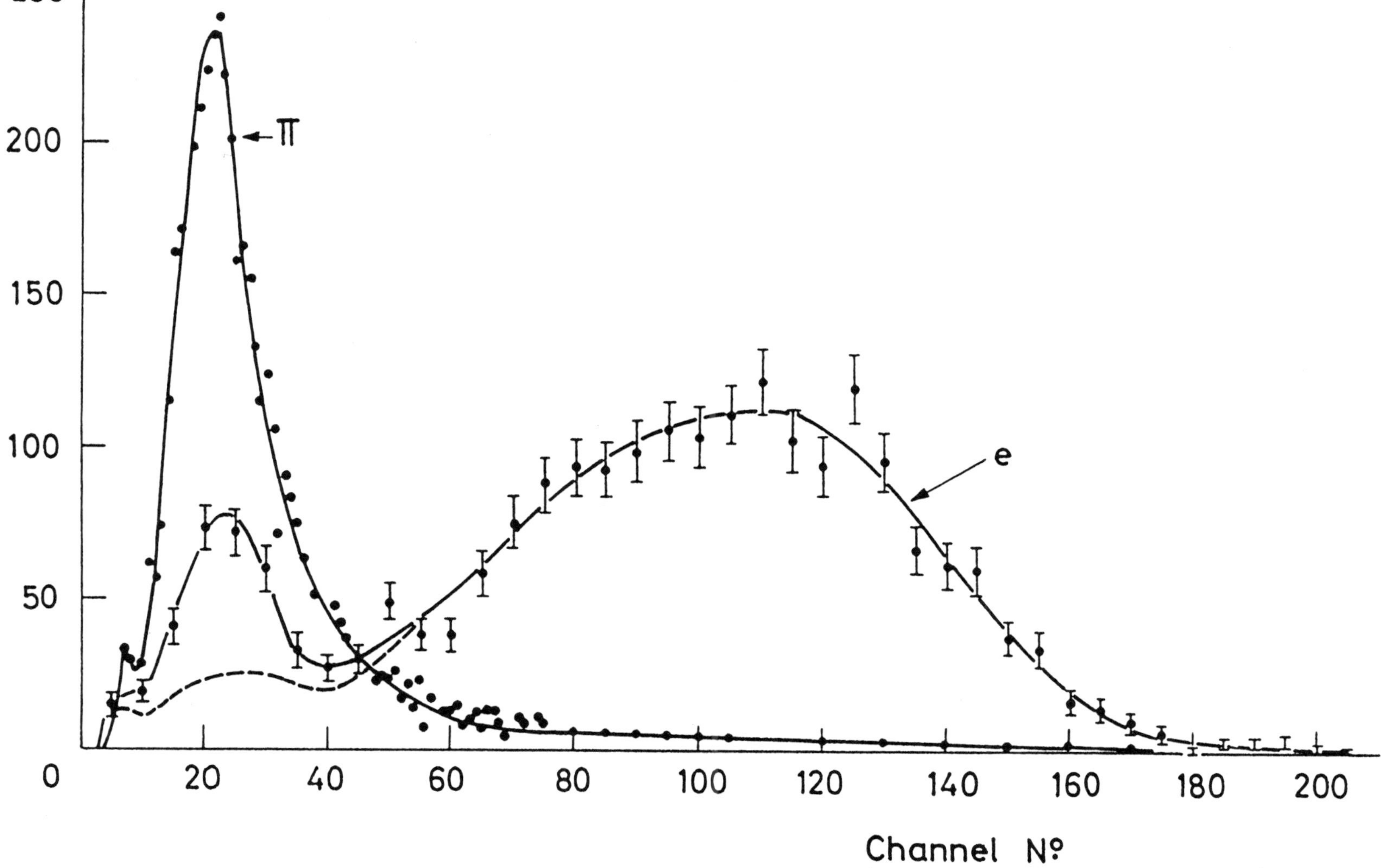

FIG. 3 π and e spectra in the Čerenkov counter

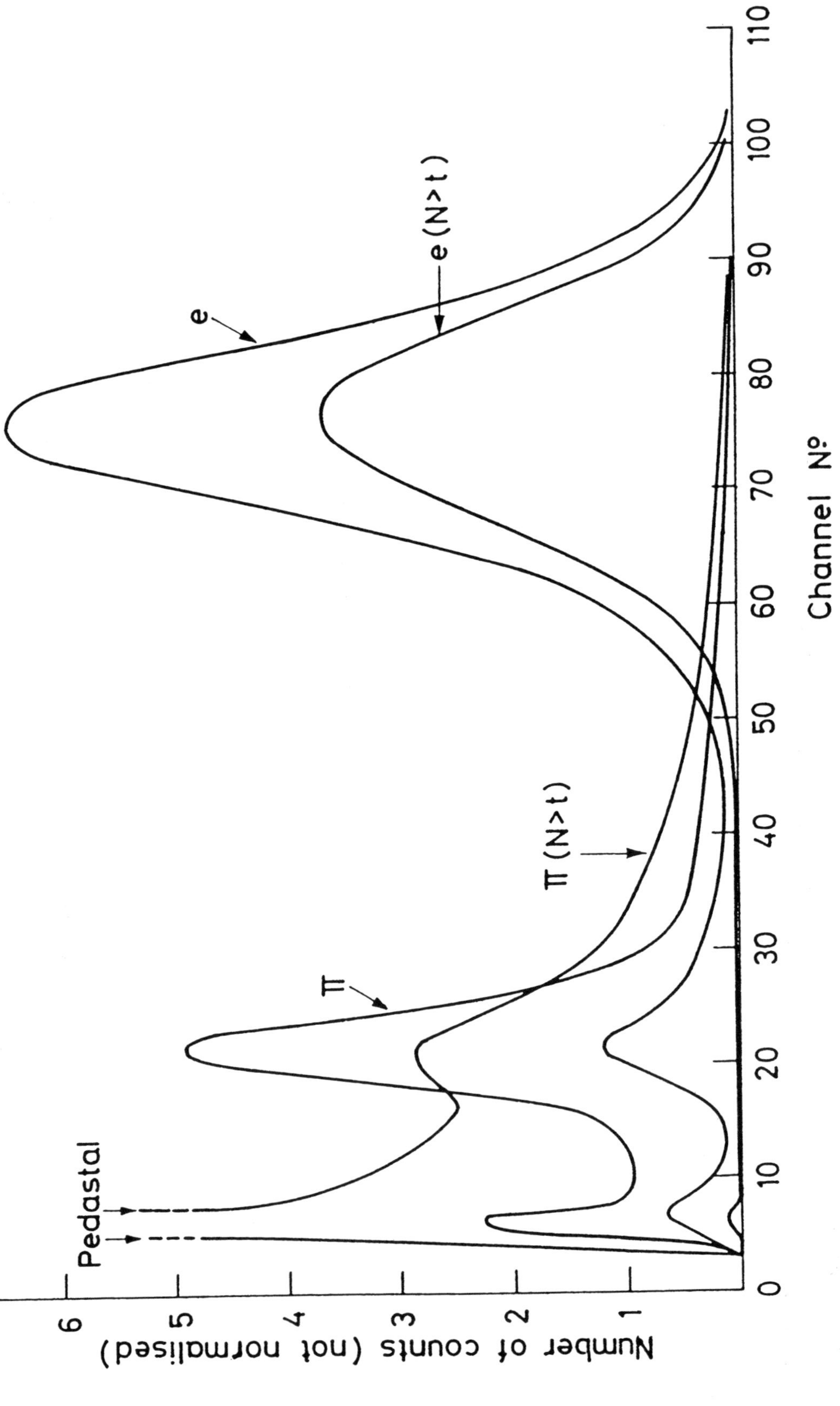

FIG. 4 Integral spectra in the Čerencov counter

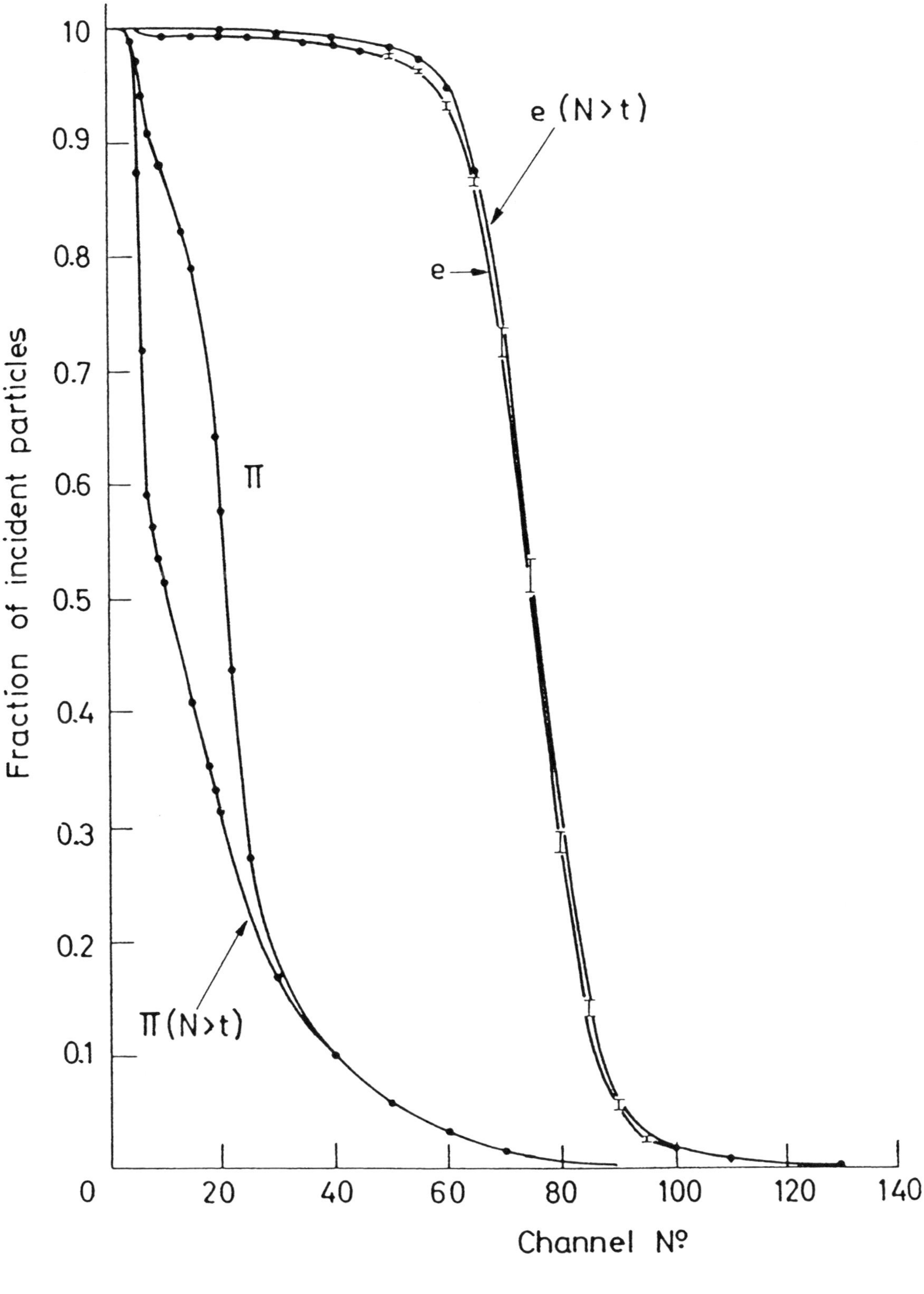

FIG. 5 Showing the minimum π efficiency as a function of electron efficiency.

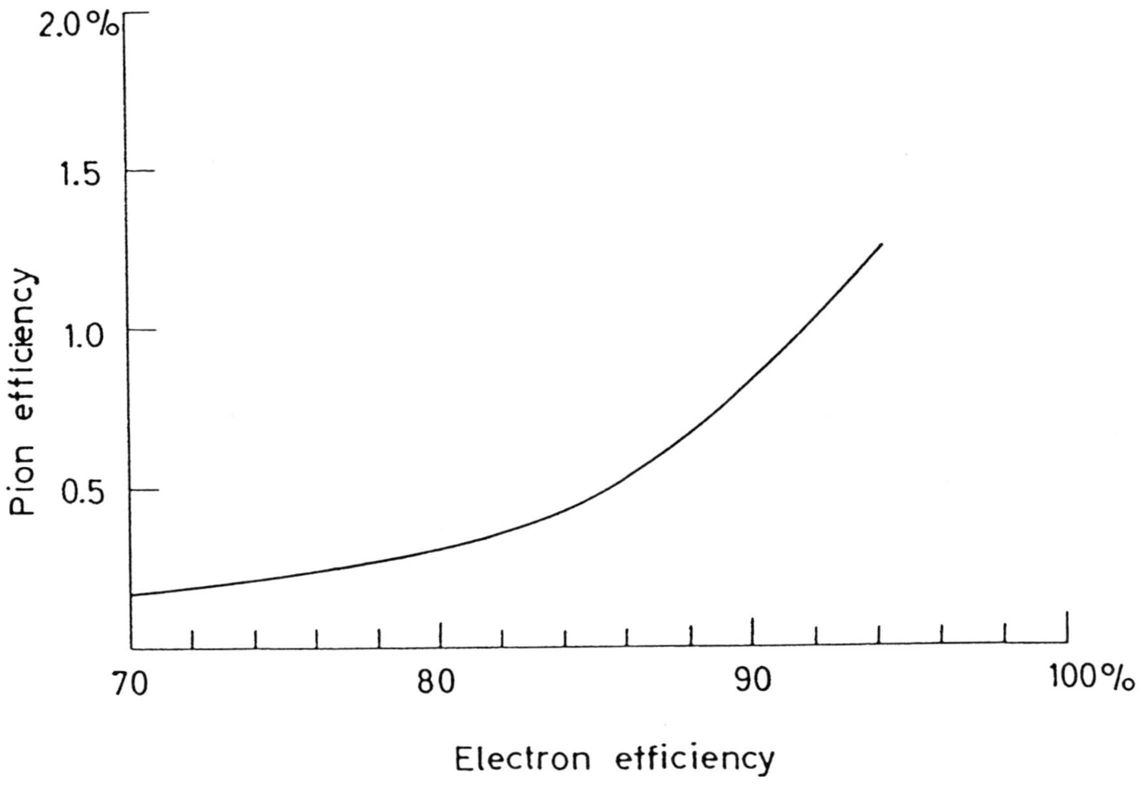

A. Buhler, T. Massam, Th. Muller and A. Zichichi

RANGE MEASUREMENTS FOR MUONS IN THE GeV REGION

From
CERN Report 64-31, 24 June 1964

1964

CERN 64- 31
NP Division

24 June, 1964

ORGANISATION EUROPÉENNE POUR LA RECHERCHE NUCLÉAIRE
CERN EUROPEAN ORGANIZATION FOR NUCLEAR RESEARCH

RANGE MEASUREMENTS FOR MUONS IN THE GeV REGION

A. Buhler*), T. Massam, Th. Muller*) and A. Zichichi

*) Centre de Recherches Nucleaires, Strasbourg, France.

CERN 64- 31
NP Division
24 June, 1964

RANGE MEASUREMENTS FOR MUONS IN THE GeV REGION

A. Buhler*), T. Massam, Th. Muller*) and A. Zichichi

*) Centre de Recherches Nucleaires, Strasbourg, France.

RANGE MEASUREMENTS FOR MUONS IN THE GeV REGION

The results of an experiment to measure the range of high-energy muons are reported. The reason for carrying out this experiment was the lack of data on muon ranges above a few hundred MeV. We needed this data for the design of an experiment on proton-antiproton annihilation into muon pairs. On the other hand, the results given in this paper represent the first experimental check of the validity of the range-energy relation for high values of γ*), i.e. $\gamma = 12 - 25$. Measurements were made of the muon range in an absorber which consisted mainly of lead, at the following momenta:

$$P = 1.240 \pm 0.019 \text{ GeV/c}$$
$$P = 1.790 \pm 0.027 \text{ GeV/c}$$
$$P = 2.480 \pm 0.037 \text{ GeV/c} .$$

The value of the momentum was determined from the currents in the bending magnets and a calibration of beam momentum as a function of bending-magnet current. This calibration was made available by the Bubble Chamber Group, who had measured τ decays[1] to obtain the K-meson momentum in the same beam. Since the range measurements and the calibration were not made simultaneously, we are obliged to quote a momentum uncertainty equal to the momentum acceptance of the beam.

The experimental set-up and the electronic block diagram are shown in Figs. 1 and 3, respectively. The apparatus consists of two main parts: the beam telescope and the range telescope.

The beam telescope

This consists of two plastic scintillators A and B, and two Vivargent-type threshold gas Čerenkov counters D_1 and D_2[2] which are filled with ethylene. A typical calibration curve of D_1 at 1.79 GeV/c

*) $\gamma = E/m$

9118/p/cm

is presented in Fig. 2, which shows the three plateaux corresponding to pions, muons and electrons. Similar calibrations were taken for both Čerenkov counters at each of the three momenta used. Two threshold Čerenkov counters are used to provide the differential velocity selection needed to detect muons with low contamination from the antiprotons, K mesons, pions and electrons. The muon electronic signature was

$$(A\,B\,D_1)\,\bar{D}_2\ ,$$

with the pressures in D_1 and D_2 set in such a way as to have a very small efficiency in D_1 for detecting pions and heavier particles, and a very high efficiency in D_2 for electron detection. This method gave a muon beam with less than 5% contamination of other particles, as may be seen from the initial slope of the integral range curves.

Table 1 shows the beam composition at the three momenta, neglecting antiprotons.

Table 1

Momentum GeV/c	e^-%	μ^-%	π^-%
1.24	5.4	3.6	91.0
1.79	1.6	2.3	96.1
2.48	1.5	1.4	97.1

The range telescope

The range telescope consisted of:

i) An eight-gap aluminium-foil spark chamber, SC, of total density 100 mg/cm^2 which allows a precise knowledge of the particle direction. This information was included so that we may check the multiple-scattering corrections which are necessary when incident and emergent particle directions are known. These results are not yet analysed.

ii) A spark chamber assembly, SC_2, consisting of a series of spark chambers of the Krienen type[3]. These chambers are $1\ m \times 1.5\ m$ with plates of 0.5 cm brass. After the first three plates of the chamber, there is a $1m \times 1.5\ m \times 2.5$ cm plastic scintillator V_1. Before the last three plates of the assembly, there is a 3 cm thick iron plate which serves to increase the total surface density to 133 cm/cm^2 lead equivalent. This assembly was an insurance in case we were not able to select muons electronically with low contamination. Pions and electrons would have a high probability of interacting in this chamber and would thus allow us to interpret correctly any initial slope of the range curve. In practice, the electronic selection of muons was so good that we did not need to use the information of SC_2.

iii) A variable thickness of lead absorber, of cross-section $1\ m \times 1\ m$.

iv) A $40\ cm \times 40\ cm \times 1$ cm plastic scintillator C. This counter was introduced to eliminate random coincidences between beam particles and intermediate energy particles which crossed the counters V outside the lead absorber. The size of this counter was chosen so that there was a negligible restriction on the multiple scattering accepted and, therefore, multiple-scattering corrections to the range could be made assuming 'infinite slice' geometry.

v) The range chamber, SC_3, which consisted of two assemblies, very similar to SC_2, which were placed in series, and gave a total of 288 gm/cm^2 lead equivalent.

Experimental method

The range was measured in two ways:

a) Electronically Coincidences $(A\,B\,D_1)\bar{D}_2\,V_2\,C\bar{V}_3$ and $(A\,B\,D_1)\bar{D}_2\,V_2\,CV_3$, for a fixed number of monitors $(AB\,D_1)\bar{D}_2$ and as a function of variable absorber thickness, defined the differential and integral range curves. These results are shown in Figs. 4

and 5. The effect of counter C may be seen from the measurement at 1.79 GeV, where it had not yet been included.

b) Using SC_3 At each energy, the variable absorber thickness was chosen so as to bring the mean range to the centre of SC_3, and the coincidence $(A\,B\,D_1)\bar{D}_2\,C\,V_2$ was used to trigger the spark-chamber system. Figure 6 shows the range curve which was determined in this way.

Calculation of the results

The two main corrections which were made to the range were the conversion of gm/cm^2 copper to gm/cm^2 lead, and the correction to the measured range for path elongation by multiple scattering. Account was also taken of the counter material in the beam and of the variation of the Čerenkov counter surface density with pressure. The lead equivalent of the chambers varies slowly with the energy. The conversion factors CX which we have used are shown as a function of material and energy in Table 2.

Table 2

Values of CX

[CX = (dE/dX material X)/(dE/dX lead)]

Material	Energy range	
	1.0 to 2.5 GeV	10 to 100 MeV
Brass and Fe	1.24	1.40
CH_2	1.77	2.20

The multiple-scattering corrections to the range were calculated according to the method of Øverås[4]. Account was taken of the fact that the final part of the slowing down took place in brass.

9118/p/cm

Table 3 shows the values of the ranges obtained by the three methods before and after the multiple scattering correction was applied.

Table 3

Momentum (GeV/c)	Integral range	Differential range	Spark-chamber range	
1.24	895	917	914	Before correction
1.79	1313	1288	1290	
2.48	1807	1791	1811	
1.24	910	933	930	After correction
1.79	1337	1315	1318	
2.48	1834	1821	1841	

The result if summarized in Table 4, which also shows what fraction of the total surface density was not lead. The uncertainty quoted in the range is assigned partly on the spread in values found in Table 3, and partly from an 8% uncertainty assigned to the lead equivalent of the brass in the spark chambers. Consequently, the three uncertainties are strongly correlated.

Table 4

Momentum (GeV/c)	Range	Fraction not lead
1.240 ± 0.019	925 ± 25	0.35
1.790 ± 0.027	1325 ± 25	0.25
2.480 ± 0.037	1830 ± 25	0.18

The uncorrected and corrected experimental results are shown in Fig. 7, compared with the theoretical expectations of Sternheimer[5)], which include the density effect.

Table 5 shows the values of the standard deviations of the assumed Gaussian distribution of the straggling, compared with the values given by Sternheimer. The experimental values were estimated from the quartiles of the observed distributions. It may be seen that the spark-chamber values are in fair agreement with the predicted values, but the values obtained from the electronic method are consistently higher. This we attribute to the fact that, in the electronic method, the thickness of the variable lead absorber is changed during the measurement, whereas, in the spark-chamber method, it is kept fixed.

Table 5

Momentum (GeV/c)	σ electronic integral range %	σ spark chambers %	σ Sternheimer %
1.24	6.2 ± 0.8	5.2 ± 0.4	3.3
1.79	5.2 ± 0.6	3.8 ± 0.3	3.6
2.48	5.3 ± 0.4	3.4 ± 0.3	3.9

σ is given as a fraction of the mean range.

In conclusion, we may say that the experimental results agree with the Sternheimer range energy data within 2% for muons with γ of up to 23. Further, there is fair agreement between the experimental and theoretical values of the range straggling, provided that the measurement is made in fixed geometry.

Acknowledgements

Dr. F. Krienen has lent us his heavy-plate spark chambers and Dr. M. Vivargent his design of the gas-threshold Čerenkov counters. To both of them we would like to express our sincere appreciation and gratitude. We also wish to thank Dr. H. Øverås for his advice on the multiple-scattering corrections, and Prof. M. Conversi for his collaboration during part of the experimental running at the proton-synchrotron. The assistance of our technicians Messrs. J. Berbiers and B. Nicolai is greatly appreciated.

9118/p/cm

REFERENCES

1) For example, Some results on 3 GeV/c K^+ interactions in hydrogen, M. Ferro-Luzzi, R. George, Y. Goldschmidt-Clermont, V.P. Henri, B. Jongejans, D.W. Leith, G.R. Lynch, F. Muller and J.M. Perreau, Proceedings of the Sienna International Conference on Elementary Particles, vol. 1, p. 182.

2) M. Vivargent, Nucl.Instr. and Meths. 22, 165 (1963).

3) H. Faissner, F. Ferrero, A. Ghani, E. Heer, F. Krienen, G. Muratori, T.B. Novey, M. Reinharz and R.A. Salmeron, Nucl.Instr. and Meths. 20, 213 (1963).

4) H. Øverås: on Small-angle multiple scattering in confined bodies, CERN report 60-18.

5) R.M. Sternheimer, Phys.Rev. 115, 137 (1959).
R.M. Sternheimer, Phys.Rev. 118, 1045 (1960).

6) R.M. Sternheimer, Phys.Rev. 117, 485 (1960).

<u>Figure Captions</u>

Fig. 1 : Experimental set-up.

D_1, D_2	:	variable pressure threshold gas Cerenkov counters.
A,B	:	plastic scintillation counters 8 cm diameter, 1 cm thick
C	:	plastic scintillation counter, $40 \times 40 \times 1$ cm^3
SC_1	:	thin-foil 8-gap spark chamber
SC_2	:	thick-plate 14-gap spark chamber
SC_3	:	thick-plate 28-gap spark chamber

View "A" shows the arrangement of the plates and incorporated scintillation counters and iron absorber. The variable thickness of lead is always increased starting from counter C towards SC_2.

Fig. 2 : Calibration curve of the threshold gas Cerenkov counter D_1. $N(A\,B\,D_1)$ is the number of triple coincidences for fixed number of monitor (A,B).

Fig. 3 : Electronic block diagram.

Scintillation counters V_1, V_2, V_3 are shown outside the spark chambers for clarity. They are marked P1 on Fig. 1, view "A".

Del	:	variable delay lines
F.D.	:	fast discriminators
3rd amp	:	fast three-stage amplifier, 2.5 nsec rise-time
scaler $A\,B\,D_1\bar{D}_2$	:	records the incoming muons
scaler int.	:	records integral range curve (coincidence unit 2
scaler diff.	:	records differential range curve (coincidence unit 3)
SC trigger	:	trigger for spark chambers, film advancement and dead-time.

Fig. 4 : Differential range curves obtained by the electronic method.

The abscissa is the thickness of the variable absorber. To this should be added 245 gm/cm^2 at 1.24 GeV/c, 242 gm/cm^2 at 1.79 GeV/c, and 241.0 gm/cm^2 at 2.48 GeV/c. These values are lead equivalent, uncorrected for multiple scattering.

9118/p/cm

Fig. 5 : Integral range curves obtained by the electronic method.

The abscissa is the thickness of the variable absorber. To this should be added 327 gm/cm^2 at 1.24 GeV/c, 325 gm/cm^2 at 1.79 GeV/c, and 323 gm/cm^2 at 2.48 GeV/c.

Fig. 6 : Differential range curve obtained from the spark chamber measurements. Multiple scattering corrections have not been made.

Fig. 7 : Mean values obtained for the mean range as a function of momentum, before and after correcting for multiple scattering.

The solid curve shows the range-momentum calculations of Sternheimer.

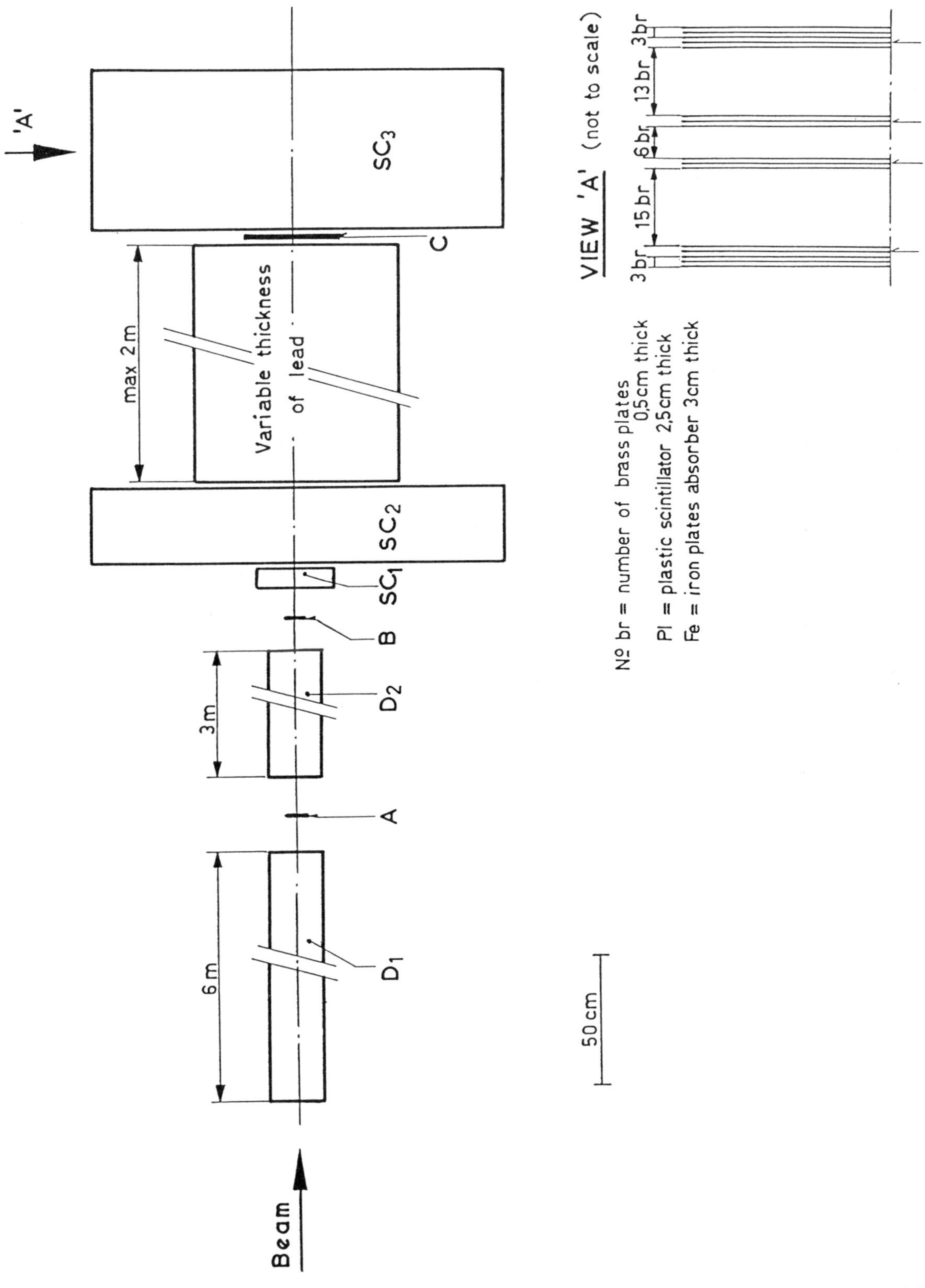
'A'
SC3
C
max 2m
Variable thickness of lead
SC2
SC1
B
3m
D2
A
6m
D1
Beam
50cm
VIEW 'A' (not to scale)
3br
13br
6br
15br
3br
Fe
Pl
Fe
Pl
Nº br = number of brass plates 0,5cm thick
Pl = plastic scintillator 2,5cm thick
Fe = iron plates absorber 3cm thick

FIG. 1

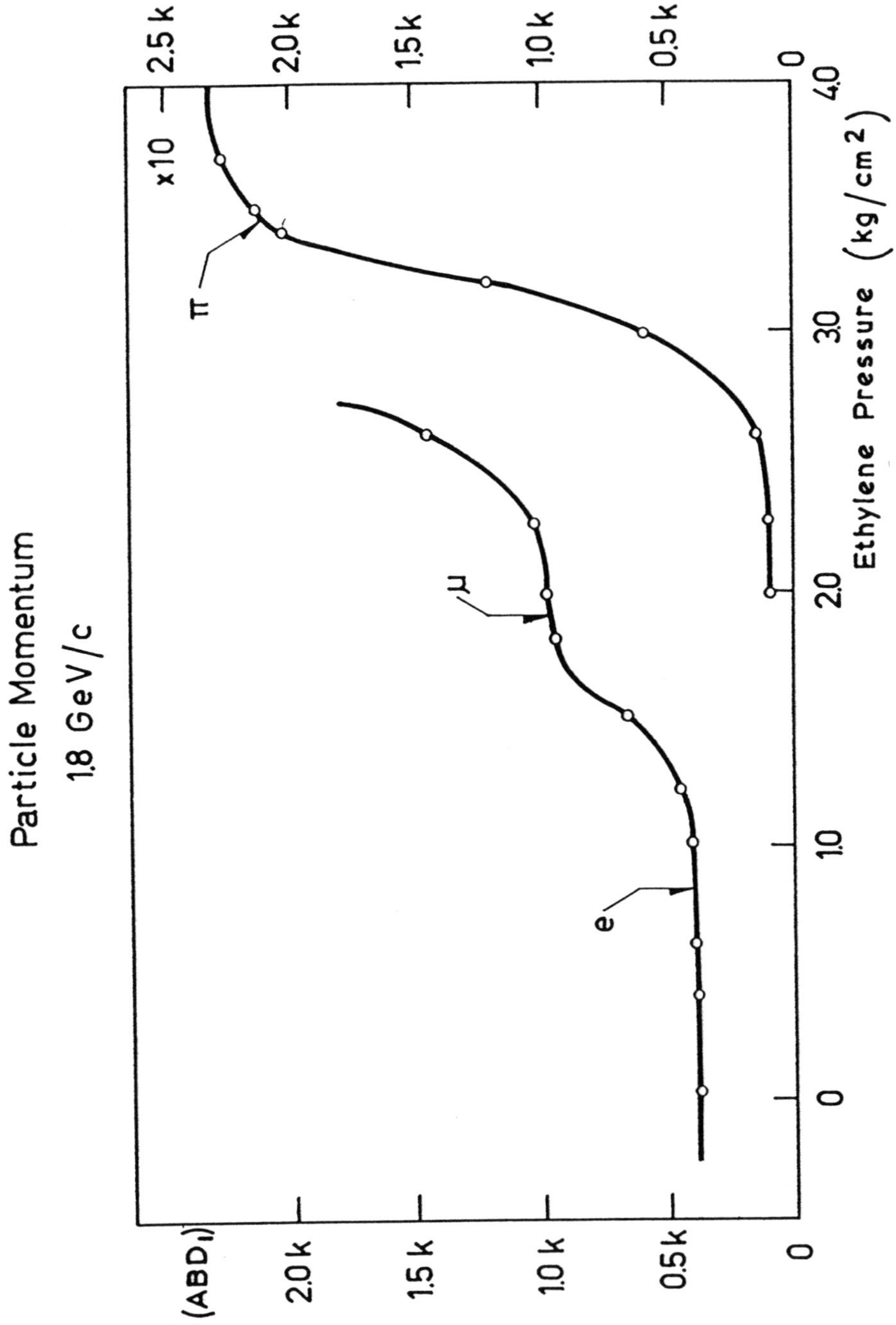

Zero pressure corresponds to atmospheric pressure

FIG. 2

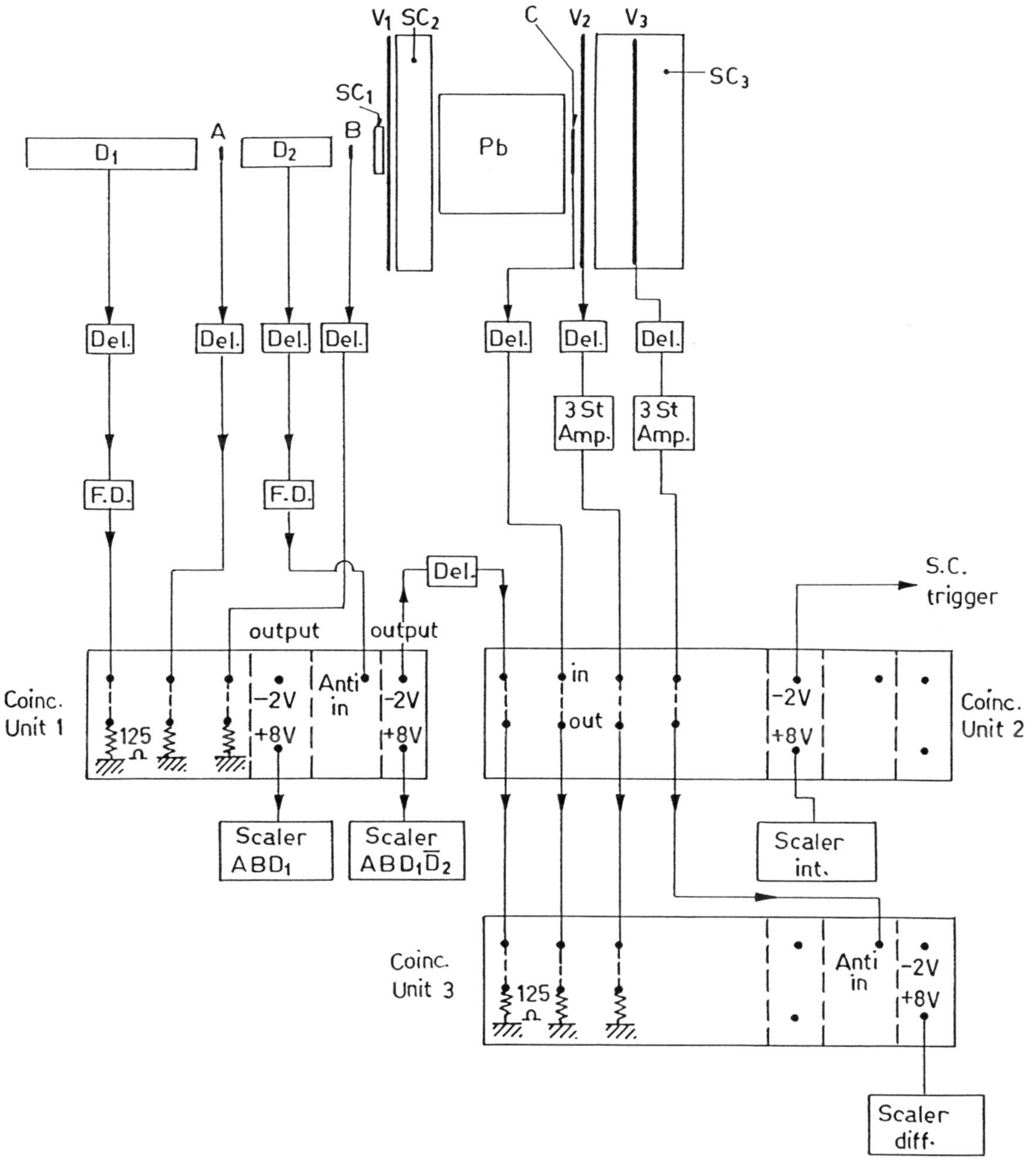

FIG.3

SIS/R/8930

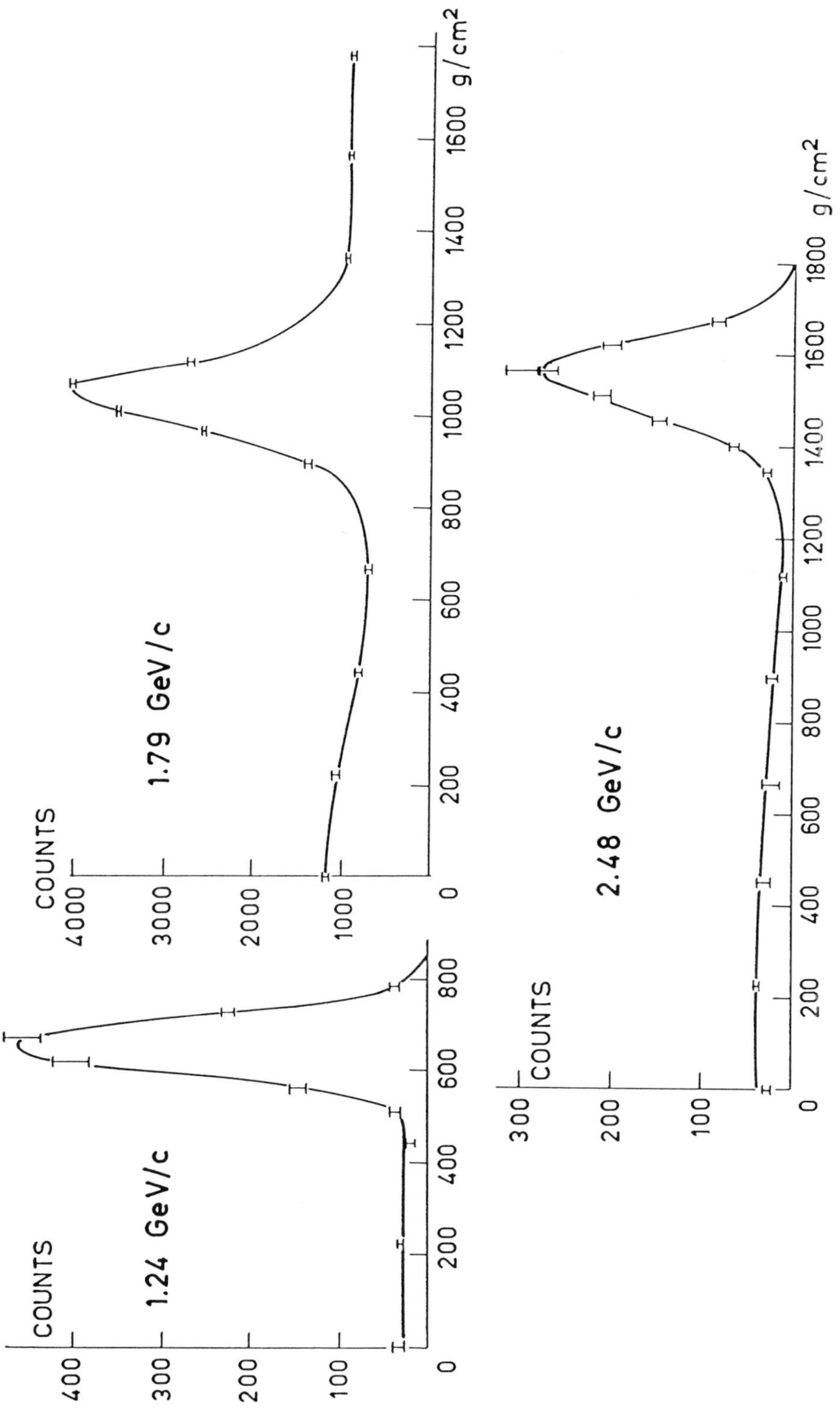

FIG.4

SIS/R/8931

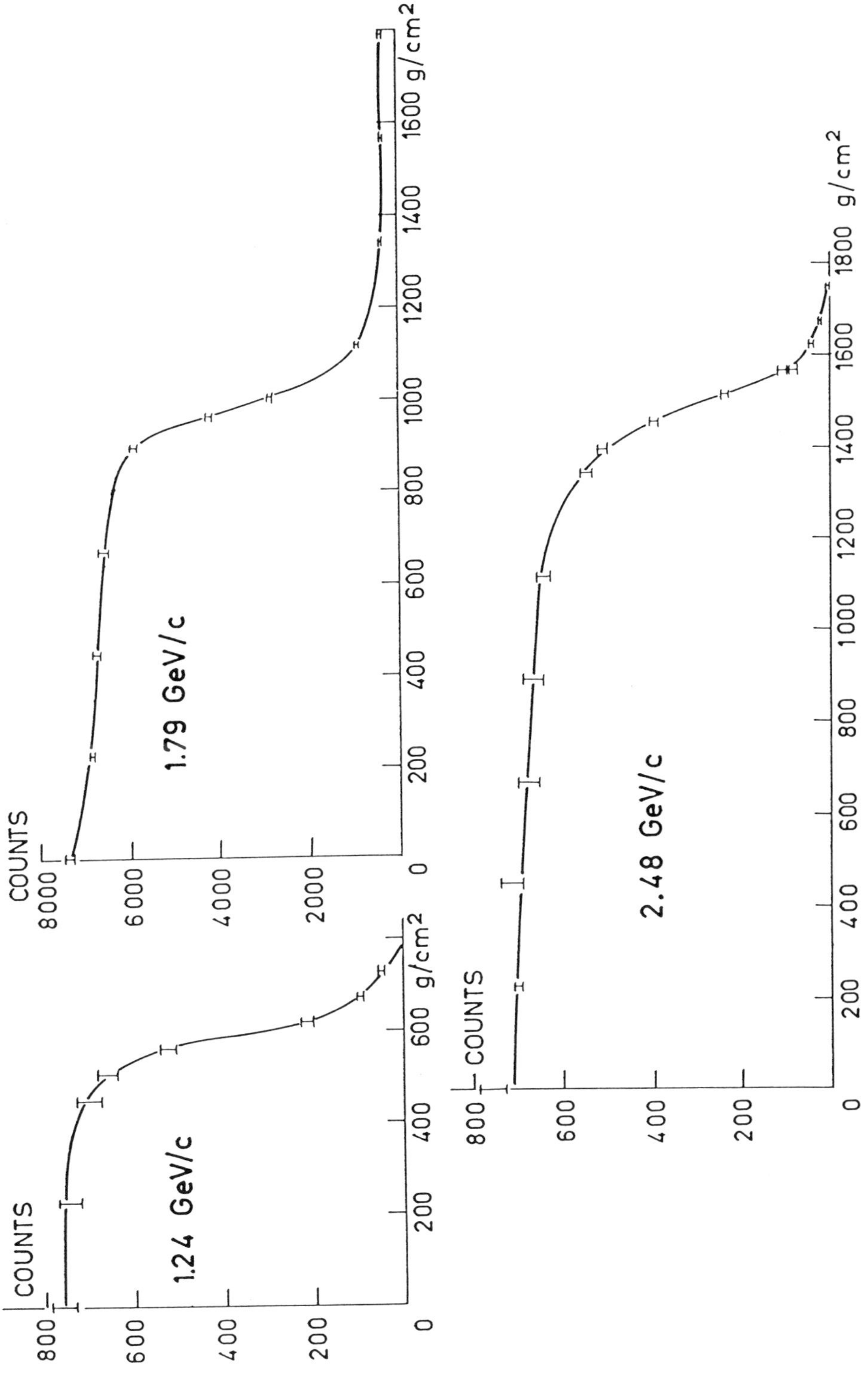
COUNTS
1.24 GeV/c
g/cm2
COUNTS
1.79 GeV/c
g/cm2
COUNTS
2.48 GeV/c
g/cm2

FIG.5

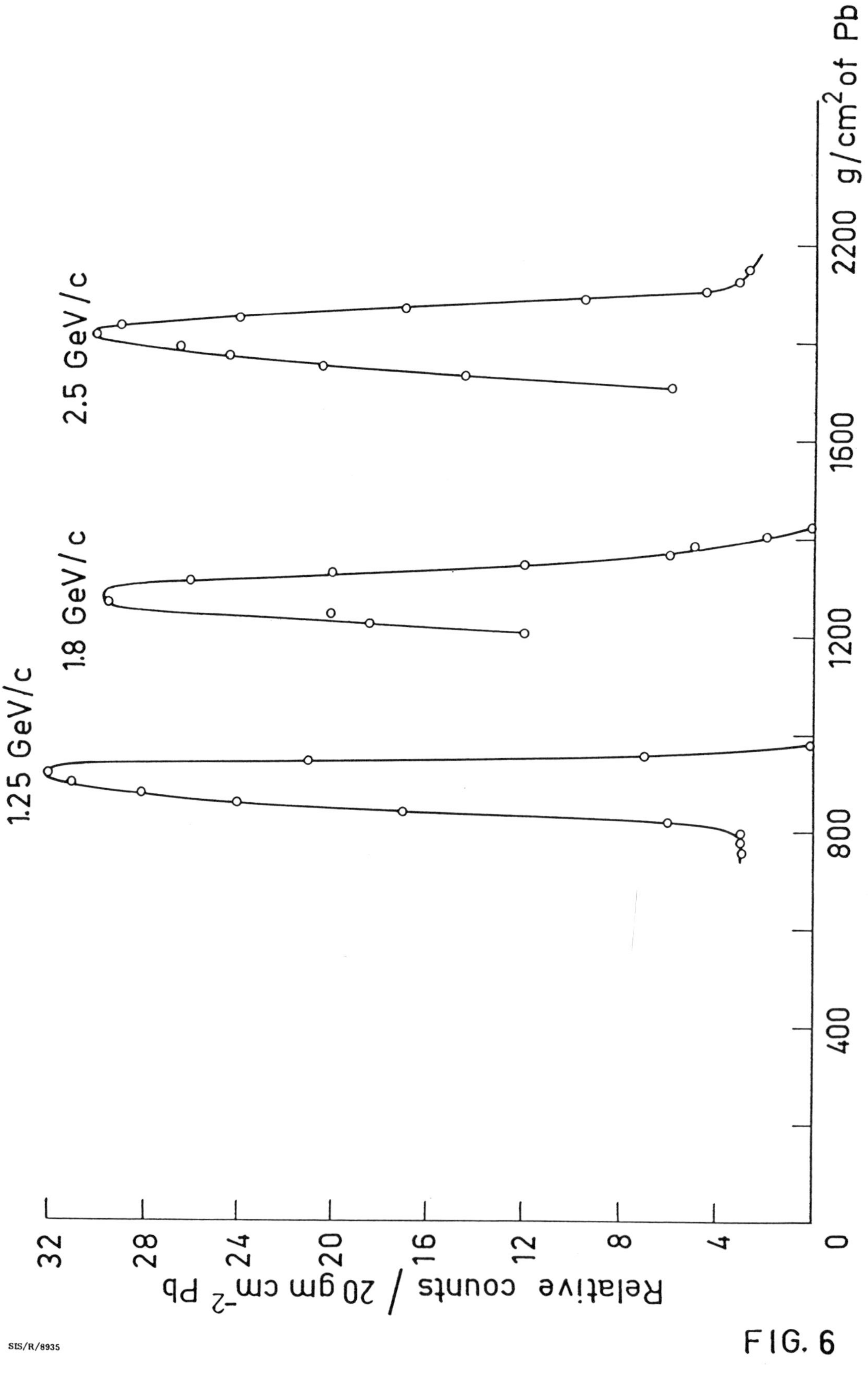

FIG. 6

SIS/R/8935

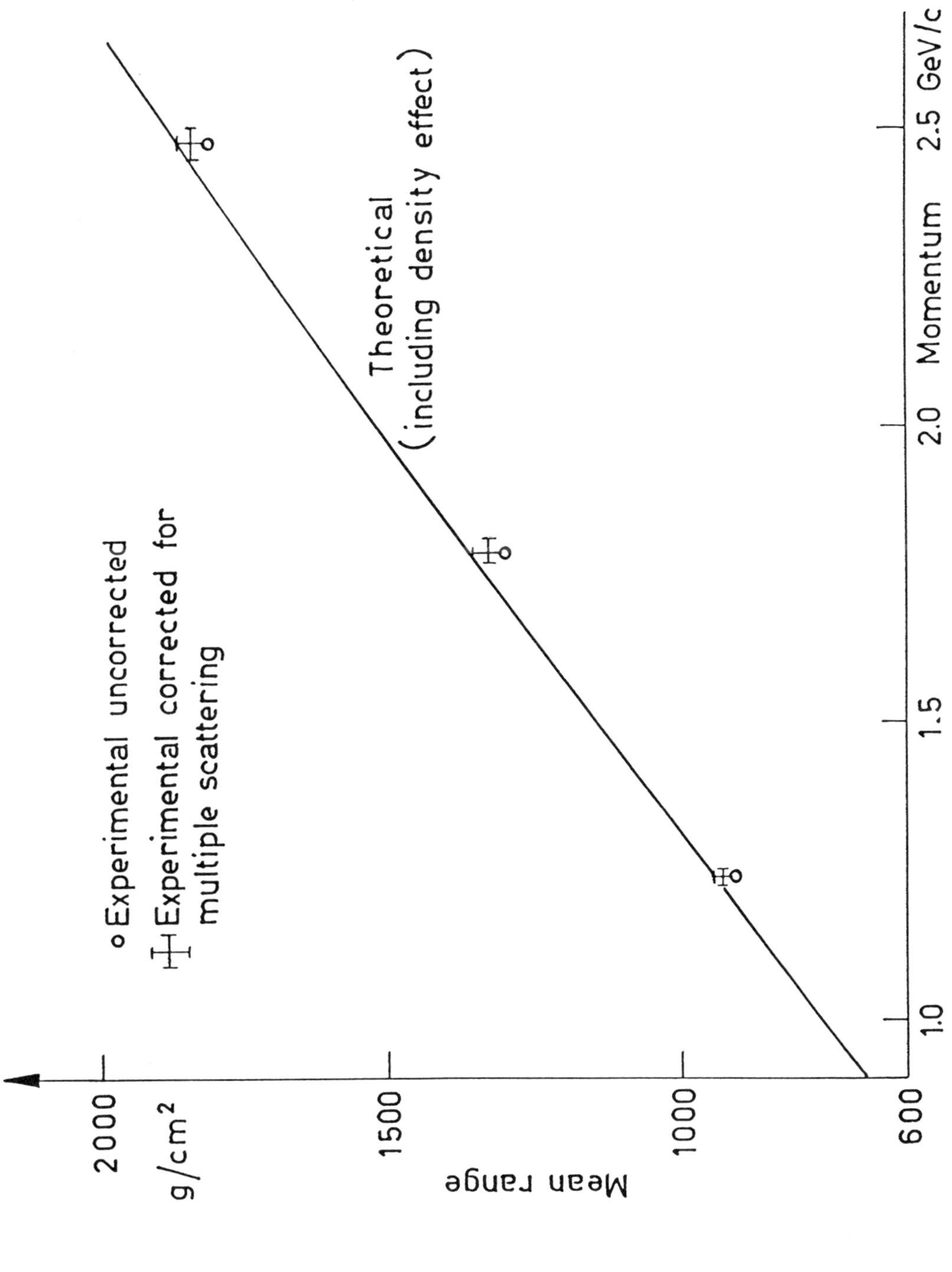
∘ Experimental uncorrected
Experimental corrected for multiple scattering
Theoretical (including density effect)
Mean range
g/cm²
2000
1500
1000
600
1.0
1.5
2.0
2.5 GeV/c
Momentum

FIG. 7

A. Buhler, T. Massam, Th. Muller and A. Zichichi

RANGE MEASUREMENTS FOR MUONS IN THE GeV REGION

From
Nuovo Cimento 35, *759 (1965)*

1965

A. Buhler, *et al.*
1° Febbraio 1965
Il Nuovo Cimento
Serie X, Vol. 35, pag. 759-767

CERN
SERVICE D'INFORMATION
SCIENTIFIQUE

Range Measurements for Muons in the GeV Region.

A. Buhler (*), T. Massam, Th. Muller (*) and A. Zichichi
CERN - Geneve

(ricevuto il 4 Luglio 1964)

Summary. — The range of highly relativistic muons has been measured. The results of the experiment are presented and compared with the theoretical expectations. The agreement between experiment and theory is estabilished to be within 2% up to a value af γ as high as 23.

1. – Introduction.

The results of an experiment to measure the range of high-energy muons are reported. The reason for carrying out this experiment was the lack of data on muon ranges above a few hundred MeV. We needed this data for the design of an experiment on proton-antiproton annihilation into muon pairs. On the other hand, the results given in this paper represent the first experimental check of the validity of the range-energy relation for high values of γ (**), *i.e.* $\gamma = 12 \div 23$. Measurements were made of the muon range in an absorber which consisted mainly of lead, at the following momenta:

$$P = 1.240 \pm 0.019 \text{ GeV/c},$$

$$P = 1.790 \pm 0.027 \text{ GeV/c},$$

$$P = 2.480 \pm 0.037 \text{ GeV/c}.$$

The value of the momentum was determined from the currents in the bending magnets and a calibration of beam momentum as a function of bending-magnet current. This calibration was made available by the Bubble Chamber Group,

(*) Centre de Recherches Nucléaires, Strasbourg.
(**) $\gamma = E/m$.

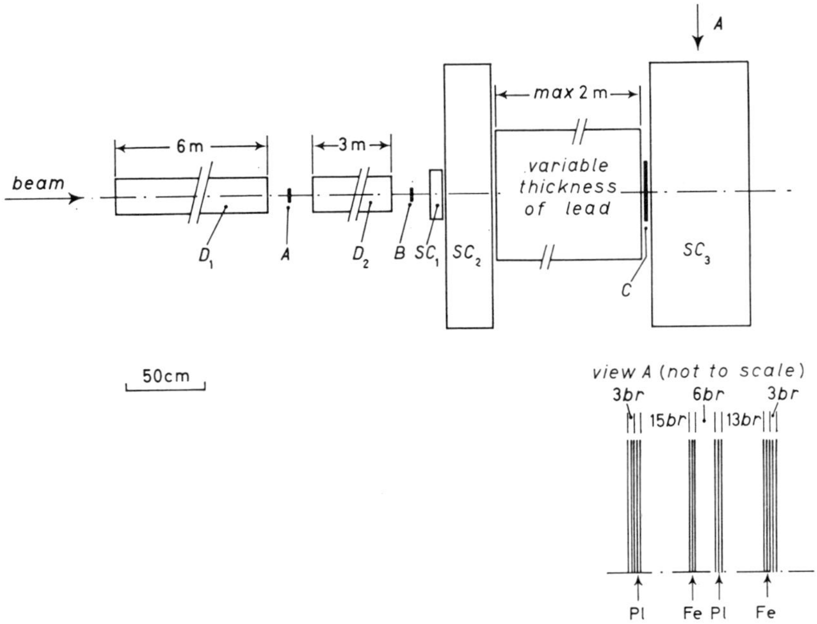

Fig. 1. – Experimental set-up. D_1, D_2: variable pressure threshold gas Čerenkov counters; A, B: plastic scintillation counters 8 cm diameter, 1 cm thick; C: plastic scintillation counter, $(40\times40\times1)\ \text{cm}^3$; SC_1: thin-foil 8-gap spark chamber; SC_2: thick-plate 14-gap spark chamber; SC_3: thick-plate 28-gap spark chamber; No. br: number of brass plates 0.5 cm thick; P_l plastic scintillator 2.5 cm thick; Fe: iron-plate absorber 3 cm thick. View A shows the arrangement of the plates and incorporated scintillation counters and iron absorber. The variable thickness of lead is always increased starting from counter C towards SC_2.

who had measured τ decays (1) to obtain the K-meson momentum in the same beam. Since the range measurements and the calibration were not made simultaneously, we are obliged to quote a momentum uncertainty equal to the momentum acceptance of the beam.

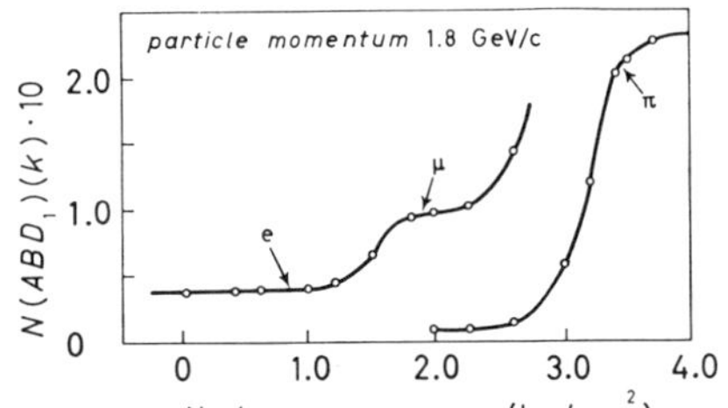

Fig. 2. – Calibration curve of the threshold gas Čerenkov counter D. $N(ABD_l)$ is the number of triple coincidences for fixed number of monitor (A, B). Particle momentum 1.8 GeV/c. Zero pressure corresponds to atmospheric pressure.

(1) For example, M. FERRO LUZZI, R. GEORGE, Y. GOLDSCHMIDT-CLERMONT, V. P. HENRI, B. JONGEJANS, D. W. LEITH, G. R. LYNCH, F. MULLER and J. M. PERREAU: *Proceedings of the Sienna International Conference on Elementary Particles*, vol. **1**, p. 182.

The experimental set-up and the electronic block diagram are shown in Fig. 1 and 3, respectively. The apparatus consists of two main parts: the beam telescope and the range telescope.

2. – The beam telescope.

This consists of two plastic scintillators A and B, and two Vivargent-type threshold gas Čerenkov counters D_1 and D_2 (2) which are filled with ethylene.

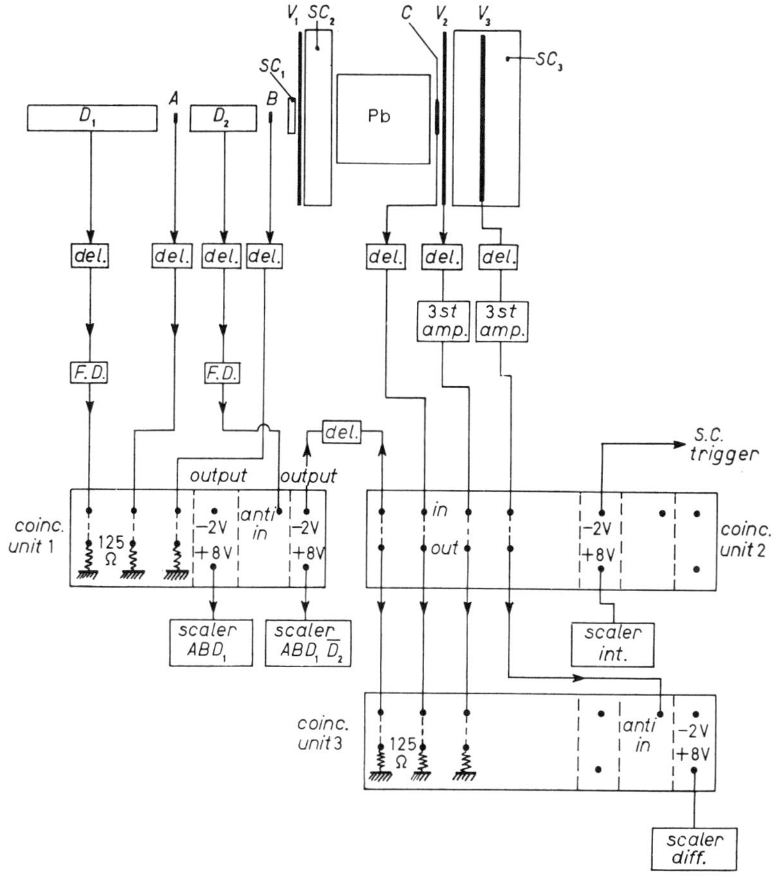

Fig. 3. – Electronic block diagram. Scintillation counters V_1, V_2, V_3 are shown outside the spark chamber for clarity. They are marked $P1$ on Fig. 1, view A. Del: variable delay lines; F.D.: fast discriminators; 3-st amp.: fast three-stage amplifier, 2.5 ns rise-time; scaler $ABD_1\bar{D}_2$: records the incoming muons; scaler int.: records integral range curve (coincidence unit 2); scaler diff.: records differential range curve (coincidence unit 3); SC trigger: trigger for spark chambers, film advancement and dead-time.

(2) M. VIVARGENT: *Nucl. Instr. and Meth.*, **22**, 165 (1963).

A typical calibration curve of D_1 at 1.79 GeV/c is presented in Fig. 3, which shows the three plateaux corresponding to pions, muons and electrons. Similar calibrations were taken for both Čerenkov counters at each of the three momenta used. Two threshold Čerenkov counters are used to provide the differential velocity selection needed to detect muons with low contamination from the antiprotons, K-mesons, pions and electrons. The muon electronic signature was

$$(ABD_1)\bar{D}_2,$$

with the pressures in D_1 and D_2 set in such a way as to have a very small efficiency in D_1 for detecting pions and heavier particles, and a very high efficiency in D_2 for electron detection. This method gave a muon beam with less than 5 % contamination of other particles, as may be seen from the initial slope of the integral range curves.

Table I shows the beam composition at the three momenta, neglecting antiprotons.

TABLE I.

Momentum GeV/c	e^- (%)	μ^- (%)	π^- (%)
1.24	5.4	3.6	91.0
1.79	1.6	2.3	96.1
2.48	1.5	1.4	97.1

3. – The range telescope.

The range telescope consisted of:

i) An eight-gap aluminium-foil spark chamber, SC, of total density 100 mg/cm² which allows a precise knowledge of the particle direction. This information was included so that we may check the multiple-scattering corrections which are necessary when incident and emergent particle directions are known. These results are not yet analysed.

ii) A spark chamber assembly, SC_2, consisting of a series of spark chambers of the Krienen type (³). These chambers are 1 m × 1.5 m with plates of 0.5 cm brass. After the first three plates of the chamber, there is a 1 m × 1.5 m × ×2.5 cm plastic scintillator V_1. Before the last three plates of the assembly, there is a 3 cm thick iron plate which serves to increase the total surface density

(³) H. FAISSNER, F. FERRERO, A. GHANI, E. HEER, F. KRIENEN, G. MURATOR. T. B. NOVEY, M. REINHARZ and R. A. SALMERON: *Nucl. Instr. and Meth.*, **20**, 213 (1963)

to 113 cm/cm^2 lead equivalent. This assembly was an insurance in case we were not able to select muons electronically with low contamination. Pions and electrons would have a high probability of interacting in this chamber and would thus allow us to interpret correctly any initial slope of the range curve. In practice, the electronic selection of muons was so good that we did not need to use the information of SC_2.

iii) A variable thickness of lead absorber, of cross-section 1 m $\times$ 1 m.

iv) A 40 cm $\times$ 40 cm $\times$ 1 cm plastic scintillator C. This counter was introduced to eliminate random coincidences between beam particles and intermediate energy particles which crossed the counters V outside the lead absorber. The size of this counter was chosen so that there was a negligible restriction on the multiple scattering accepted and, therefore, multiple-scattering corrections to the range could be made assuming « infinite slice » geometry.

v) The range chamber, SC_3, which consisted of two assemblies, very similar to SC_2, which were placed in series, and gave a total of 288 g/cm^2 lead equivalent.

4. – Experimental method.

The range was measured in two ways:

a) *Electronically.* Coincidences $(ABD_1)\,\overline{D}_2\,V_2\,CV_3$ and $(ABD_1)\overline{D}_2V_2C\overline{V}_3$ for a fixed number of monitors $(ABD_1)\overline{D}_2$ and as a function of variable absorber thickness, defined the differential and integral range curves. These results are shown in Fig. 4 and 5. The

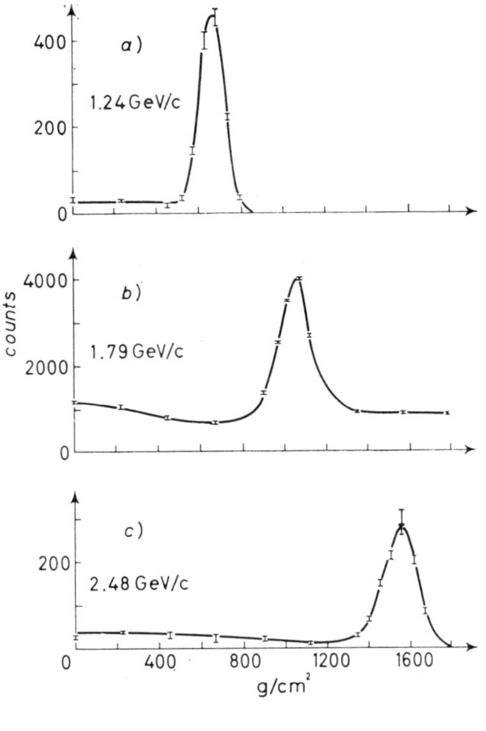

Fig. 4. – Differential range curves obtained by the electronic method. The abscissa is the thickness of the variable absorber. To this should be added 245 g/cm^2 at 1.24 GeV/c, 242 g/cm^2 at 1.79 GeV/c, and 241 g/cm^2 at 2.48 GeV/c. These values are lead equivalent, uncorrected for multiple scattering.

effect of counter C may be seen from the measurement at 1.79 GeV, where it had not yet been included.

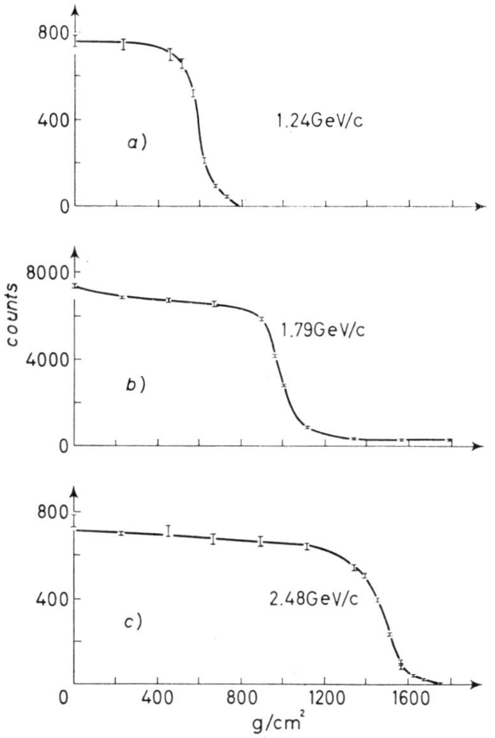

Fig. 5. – Integral range curves obtained by the electronic method. The abscissa is the thickness of the variable absorber. To this should be added 327 g/cm² at 1.24 GeV/c, 325 g/cm² at 1.79 GeV/c, and 323 g/cm² at 3.48 GeV/c.

b) Using SC_3. At each energy, the variable absorber thickness was chosen so as to bring the mean range to the centre of SC_3, and the coincidence $(ABD_1)\bar{D}_2CV_2$ was used to trigger the spark-chamber system. Figure 6 shows the range curve which was determined in this way.

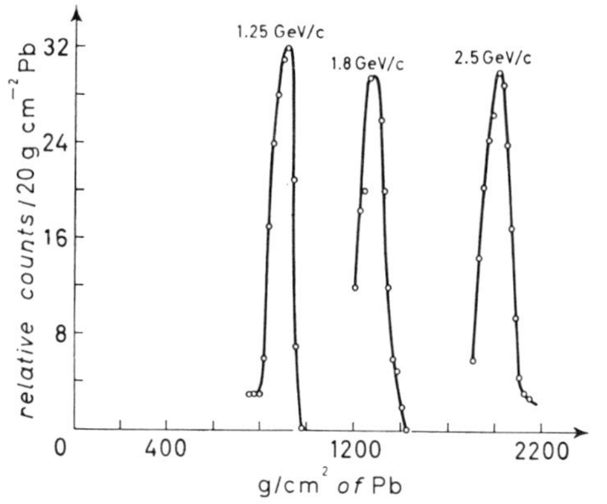

Fig. 6. – Differential range curve obtained from the spark chamber measurements. Multiple scattering corrections have not been made.

5. – Calculation of the results.

The two corrections which were made to the range were the conversion of g/cm² copper to g/cm² lead, and the correction to the measured range for path elongation by multiple scattering. Account was also taken of the counter material in the beam and of the variation of the Čerenkov counter surface density with pressure. The lead equivalent of the chambers varies slowly with the energy. The conversion factors CX which we have used are shown as a function of material and energy in Table II.

TABLE II. – *Values of CX.* [$CX = (dE/d\chi$ material $X)/(dE/d\chi$ lead)].

Material	Energy range	
	1.0 to 2.5 GeV	10 to 100 MeV
Brass and Fe	1.24	1.40
CH_2	1.77	1.20

The multiple-scattering corrections to the range were calculated according to the method of Øverås (4). Account was taken of the fact that the final part of the slowing-down took place in brass.

Table III shows the values of the ranges obtained by the three methods before and after the multiple scattering correction was applied.

TABLE III.

Momentum (GeV/c)	Integral range	Differential range	Spark-chamber range	
1.24	895	917	914	Before correction
1.79	1313	1288	1290	
2.48	1807	1791	1811	
1.24	910	933	930	After correction
1.79	1337	1315	1318	
2.48	1834	1821	1841	

The result is summarized in Table IV, which also shows what fraction of the total surface density was not lead. The uncertainty quoted in the range is assigned partly on the spread in values found in Table III, and partly from an 8% uncertainty assigned to the lead equivalent of the brass in the spark chambers. Consequently, the three uncertainties are strongly correlated.

TABLE IV.

Momentum (GeV/c)	Range	Fraction not lead
1.240 ± 0.019	925 ± 25	0.35
1.790 ± 0.027	1325 ± 25	0.25
2.480 ± 0.037	1830 ± 25	0.18

(4) H. ØVERÅS: *On small-angle multiple scattering in confined bodies*, CERN report 60-18.

765

The uncorrected and corrected experimental results are shown in Fig. 7, compared with the theoretical expectations of STERNHEIMER ([5]), which include the density effect.

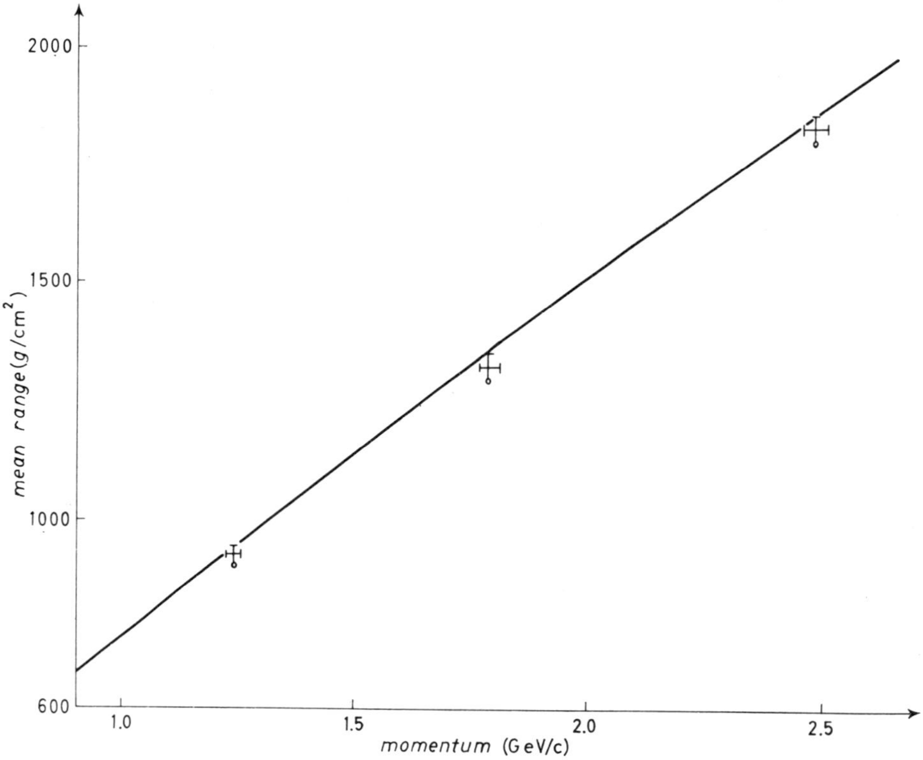

Fig. 7. – Mean values obtained for the mean range as a function of momentum, before and after correcting for multiple scattering. The solid curve shows the theoretical range-momentum calculations of Sternheimer (including density effect). ○ Experimental uncorrected; + experimental corrected for multiple scattering.

Table V shows the values of the standard deviations of the assumed Gaussian distribution of the straggling, compared with the values given by STERNHEIMER ([6]). The experimental values were estimated from the quartiles of the observed distributions. It may be seen that the spark-chamber values are in fair agreement with the predicted values, but the values obtained from the electronic method are consistently higher. This we attribute to the fact that,

([5]) R. M. STERNHEIMER: *Phys. Rev.*, **115**, 137 (1959); **118**, 1045 (1960).
([6]) R. M. STERNHEIMER: *Phys. Rev.*, **117**, 485 (1960).

TABLE V.

Momentum (GeV/c)	σ electronic integral range (%)	σ spark chambers (%)	σ Sternheimer (%)
1.24	6.2 ± 0.8	5.2 ± 0.4	3.3
1.79	5.2 ± 0.6	3.8 ± 0.3	3.6
2.48	5.3 ± 0.4	3.4 ± 0.3	3.9

σ is given as a fraction of the mean range.

in the electronic method, the thickness of the variable lead absorber is changed during the measurement, whereas, in the spark-chamber method, it is kept fixed.

In conclusion, we may say that the experimental results agree with the Sternheimer range-energy data within 2% for muons with γ of up 23. Further, there is fair agreement between the experimental and theoretical values of the range straggling, provided that the measurement is made in fixed geometry.

* * *

We wish to thank Prof. M. CONVERSI for his collaboration during part of the experimental running at the proton-synchrotron and Dr. H. ØVERÅS for his advice on the multiple-scattering corrections. Dr. F. KRIENEN has lent us the heavy-plate spark chambers and Dr. M. VIVARGENT his design of the gas-threshold Čerenkov counters. To both of them we would like to express our sincere appreciation and gratitude. The assistance of our technicians Messrs. J. BERBIERS and B. NICOLAI is greatly appreciated.

RIASSUNTO

I risultati sperimentali delle misure di « range » per muoni altamente relativistici sono presentati e discussi in relazione alle previsioni teoriche. L'accordo tra teoria ed esperienza è stabilito al 2% fino al più alto valore di γ misurato e cioè $\gamma = 23$.

767

T. Massam, Th. Muller, M. Schneegans and A. Zichichi

A NEW ELECTRON DETECTOR WITH HIGH REJECTION POWER AGAINST PIONS

From
Nuovo Cimento 39, 464 (1965)

1965

T. MASSAM, *et al.*
16 Settembre 1965
Il Nuovo Cimento
Serie X, Vol. **39** pag. 464-473

CERN
SERVICE D'INFORMATIONS
SCIENTIFIQUE

A New Electron Detector with High Rejection Power against Pions.

T. MASSAM, TH. MULLER (*), M. SCHNEEGANS (*) and A. ZICHICHI
CERN - Geneva

(ricevuto il 21 Maggio 1965)

Summary. — An electron detector, which consists of five elements, each one being made of a lead layer followed by a plastic scintillation counter and a two-gap spark chamber, is described. The rejection power of this new detector against pions is of the order of $4\cdot10^{-4}$, the efficiency for electron detection varies from 75% to 85%, and the energy resolution can be as good as 10%, in the energy range 1.1 GeV to 2.5 GeV.

1. – Introduction.

The detection of electrons and the simultaneous rejection of pions with a high rejection power is a problem which is of great interest for high-energy physics and which has received much attention in recent years.

Two types of technique have been used so far (1-5): visual and nonvisual. The visual techniques essentially use heavy-plate spark chambers, while the

---

(*) Institut de Recherches Nucléaires et Centre National de la Recherche Scientifique, Strasbourg.

(1) H. FAISSNER, F. FERRERO, A. GHANI, F. KRIENEN, T. B. NOVEY and M. REINHARZ: *Proceedings of the 1962 Conference on Instrumentation for High-Energy Physics* (*CERN*), p. 161.

(2) G. BACKENSTOSS, B. D. HYAMS, G. KNOP and U. STIERLIN: *Nucl. Instr. and Meth.*, **21**, 155 (1963); **20**, 294 (1963).

(3) R. A. ZDANIS, L. MADANSKY, R. W. KRAEMER, S. HERTZBACH and R. STRAND: preprint.

(4) C. A. HEUSCH and C. Y. PRESCOTT: CTSL-41.

(5) H. RUDERMAN, R. GOMEZ and A. V. TOLLESTRUP: CTSL-31.

nonvisual use either lead-glass Čerenkov counters or sandwich-type counters, which consists of layers of lead and plastic scintillator or of lead and plexiglass. The results obtained with both techniques barely reach the value of 1% for the rejection power against pions (*). Further, the visual technique provides no fast signal and so can only be used under very special conditions. The main limitation of the nonvisual technique is that the detector is not able to distinguish the shower originating from neutral pions, produced by the charge-exchange process, from that generated by an electron. In some earlier work (6), we attempted to improve the pion rejection by placing 1 cm of lead and a scintillation counter in front of a lead-glass Čerenkov counter and thus reached 0.3% rejection power.

The possibility of identifying the charge-exchange process by means of the charge-exchange scattering angle allows the elimination of those events in which the angle between the incident particle and the mean direction of the shower is greater than some minimum observable value.

The detector described here is a combination of both techniques. It provides a fast electronic signal, allows the identification of the charge-exchange process and is very compact along the direction of motion of the detected particle.

2. – Description.

Figure 1 shows the structure of the detector. The chambers K are thin-plate spark chambers, of total surface density 0.1 g/cm² of aluminium, which define the direction of the incident electron. K consists of a total of eight gaps, each of 8 mm spacing. The first layer of the sandwich consists of 1 cm of lead, followed by 1.5 cm of scintillator and then two spark-chamber gaps, S, with 7 mm spacing. The four subsequent layers are similar except that they contain 0.5 cm of lead. The whole assembly is mounted on a 2 cm thick aluminium back-plate. The lateral dimensions chosen for the detector described here are (35×35) cm² for the first set of chambers in K, (43×43) cm² for the second set of chambers in K, (60×60) cm² for the spark chambers within the sandwich, and (70×70) cm² for the scintillators. Each of the counters in the sandwich was viewed by two photomultipliers, the signals of which were added

(*) We define the rejection power against pions to be the ratio R:

$$\mathrm{R}=\frac{\text{number of pions which simulate electrons}}{\text{total number of pions}}.$$

This ratio R is also equivalent to the efficiency of the detector for pions.

(6) T. Massam, Th. Muller and A. Zichichi: CERN-63-25. Also, *Proceedings of the Sienna International Conference on Elementary Particles* (1964), p. 488.

linearly. All photomultiplier gains were equalized by taking discriminator curves of the output of each photomultiplier when an electronically defined beam of muons was allowed to pass through the sandwich. The counter M is an additional facility which is used to veto multiple particles in the calibration or in an experiment. The counter which was used in the calibrations was 28 cm in diameter and 1.5 cm thick.

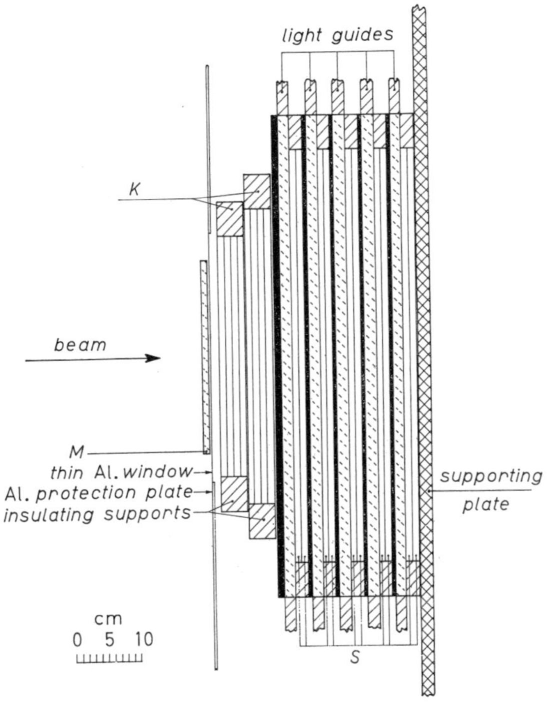

Fig. 1. – The structure of the detector. Not shown are the mirrors which are inserted between the light guides to allow 90° stereo-views of the spark chambers, and also the cylindrical lens which is placed in front of the chamber-mirror system. K and S are thin-plate spark chambers and M is an auxiliary counter which is used in conjunction with the sandwich. [cross-hatched] anticorodal (Al); [dashed] plastic scintillator; [hatched] plexiglass; [black] lead.

3. – Calibration.

Figure 2 shows the electronic set-up which was used to calibrate the detector. D_1 and D_2 are ethylene-filled gas Čerenkov counters (7) and A and B, are 8 cm diameter beam-defining plastic scintillation counters.

S_1 to S_5 are the five counters in the sandwich. Electrons were defined by the coincidence ABD_1, muons $ABD_1\bar{D}_2$ and pions by $AB\bar{D}_1\bar{D}_2$. The effects of $\bar{D}_1$ and $\bar{D}_2$ were measured separately to check that the contamination of electrons in the electronically selected pion beam was less than 10^{-4}.

The definition of a single electron by the sandwich was

$$M\cdot(\bar{M} > t_1)\cdot(S_1 > t_2)\cdot S_2\cdot S_3\cdot S_4\cdot S_5\cdot\left[\left(\sum_{i=1}^{5} S_i\right) > t_3\right].$$

(7) M. Vivargent, G. von Dardel, R. Mermod, G. Weber and K. Winter: *Nucl. Instr. and Meth.*, **22**, 165 (1963).

$\bar{M}$ threshold, t_1, was set to reject double particles while accepting single particles with high efficiency. The discrimination level of S_1 was set high to ensure that the shower started in the first slice of lead. This reduced the pion trigger rate caused by charge exchange, since only those pions which suffered charge

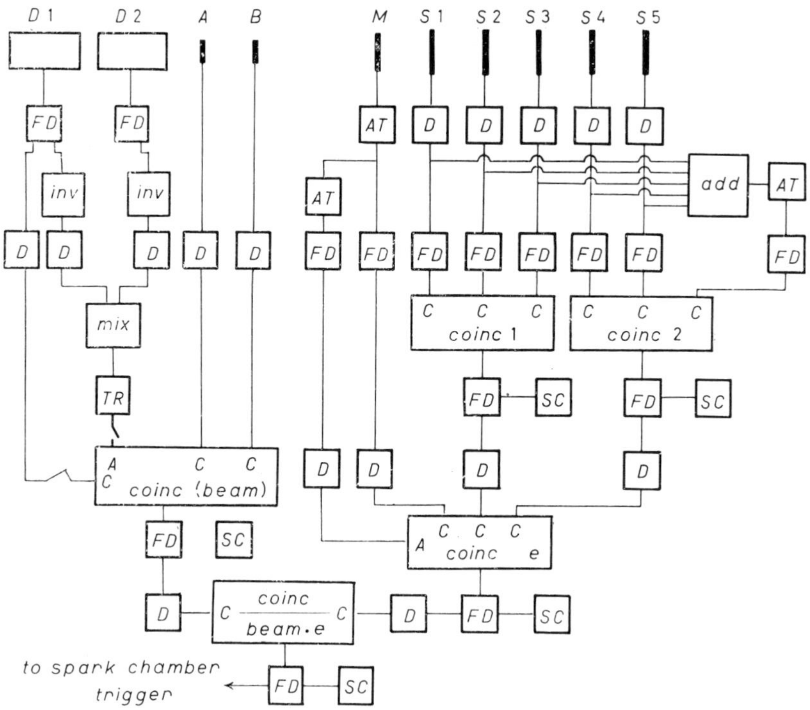

Fig. 2. – The electronic set-up: *D*, delay; *AT*, attenuator; *FD*, fast discriminator; *INV*, inverter; *ADD*, linear adding circuit; *MIX*, mixer; *TR*, trigger circuit; *SC*, scaler; *COINC*, coincidence unit; *C*, coincidence input; *A*, anticoincidence input

exchange in the first part of the first slice of lead had an appreciable probability of initiating a shower before S_1. The thickness of this first slice of lead was chosen experimentally as a compromise between high pulse height in S_1 for electrons (high electron efficiency) and low thickness of lead (low pion charge exchange). The discrimination level t_3 of $\sum S_i$ is somewhat arbitrary and is a compromise between electron efficiency and acceptable pion trigger rate. To a first approximation, the final results of the analysis imply that the final pion efficiency is not dependent on the electronic efficiency. The conditions under which we worked were established at 1.1 GeV/c, with the discrimination level of S_1 set to give 85 % electron efficiency and the discrimination level of $\sum S_i$ set to reduce this to 75 % electron efficiency.

A coincidence between an electron or pion in the beam and an « electron » defined by the sandwich triggered the spark-chamber system. The pulses from the counters S_1 to S_5 and also the sum of these, $\sum S_i$, were displayed on an oscilloscope and photographed to monitor the performance of the electronics.

4. – Performance.

Figure 3*a* shows a typical 1.8 GeV electron in the spark chambers. Figures 3*b* and 3*c* show typical pion events. From these we see that the characteristic of an electron shower is the dense central cone. Pions may be accepted electronically if they undergo inelastic interaction as in Fig. 3*b*, or charge exchange as in Fig. 3*c*. Notice that the former is characterized by a few tracks at large angles, and that the latter shows clearly the asymmetric nature of the shower.

A simple, nonsubjective method of scanning the events, which rejects the pion events with high efficiency, is to count the sparks produced within a narrow cone of half angle about 6°. The apex of the cone was positioned on the centre of the first lead plate and the axis was placed along the track in chamber K. This measurement was made in two orthogonal views of the chambers and the result were plotted as a scatter diagram of the number of sparks counted in the direct view against the number counted in the reflected view. A set of results for electrons and pions at 1.8 GeV/c is shown in Fig. 4*a* and 4*b*.

The value of 6° for the half angle of the cone was chosen so as to give the lowest pion efficiency for a given electron efficiency, but was found to be not very critical.

In our analysis we imposed the condition that the sum of the numbers of sparks counted in the cone in the two views must be greater than some limit which gave 90 % efficiency for the electrons. The variation of this limit with energy is shown in Fig. 5.

Figure 6 summarizes the results. Figure 6*a* shows the variation, with energy, of the electronic efficiency for pions and electrons. The behaviour of the electron curve is simply the effect of the energy resolution of the system, while the pion curve may be understood in terms of the increased energy of the charge-exchange products of the increased multiplicity in inelastic events, together with the increase in specific ionization of the incident pion and the corresponding increase in the energy of the knock-on electrons produced before or in the lead plates.

Figure 6*b* shows the variation with energy of the fraction of the electronically selected pion events which are also accepted as « electrons » in the spark-chamber analysis. The decrease with energy in this curve shows that the

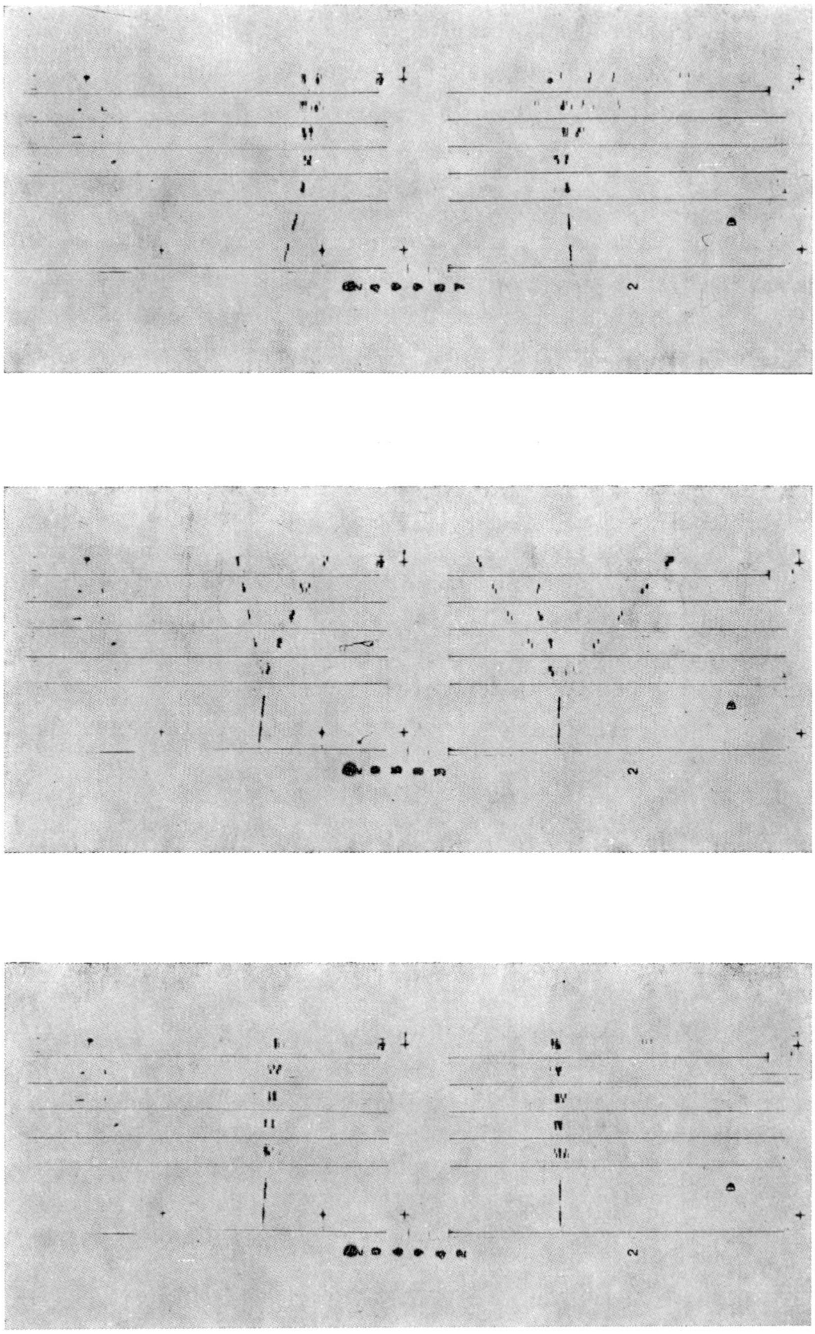

c) *b*) *a*)

Fig. 3. – Typical events observed in the spark chamber: *a*) electron shower; *b*) pion inelastic interaction; *c*) pion charge exchange giving an asymmetric shower.

7033

extra events accepted electronically, as the energy is increased, are recognized and rejected as spurious in the spark-chamber analysis. This compensation

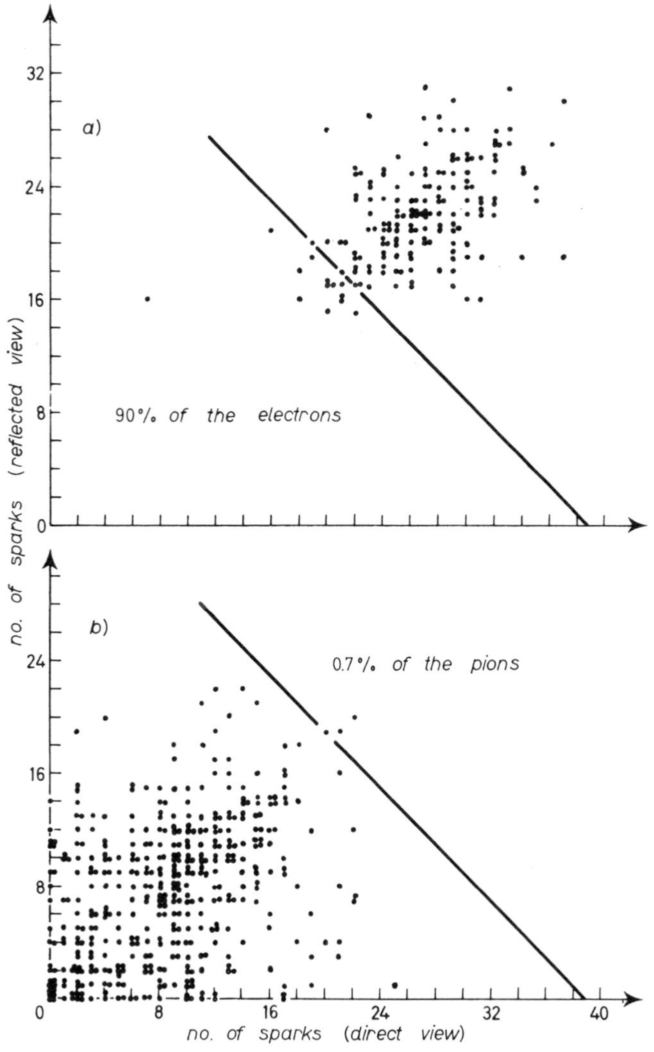

Fig. 4. – The results of counting sparks inside the standard cone in the direct and reflected views for *a*) electrons and *b*) pions. The 90% electron efficiency limit is drawn in both diagrams.

is aided by the fact that at higher energies we are able to set a higher spark-count threshold. This result is the basis of our statement, earlier in this paper, that the results are approximately independent of the electronic discrimination level: we expect the effect of varying the discrimination level at fixed energy to be similar to the effect of varying the energy at fixed discrimination level.

Figure 6c shows the over-all efficiency obtained by multiplying together the curves of Fig. 6a and 6b. It shows that for an electron efficiency which varies from 75 % at 1.1 GeV/c to 85 % at 2.5 GeV/c, the pion efficiency is effectively constant at $4 \cdot 10^{-4}$. Similar results were obtained for a second telescope of identical design.

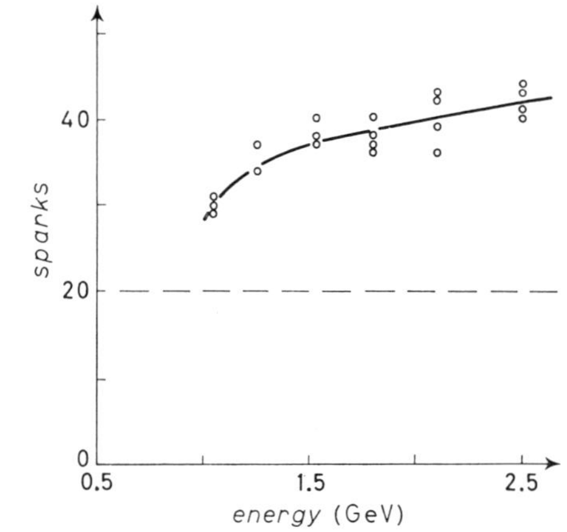

Fig. 5. – The variation of the 90% electron efficiency limit with energy. The horizontal line at 20 sparks indicates the spark count obtained from a noninteracting pion. The points were taken over a period of five months.

So far, we have described only the pion rejection properties of the detector, which were of primary interest in its design. It is relevant also to include here our knowledge of its energy resolution. This was measured by taking discrimination curves of the sum of all pulse heights. The discrimination level t_2 of the first counter S_1 was set at a low value, but otherwise the electronic system was the same as described above. Figure 7a shows the results which were obtained. The solid curves are the probability integral, adjusted to fit the data. The resolution, though not very good, was

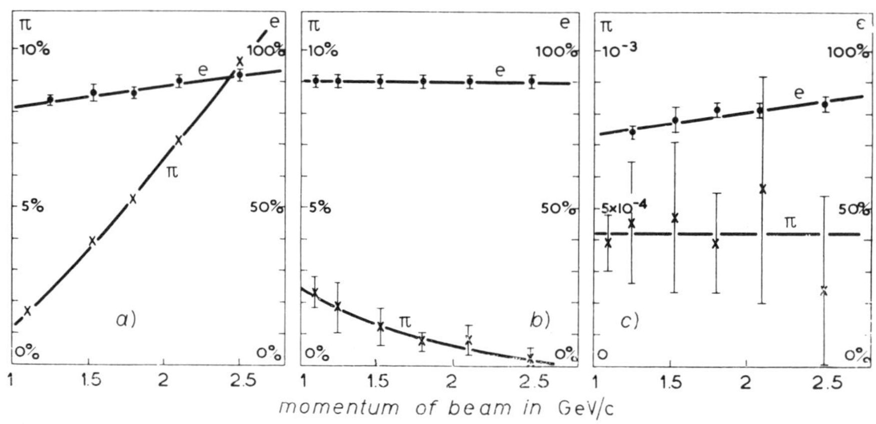

Fig. 6. – a) The variation with energy of the electronic efficiency of pions and electrons; b) the variation with energy of the fraction of electronically selected pion events which are also accepted as « electrons » in the spark-chamber analysis; c) the over-all efficiency for pions and electrons as a function of energy.

7035

sufficient for our needs, but was limited by the length of the detector which could be installed in our experimental arrangement.

To investigate the utility of increasing the number of layers, we placed our second electron detector after the first in the beam and arranged for the sum of its pulse heights to be added linearly to that of the first.

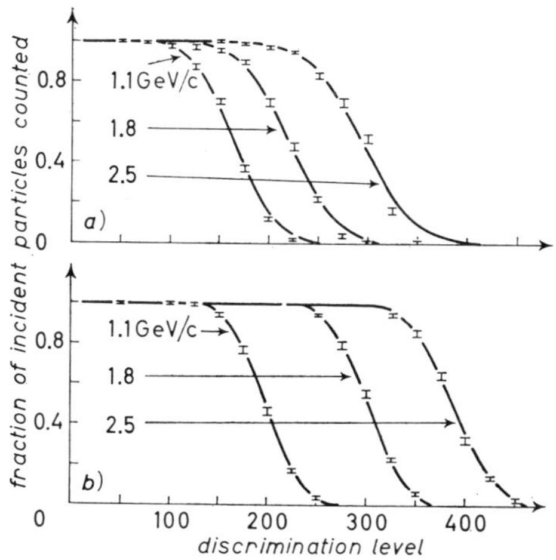

Fig. 7. – Discrimination curves of $\sum S_i$ for electrons at 1.25, 1.80 and 2.50 GeV/c for *a*) one telescope and for *b*) two telescopes placed in series in the beam and their outputs added linearly.

The pulse heights of the two detectors were equalized. Figure 7*b* shows results obtained in this way. The pulse heights and the resolution are given in Table I, where it is seen that the resolution with the two chambers in series compared well with values which may be obtained with a total absorption lead-glass Čerenkov counter. A typical figure (8) for the latter instrument is

$$\frac{\text{standard deviation}}{\text{mean pulse height}} = 0.075$$

at 1.8 GeV/c, while our detector could be modified to give a value of 0.10.

TABLE I.

	One telescope			Two telescopes added together		
Energy	1.25	1.80	2.50	1.25	1.80	2.50
Mean discriminator level	165	220	293	198	302	388
Standard deviation / Mean pulse height	0.21	0.18	0.15	0.15	0.10	0.093

5. – Conclusions.

We have shown in the present paper how it is possible to construct an electron detector with a rejection power against pions as high a $4\cdot10^{-4}$ in the

(8) This is a result which we have obtained using the instrument described by G. GATTI, G. GIACOMELLI, W. A. LOWE, W. C. MIDDELKOOP and T. YAMAGATA: *Rev. Sci. Instr.*, **32**, 848 (1961).

7036

energy range 1.1 to 2.5 GeV. This is, to our knowldege, an order of magnitude better than the rejection achieved by previous instruments. The energy resolution of this detector can be made as good as that of a lead-glass Čerenkov counter with the advantage of shorter length and greater flexibility in the construction.

* * *

We gratefully acknowledge the interest and support given to the present work by Prof. P. PREISWERK. The help of our technicians Messrs. J. BERBIERS, K. LEY and B. NICOLAI is highly appreciated. We would also like to express our appreciation for the diligent work done by our scanners: Mrs. R. BLANC, Misses A. BENEVENTANO, C. WELLS, M. LUDSTROM and J. VERATRINI.

RIASSUNTO

Si descrivono i parametri costruttivi ed i risultati ottenuti con un sistema che consiste di cinque elementi, ciascuno dei quali consta di uno strato di piombo seguito da uno scintillatore plastico e da una camera a scintilla a doppio gap, e che possiede un potere di reiezione contro i pioni dell'ordine di $4\cdot10^{-4}$, con una efficienza per rivelazione di elettroni che varia tra 75% e 85% nell'intervallo di energia da 1.1 GeV a 2.5 GeV. La risoluzione in energia di questo apparato può raggiungere il 10%.

7037

T. MASSAM, *et al.*
16 Settembre 1965
Il Nuovo Cimento
Serie X, Vol. 39, pag. 464-473

D. Bollini, A. Buhler-Broglin, P. Dalpiaz, T. Massam, F. Navach,
F.L. Navarria, M.A. Schneegans et A. Zichichi

UN GRAND DÉTECTEUR E.M. À HAUTE RÉJECTION DES PIONS

From
Revue de Physique Appliquée 4, *108 (1969)*

1969

REVUE DE PHYSIQUE APPLIQUÉE TOME 4, JUIN 1969, PAGE 108.

UN GRAND DÉTECTEUR E.M. A HAUTE RÉJECTION DES PIONS

D. BOLLINI, A. BUHLER-BROGLIN, P. DALPIAZ, T. MASSAM, F. NAVACH, F. L. NAVARRIA, M. A. SCHNEEGANS et A. ZICHICHI,

C.E.R.N., Geneva, Switzerland,
Istituto di Fisica dell'Università di Bologna, Italie,
Istituto Nazionale di Fisica Nucleare, Sezione di Bologna, Italie,
Centre de Recherches Nucléaires, Strasbourg, France.

Résumé. — Un grand détecteur de gerbes électromagnétiques comprenant neuf couches — chacune faite de plomb, d'une chambre à étincelles et d'un compteur à scintillations — est décrit. La réjection contre les pions est de 6×10^{-4}, l'efficacité de détection des électrons varie de 68 % à 80 % et la résolution en énergie de 15 % à 10 % dans la région d'énergie de 0,4 GeV à 1,1 GeV.

Abstract. — A large electromagnetic shower detector which consists of nine layers, each one made of lead, spark-chamber and scintillation counter, is described. The rejection power against pions is of 6×10^{-4}, the efficiency for electron detection varies from 68 % to 80 % and the energy resolution from 15 % to 10 % in the energy range 0.4 GeV to 1.1 GeV.

I. Introduction. — En 1964, la collaboration Bologne-C.E.R.N.-Strasbourg a construit et étudié un détecteur [1] constitué de couches successives de plomb, de chambres à étincelles et de scintillateurs en « sandwich », capable de distinguer les électrons des pions à des énergies allant de 1,1 à 2,5 GeV. Le détecteur décrit ici est une version agrandie, réalisée en 1967, capable de détecter des électrons ou des

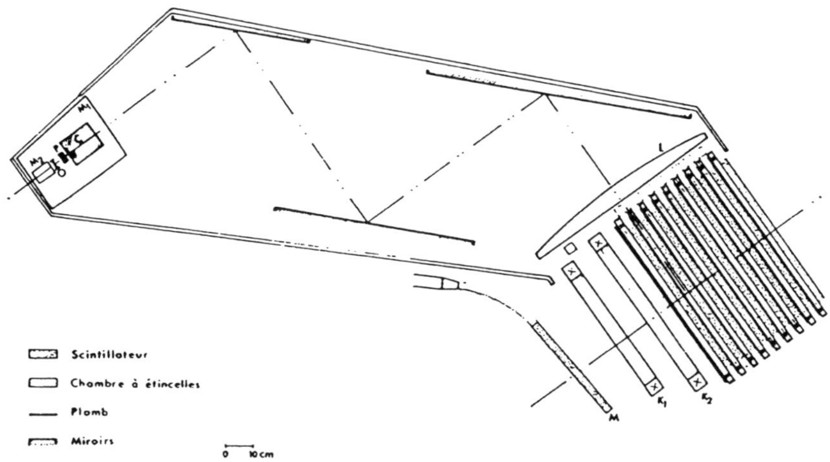

FIG. 1. — Section verticale du détecteur : devant les neuf couches du « sandwich », on voit les chambres à étincelles à feuilles minces K_1 et K_2 et le compteur M. La lentille L, les grands miroirs, les miroirs à 45° M_1 et M_2, l'objectif O amènent l'image des chambres sur le film F. Les croix sur les chambres à étincelles sont des marques fiducielles.

rayons γ à des énergies allant de quelques centaines de MeV à plusieurs GeV en présence d'un important fond de pions. Ce détecteur combine la méthode électronique utilisant les impulsions produites par des compteurs à scintillation pour réaliser une discrimination électron-pion et la méthode optique qui consiste à visualiser les gerbes dans des chambres à étincelles et permet ainsi d'améliorer la réjection des pions. Deux tels détecteurs ont été utilisés dans plusieurs expériences [2, 3, 4].

II. **Description.** — La figure 1 représente une section verticale du détecteur. Il consiste essentiellement en un « sandwich » compact de neuf couches triples comprenant chacune une plaque de plomb, une chambre à étincelles et un compteur à scintillation. Les dimensions du parallélépipède sensible sont : 60×120 cm² de surface et 50 cm de profondeur.

L'épaisseur totale de plomb correspond à 8,6 longueurs de radiation. La première plaque a 1 cm d'épaisseur assurant, avec une probabilité supérieure à 80 %, la multiplication de l'électron incident.

Les neuf chambres à étincelles identiques ont chacune deux intervalles de 0,85 cm de large et leurs dimensions sont : $70 \times 130 \times 2$ cm³. Leurs plaques sont des feuilles d'aluminium de 1 mm d'épaisseur tenues par un cadre de plexiglas. Une barre centrale de 0,7 cm de large sépare les chambres en deux moitiés optiquement indépendantes. Les chambres sont remplies d'hénogal (30 % He, 70 % Ne) et une haute tension de 9 kV est appliquée sur le plateau central. Un champ de balayage de 30 V donne 95 % d'efficacité pour un temps de sensibilité de 1 μs.

Les neuf compteurs à scintillation identiques se composent chacun d'une plaque de scintillateur plastique de $60 \times 120 \times 2,5$ cm³, vue par 4 photomultiplicateurs 56 AVP à travers deux guides de lumière rectangulaires de $60 \times 65 \times 2,5$ cm³ prolongeant le scintillateur à ses deux extrémités plus étroites.

On peut voir (*fig.* 1), en avant du « sandwich », les deux chambres cinématiques à feuilles minces (25 μm), comportant chacune 6 intervalles de 0,8 cm et permettant de reconstruire la direction des particules chargées incidentes avec une précision de $\pm$ 1°. Devant ces chambres, un compteur à scintillation (M) garantit une particule incidente chargée et réduit la multiplicité de ces particules. Lorsque le détecteur est utilisé comme spectromètre de rayons γ, un convertisseur de plomb est ajouté.

Le système optique permettant de photographier avec une seule caméra les deux moitiés optiquement indépendantes des chambres est visible sur la figure 1.

Chaque moitié du système transmet l'image de deux vues orthogonales des chambres.

La logique électronique adoptée après différentes études consiste à demander une certaine multiplication dans la première plaque de plomb ainsi qu'une énergie totale laissée par la particule dans le détecteur supérieure à une certaine valeur. L'impulsion discriminée du premier compteur ainsi que la somme linéaire discriminée des neuf compteurs sont envoyées dans la coïncidence finale qui déclenche la logique des chambres à étincelles.

III. **Calibrations.** — Calibrer le détecteur consiste à le placer dans un faisceau direct d'électrons ou de pions et à mesurer sa réponse en hauteur d'impulsion et en efficacité, à différents moments de la particule incidente. On peut ainsi déterminer les conditions électroniques pour réaliser le meilleur compromis entre

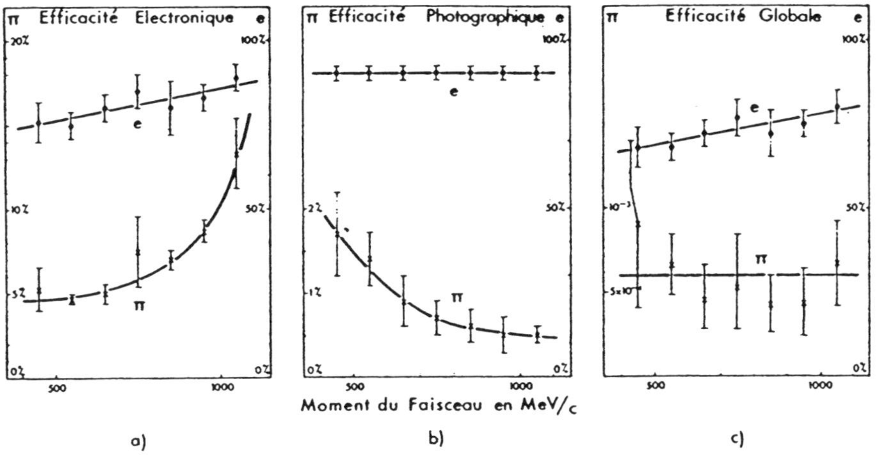

FIG. 2. — Efficacités de détection « électronique », « photographique » et globale pour les électrons et les pions en fonction du moment de la particule incidente.

haute efficacité de détection pour les électrons et bonne réjection des pions. Ceci est la « calibration électronique ». Dans une deuxième étape, on prend des photos des chambres à étincelles dans les conditions électroniques déterminées de manière à définir des critères pour une réjection supplémentaire des pions. C'est la « calibration photographique ». Ces calibrations ont été effectuées en de nombreux points de la surface du détecteur de manière à connaître l'efficacité moyenne de détection et à étudier son uniformité et les effets de bord.

IV. **Performances.** — Les résultats des « calibrations électroniques » et des « calibrations photographiques » sont représentés respectivement sur les figures 2 a et 2 b en fonction du moment de la particule incidente.

La figure 2 c montre les efficacités totales pour les pions et les électrons, obtenues en multipliant les efficacités « électroniques » et « photographiques » à différents moments. Les valeurs citées sont des moyennes sur différentes positions d'impact du faisceau et sur différentes périodes de temps des efficacités de détection et les erreurs représentées contiennent les fluctuations avec ces deux facteurs. On voit que l'efficacité de réjection totale pour les pions ($\sim 5 \times 10^{-4}$) ne varie pas sensiblement avec le moment, pendant que l'efficacité pour les électrons augmente de 68 % à 450 MeV/c à 80 % à 1050 MeV/c.

La résolution en énergie du détecteur, obtenue en combinant l'information des hauteurs d'impulsions et celle du nombre d'étincelles présentes dans la gerbe, est de 15 % à 400 MeV et de 10 % à 1100 MeV.

BIBLIOGRAPHIE

[1] MASSAM (T.) *et al.*, *Nuovo Cim.*, 1965, **39**, 464.
[2] BOLLINI (D.) *et al.*, *Nuovo Cim.*, 1968, **56** A, 1173.
[3] BOLLINI (D.) *et al.*, *Nuovo Cim.*, 1968, **57** A, 404.
[4] BOLLINI (D.) *et al.*, *Nuovo Cim.*, 1968, **58** A, 289.

M. Basile, J. Berbiers, D. Bollini, A. Buhler-Broglin, P. Dalpiaz, P.L. Frabetti,
T. Massam, F. Navach, F.L. Navarria, M.A. Schneegans and A. Zichichi

A LARGE ELECTROMAGNETIC SHOWER DETECTOR WITH HIGH REJECTION POWER AGAINST PIONS

From
Nuclear Instruments and Methods 101, *433 (1972)*

1972

NUCLEAR INSTRUMENTS AND METHODS 101 (1972) 433–454; © NORTH-HOLLAND PUBLISHING CO.

A LARGE ELECTROMAGNETIC SHOWER DETECTOR WITH HIGH REJECTION POWER AGAINST PIONS

M. BASILE, J. BERBIERS, D. BOLLINI, A. BUHLER-BROGLIN, P. DALPIAZ, P. L. FRABETTI, T. MASSAM, F. NAVACH, F. L. NAVARRIA, M. A. SCHNEEGANS and A. ZICHICHI

CERN, Geneva, Switzerland; Istituto di Fisica dell'Università di Bologna, Italy; Istituto Nazionale di Fisica Nucleare, Sezione di Bologna, Italy; Centre National de Recherches Scientifiques, Strasbourg, France

Received 17 January 1972

A large electromagnetic shower detector for identification and energy measurements of γ-rays (between 150 and 1600 MeV) and electrons (between 400 and 1100 MeV), in the presence of high pion background, is described.

The detector is based on the principle of simultaneous measurement of the spatial development of the electromagnetic cascade and of its energy release. It consists of 1) two six-gap thin-plate spark chambers for the reconstruction of the incoming particle trajectories; and 2) nine elements, each made of a lead foil, a spark chamber, and a plastic scintillator, all sandwiched together; here the shower development is studied. When used for γ-detection, a 0.5 cm Pb foil is placed in front of the thin-plate spark chambers, in order to allow the detection of the γ-conversion process and the identification of the γ-direction. The dimensions of the detector are $60 \times 120\,\text{cm}^2$ front face, and 50 cm depth along the electromagnetic shower development.

A pion rejection power of the order of 5×10^{-4} between 400 and 1100 MeV, for electron efficiencies varying from 70% to 80%, is obtained. The pion rejection efficiency in the γ-case is highly improved by the anticoincidence efficiency factor, while the γ-detection efficiency depends on the precision required in the reconstruction of the γ-ray direction. The γ-ray and electron energy resolution is about $\pm 15\%$.

1. Introduction

The purpose of this paper is to describe a large electromagnetic shower detector ($60 \times 120 \times 50$ cm^3) consisting essentially of a "Sandwich" of nine slices, each composed of a lead sheet, a two-gap optical spark-chamber, and a plastic scintillator counter. Clearly recognizable showers are produced in the detector, which provides signals for the electronic identification of the electromagnetic events, and consequently the rejection of most hadronic events from the trigger. This system is more powerful than lead-glass Cherenkov counters and lead-scintillator or lead-plexiglass counters for the rejection of pions, because the details of the shower development can be detected with fast electronic signals at any stage of the shower development along the nine slices, and thus the best conditions for fast hadronic rejection can be obtained. For example, an order of magnitude in π-rejection is gained by requiring suitable pulse height conditions after two radiation lengths of lead and adding up linearly all the counter signals. Moreover, the finer details of the shower can be studied off-line in the spark chamber pictures, thus providing afurt her rejection of the hadronic events.

Two thin-foil spark chambers placed in front of the Sandwich allow the reconstruction of the trajectories of the incident particles and, together with the Sandwich information, permit the identification of the nature of the particle by optical inspection. The electron's direction can be measured to $\pm 0.3°$, while the precision for a γ-ray reconstructed from the shower is much poorer. However, an accuracy of $\pm 1°$ can be obtained if the γ-ray is converted in a lead sheet preceding the thin-foil spark chambers, and if the tracks are measured only when they face the shower developed in the Sandwich.

This electromagnetic shower detector is based on the same principle as that of an electron detector designed by some of us in 1964[1]) to recognize electrons in the presence of a high pion background at energies between 1.1 and 2.5 GeV. The new detector, constructed in 1967, is able to identify γ-rays or electrons in the presence of a high pion background, at energies ranging from a few hundred MeV to several GeV. By combining the electronic and optical methods a pion rejection power of the order of 5×10^{-4} is obtained between 400 and 1100 MeV, corresponding to an efficiency for the electron detection between 70% and 80%. The performance improves at higher energy, but an interesting feature of the instrument is its behaviour at an energy as low as 400 MeV.

The combination of the pulse height and spark counting information yields a measurement of the energy of electrons or γ-rays with an accuracy better than $\pm 15\%$.

A brief description of the most interesting features

Fig. 1. Two views of the two electromagnetic shower detectors as used in experiments. The beam is coming from the left and the target can be placed at the centre of the turntable. The sensitive parts of the shower detectors are contained in the two strong metallic cages seen on the right. The aluminium-covered extensions contain the optics for the spark chambers. The γ-ray converters and the phototubes of counters M are visible in front of the sensitive part of the detectors.

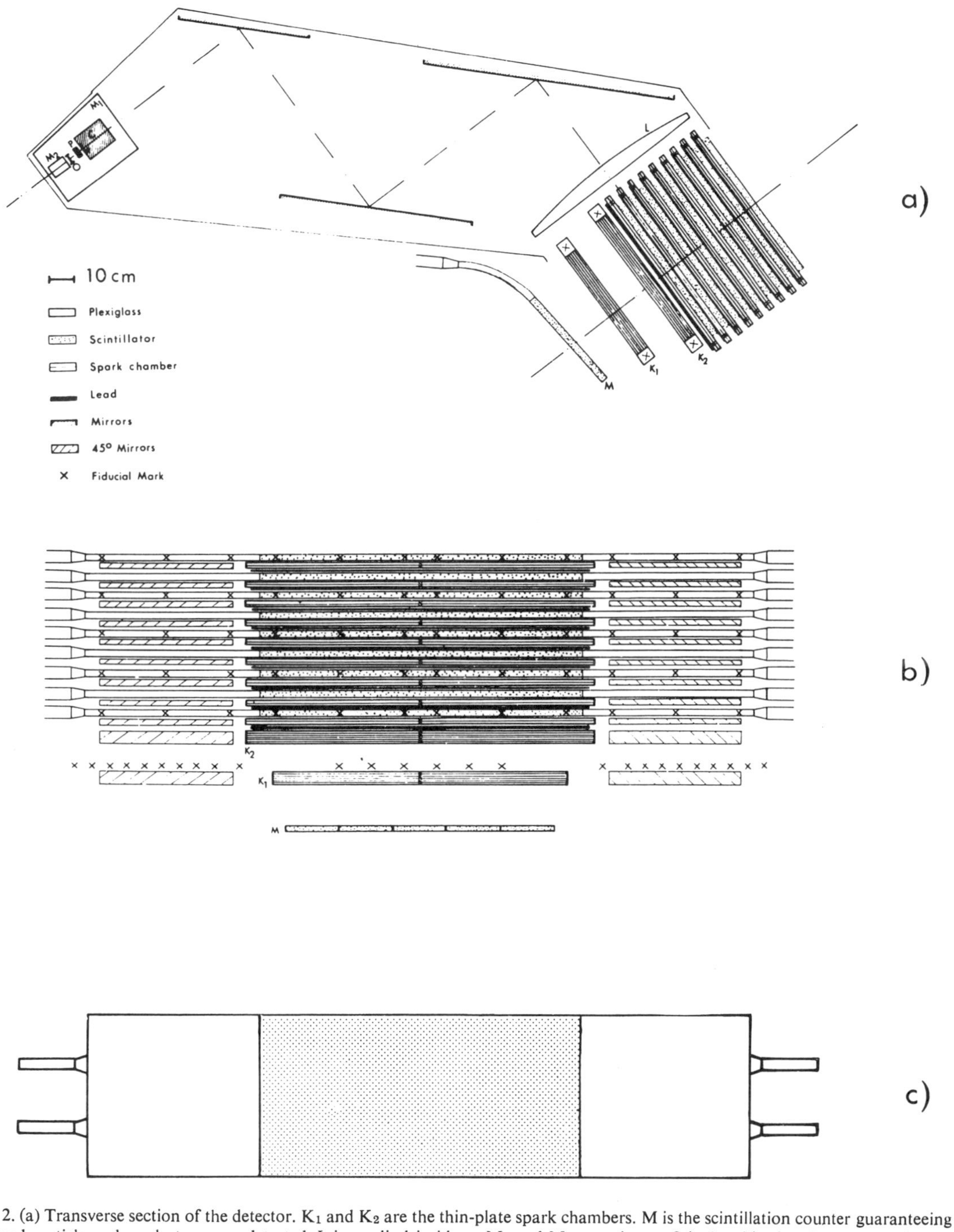

Fig. 2. (a) Transverse section of the detector. K_1 and K_2 are the thin-plate spark chambers. M is the scintillation counter guaranteeing charged particles, when electrons are detected. L is a cylindrical lens, M_1 and M_2 are mirrors, O is the objective, P a system of three mirrors, and F the film in its container C. The crosses near the spark chambers are fiducial marks. (b) Longitudinal section of the detector. Notice the plexiglass bar splitting the chambers in the middle. The 45° mirrors inserted between the light guides of the counters yield the reflected views of the left and the right halves of the chambers. The splitting into five parts of counter M for electron detection is shown. (c) Structure of one of the nine identical plastic scintillator counters of the shower detector.

of this new detector has already been published elsewhere[2]). Two such detectors have been used with the Bologna–CERN neutron missing-mass spectrometer[3]) in a series of fourteen experiments at the CERN Proton Synchrotron during the period 1967–1970[4]). We mention for example its use as electron detectors, in the study of the e^+e^- decay mode of the vector mesons ω and ϕ[5,6]); as γ-ray detectors, in the study of the $\gamma\gamma$ decay mode of the X^0-meson[7]) and of the $\eta\gamma$ decay mode of the ϕ-meson[8]); and as a γ-ray trigger in the study of the neutral A_2 structure[9]). Fig. 1 shows the electromagnetic shower detectors as they were used in these experiments. Both shower detectors were mounted in a frame with two perpendicular axes of rotation, allowing the data-taking to be optimized and also all regions of the detectors to be calibrated with electron, pion, or γ-ray beams at any time.

2. Description of the apparatus

2.1. Shower detector

The structure of the shower detector is shown in fig. 2. Each of the nine slices of the Sandwich consists of a layer of lead followed by a two-gap spark chamber and a scintillation counter. Two six-gap thin-plate spark chambers are placed in front of the Sandwich.

2.1.1. *Spark chambers*

The nine identical spark chambers of the Sandwich have external dimensions of $70 \times 130 \times 2$ cm^3, and consist of two 0.85 cm wide gaps with the high voltage applied on the central plate. The plates are made of 1 mm thick sheets of aluminium fixed on a plexiglass frame. In front of each chamber there is a lead plate of dimensions 130×70 cm^2, laminated with two thin iron foils to improve rigidity. The first lead plate is one cm thick; this gives to an electron more than 70% probability of starting a shower in the energy range considered. The other eight lead plates each have a thickness of 0.5 cm; the nine lead plates provide a total thickness of 5 cm. To improve the rigidity of the chambers, these were divided into two halves by a 0.7 cm wide plexiglass bar, blackened to avoid any reflection and pierced by holes to allow gas circulation. The electrical capacity of each two-gap chamber is 1600 pF, and a high voltage of 9 kV was applied through a storage capacity of 9000 pF for each chamber. Henogal (He 30%, Ne 70%) was used for all chambers with a slight, automatically regulated, overpressure.

Fig. 3 shows the efficiency, for single tracks, of the spark chambers as a function of the triggering delay for various clearing fields. For a 0 V clearing field, the curve drops at 60 μs. A clearing field of 30 V is convenient to use, since it yields 95% efficiency at the triggering delays used in the experiments, and corresponds to 1 μs sensitivity time of the chambers.

The two kinematic chambers each have six gaps of 0.8 cm, and their dimensions are $55 \times 110 \times 5.0$ cm^3 for the first chamber and $64 \times 130 \times 5.0$ cm^3 for the second one. Their walls are made of thin (25 μm) foils of aluminium, the total thickness of material (0.035 cm) corresponding to a probability smaller than 10^{-3} that an electron will start a shower in the kinematic cham-

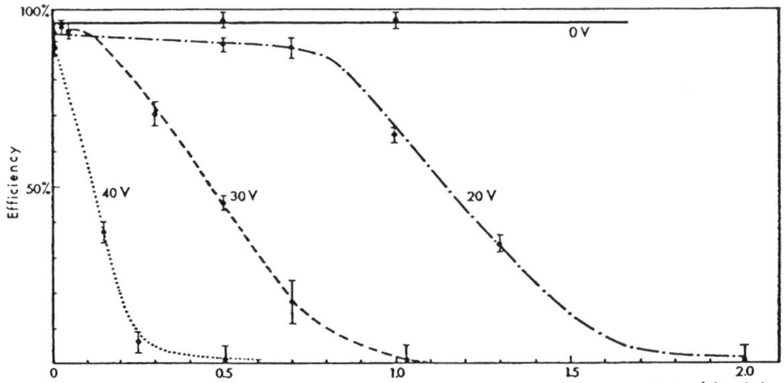

Fig. 3. Efficiency of the nine two-gap spark chambers as a function of the triggering delay for various values of the clearing field. The delay in the abcissa is an extra delay added to the fixed electronic delay ($\approx$ 500 ns) between the event and the final trigger. The efficiency, defined as the mean efficiency of the nine two-gap spark chambers when traversed by straight cosmic-ray tracks, is given by $N_s/(18 \times N_p)$, where N_s is the total number of sparks counted in N_p pictures considered.

bers. A clearing field of 45 V could be used in these chambers without reducing the efficiency below 98%.

2.1.2. *Scintillation counters*

Each of the nine identical counters (fig. 2c) of the Sandwich is composed of:

- one plastic scintillator NE 102A of dimensions: $60 \times 120 \times 2.5$ cm^3;
- two rectangular plexiglass light-guides, $60 \times 65 \times 2.5$ cm^3 at the smaller sides of the scintillator;
- four 56 AVP phototubes, two on each rectangular light-guide, with intermediate conical light-guides.

The length of the rectangular light-guide was chosen so that the variation in the illumination of its end surface with the position of the event in the scintillator was small. Simple surface matching of a fraction of this surface to the phototubes could then be made with small conical light-guides. These guides were made longer on every second counter in order to shift the photo-multipliers and give a more compact structure to the Sandwich. This optical design of the counters gives a pulse-height variation over the whole surface of scintillator smaller than $\pm 5\%$.

When the detector is used for electrons, a thick (2.5 cm) plastic scintillator counter (counter M) is put in front of the kinematic chambers. It guarantees that a charged particle enters the thin-plate chambers and allows high multiplicity events to be rejected. Counter M consists of a sheet of plastic scintillator NE 102A, of total dimensions $100 \times 40 \times 2.5$ cm^3, divided into five parts, each of dimensions 20×40 cm^2 and each viewed by a 56 AVP photomultiplier through a bent plexiglass light guide as shown in figs. 2a and 2b. The division of the counter into five parts is to achieve a better efficiency and a more uniform pulse-height spectrum, and also to allow rough kinematical selection in the final electronic trigger.

When the detector is used as a γ-ray spectrometer, a lead-plate converter of dimensions $82 \times 42 \times 0.52$ cm^3, followed by a scintillation counter, is placed 16 cm in front of the kinematic chambers. In this case, the counter has only the purpose of guaranteeing the γ-conversion, and is 1 cm thick; the plastic scintillator sheet of 78×39 cm^2 is split into three, 13 cm wide, horizontal strips, each seen by two 56 AVP photomultipliers at opposite ends.

2.2. COMPLEMENTARY APPARATUS

2.2.1. *Spark chamber optics*

The optics is designed in such a way as to allow one camera to photograph two orthogonal views of each of the two optically independent halves of the spark chambers. For this, two identical optical systems transmit, for each half, the image of a "direct view" of the chambers and that of a "reflected view"; the latter is obtained by narrow (2 cm) mirrors inserted at 45° between the plexiglass light-guides of the counters as shown in fig. 2b.

Fig. 2a shows the optical system for one half of the spark chambers. Two large plexiglass, plane-convex, cylindrical lenses L (0.75×1.25 m^2), with 4 m focal distance, are mounted on a metallic frame and are placed in front of the two halves of the chambers. They are mounted with the plane side facing the optical faces of the chambers and parallel to them, i.e. parallel to the plane of fig. 2b. Three large parallel mirrors followed by a 45° mirror M_1 bring the image of the two "direct" and the two "reflected" views to the median plane of the detector, where one camera photographs the complete system. M_2 indicates two small mirrors in the median plane, each at 45° to the optical axis of the corresponding half of the system. These two optical axes are thus made parallel and pass through two independent objectives O ("Schneider–Kreuznach" type "Componon Normal") with 60 mm focal length. Just after O, a system of three mirrors P brings the optical axes near to each other in order to have the two images on the film F. The whole optical system is enclosed in a 0.5 mm thick aluminium enclosure.

The film is mounted on 30 m rolls in a modified Beattie–Coleman magazine (model CS-48, type 10356). A special guiding system had to be designed to locate the film within ± 0.1 mm, in both its lateral position and its distance from the objective. The film used was the 35 mm wide "Adox" track chamber perforated film (SCF Super Type A), with emulsion No. 43015 or No. 43025 on the inside. Perforated film was used to facilitate automatic scanning on the CRT flying-spot digitizer "Luciole 66"[10]).

With this optical system, the final image (shown on fig. 4) is formed of four quarters: the "direct" and "reflected" view of each half of the spark chambers. A demagnification factor of 58.5, together with the dimensions of the chamber system, allows the images of the two optical systems to be positioned one under the other in the usable width of the film (25 mm). This disposition is very economic in film length – 2.5 cm per total image – and this allows 1200 pictures to be taken on one roll.

The orientation of the film is chosen so that the sparks are perpendicular to the film advance and thus to possible scratches on the film. This point is important in a shower detector where faint thin sparks

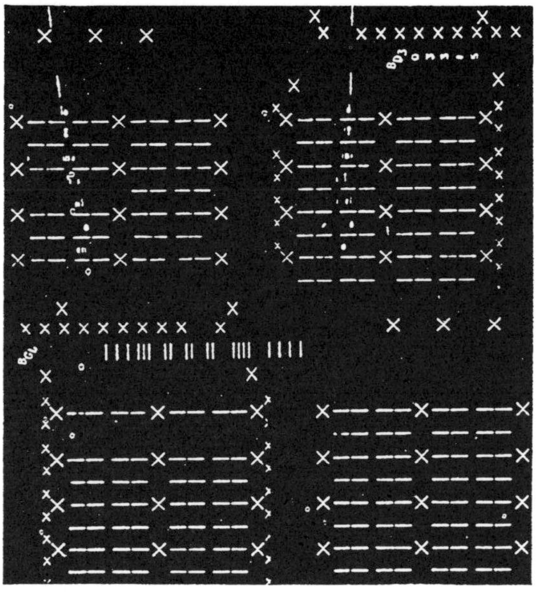

Fig. 4. Final image of the spark chambers on the film. It is formed of four quarters: the "direct" and the "reflected" views of each half of the chambers. The fiducial marks, as well as a decimal and a binary coded numerator and an indication of which quarter one is looking at (i.e. BG_4 means "bas gauche", quarter 4), are visible. A 750 MeV electron is shown.

must be clearly interpreted either visually or by an automatic scanning device.

2.2.2. *References for reconstruction*

The reconstruction in space of the tracks of the particles observed in the detector is done in two steps:

- first the tracks are reconstructed in the so-called Sandwich system, which is defined with respect to the detector, independently from its real position in space;
- then the tracks are transformed in the laboratory system.

Internal fiducial marks and well-placed external reference points are necessary. The external reference points are short cones of brass positioned so that they can easily be measured when the detector is in data-taking position. The internal fiducial marks, fixed near the chambers and photographed at the same time as the sparks, must have luminosity comparable to spark luminosity and good homogeneity of brightness over their surface, in order to facilitate accurate reading by "Luciole 66"[10]).

Electroluminescent panels made by "Thurn" (Reg. Trade Mark: "Panelume") are used. They consist of specially prepared phosphors of the zinc sulphide group, incorporated in a thin layer of high dielectric strength. The luminescent panels of 3×4 cm^2 are glued in sets of three on oxidized black Anticorrodal bars and covered with glass on which five layers of photographically deposited emulsion define the crosses. The panels were operated at 220 V and 500 Hz frequency, when new. After four years of use, a frequency of 5000 Hz was required to obtain the desired brightness.

Figs. 2 and 4 show the configuration of the fiducial marks for "direct" and "reflected" views. The bars of three crosses for the direct view are fixed in front of every second counter. The high number of marks is justified by the various checks that it is necessary to carry out on possible distortions or differences of magnification of the optical systems, and also allows the reconstruction of the true optical position of the objective.

The position of all fiducial marks plus the external brass reference points were measured in an arbitrary coordinate system, using two theodolites. The lens and mirror system were dismounted for these measurements. The Sandwich system could then be defined with three fiducial marks, and the coordinates of all the other marks and external reference points were calculated in this system.

When the detector is in position for an experiment, four non-coplanar external reference points are measured in the laboratory system. The knowledge of the coordinates of these points in the Sandwich system and in the laboratory system allows the determination of the transformation from one system to the other.

2.2.3. *Mechanics*

The nine slices of the Sandwich and the second kinematic chamber are contained in a strong metallic frame of dimensions $70 \times 74 \times 400$ cm^3, which also encloses the phototubes and their supply chains (see fig. 5). The first kinematic chamber is fixed outside the frame. The lead, spark chambers, and counters rest on the bottom of the frame and are kept in place by metallic combs. In addition, the photomultipliers and their magnetic shielding are supported by collars fixed on the frame. The spaces between the layers are made as small as possible in order to keep the Sandwich compact (49 cm for the nine slices).

Two strong arms at each end of the frame allow the heavy detector (two tons) to rotate around a horizontal axis. The frame is suspended by two 1.2 cm diameter cables at points furthest from the axis of rotation. This rotation permits the detector to be placed at any angle θ_V in the vertical plane. Furthermore, the distance from the detector to the axis of rotation can be adjusted.

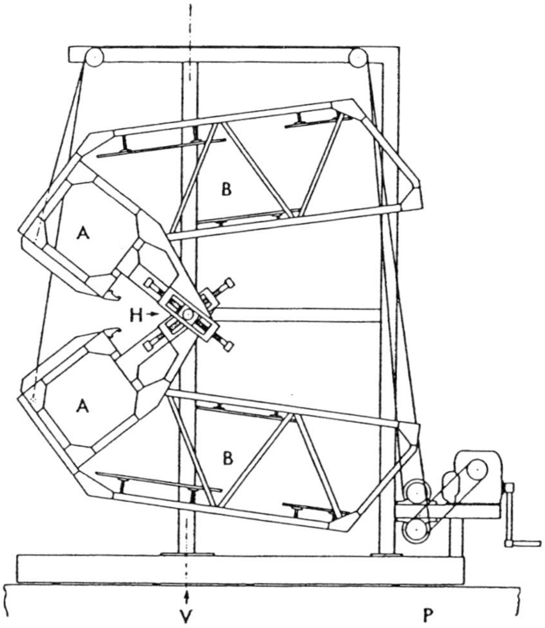

Fig. 5. Mechanics of the system of two detectors. The lead, the spark chambers, and the counters of the Sandwich are fixed inside the strong metallic frame (A). A lighter frame (B) encloses the optics of the spark chambers. Each detector can rotate around the common horizontal axis (H), and the whole system of two detectors can rotate on the platform (P) around the vertical axis (V).

When two such detectors are used in experiments, they are mounted on a turntable (see fig. 1) which allows any angle θ_H in the horizontal plane to be reached for both detectors. This is essential for calibration purposes. The position $\theta_H=0°, \theta_V=0°$ corresponds to the direct beam hitting the center of a detector. A liquid H_2 target can be mounted at the intersection point of the horizontal and vertical axes.

2.3. Electronics

The electronic logic is designed to accept events of electromagnetic nature with high efficiency. A shower is required to release a minimum energy in the Sandwich by discriminating on the linear sum of the pulses of all counters. In the case of γ-ray detection, only this condition is required in the electronic logic.

In the case of electron detection, in which the efficiency for pions must be kept as low as possible, several fast electronic logics were studied in order to have the best electron-pion discrimination. Only the logic most suitable to our purpose is described here, the variants being discussed in section 3.1.2. The principle of this logic is to require, in addition to a minimum energy release, some multiplication of the primary electron to take place in the first slice of lead (1.72 radiation lengths). This is obtained by discriminating the pulse of counter 1.

The diagram of the electronic logic used for electron detection is shown on fig. 6. The pulses from the four phototubes of each counter are equalized and passively mixed. The counter 1 pulse is split into three channels: one is sent to an encoding system, the second enters a discriminator, and the third is mixed in a fast linear mixer with the pulses from the other eight counters. The output pulse (Σ) of this mixer is split into two channels, of which one is sent to the encoding system; the other, after pulse-height discrimination, is put in coincidence (S) with the discriminated pulse of counter 1 and with the beam telescope signal (BS). A further coincidence (E) between (S) and the signal from the M

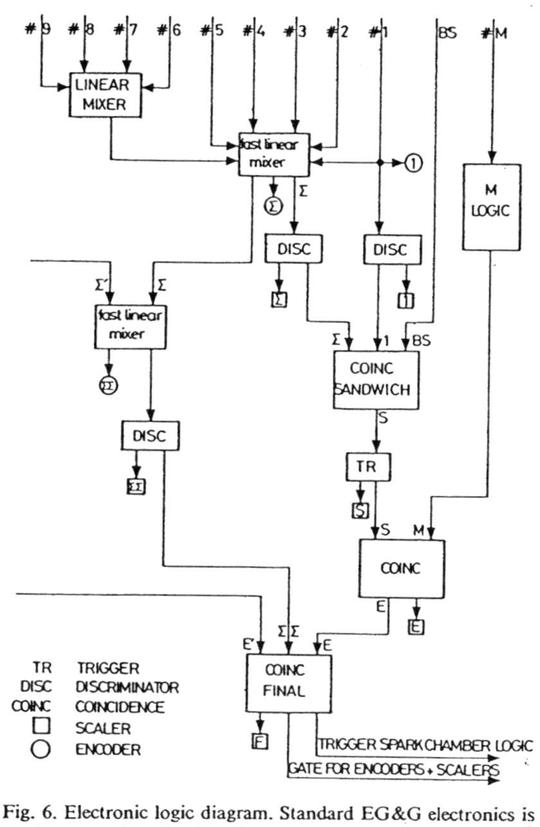

Fig. 6. Electronic logic diagram. Standard EG&G electronics is used, except for the fast linear mixers which is a modified CERN unit. In this diagram, "trigger" refers to shapers with a fixed trigger level, and "discriminators" to shapers with a critical threshold level which has to be optimized in a calibration.

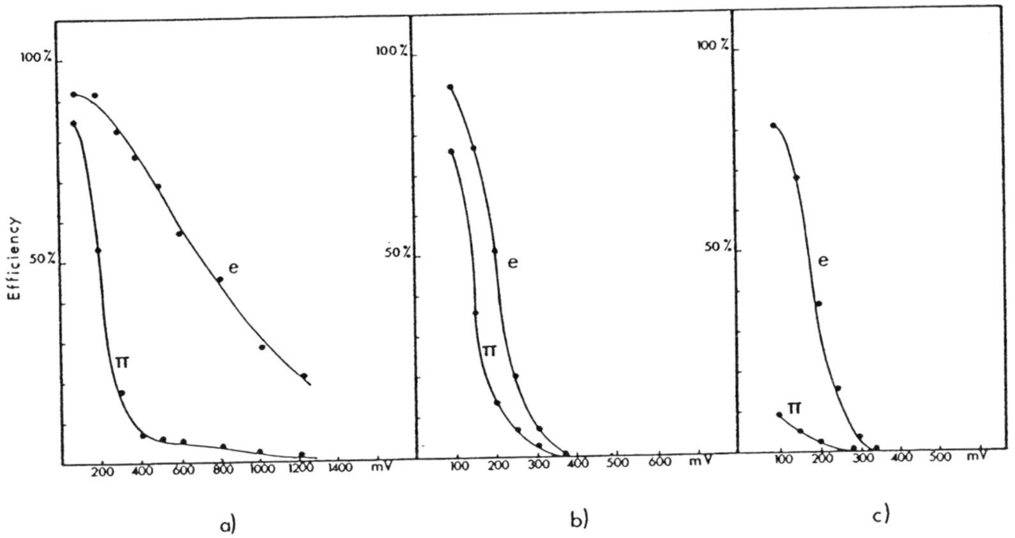

Fig. 7. Discrimination curves at 400 MeV for electrons and pions varying: (a) (1) threshold, the efficiency is given by [(1)·BS]/BS; (b) (Σ) threshold, the efficiency is given by [(Σ)·BS]/BS; (c) (Σ) threshold with (1) at 300 mV in coincidence, the efficiency is given by [(Σ)·(1)·BS]/BS.

counter logic guarantees a charged particle in the thin-foil chambers. In the case of electron detection, counter M logic also allows the rejection of events when two of the five counters are struck simultaneously, and the discrimination between single and double charged particles in each counter. The efficiency obtained is 92% for single particles, 30% for double ones (i.e. 70% rejection), and the rejection for three or more particles is excellent.

The signal (E) triggers the logic of the spark chambers and of the film and numerator advance. When two detectors are used in experiments, the signal (E′) together with E go into the coincidence (FINAL). An additional electronic condition may be introduced in the final trigger: an output of the (Σ) mixer can be sent in a fast linear mixer together with the corresponding pulse (Σ') of the second detector, and the resulting signal ($\Sigma\Sigma$) may be required to be greater than the mean pulse height expected for a certain electromagnetic energy release. This condition can yield a substantial decrease in the trigger rate of an experiment.

3. Calibrations as an electron detector

The calibration of the shower detector as an electron detector consists in measuring its response when electrons or pions* of known momentum are sent into it, in order to know its detection efficiency for electrons and to find the best conditions for a good electron–pion discrimination. Furthermore, the response of the detector has to be measured as a function of the electron energy.

3.1. Electron–pion discrimination

This calibration was done in two steps:

1) First, the discrimination levels of (1) and (Σ) discriminators were set to have the best compromise, at all energies and in all regions of the detector, between high pion rejection and high electron efficiency. This step, called "electronic calibration", was done using the standard electronic circuit described in section 2.3. The attempts to improve the electron-pion discrimination will be discussed in section 3.1.2.
2) The second step consisted in taking spark-chamber pictures of electrons and pions in the standard electronic conditions, in order to determine criteria for a further pion rejection. This step was called "picture calibration". Owing to possible drifts in the pulse heights and changes in efficiency of the spark-chambers, these calibrations had to be repeated periodically.

* Electrons or pions are selected by means of an ethylene Cherenkov counter. Its output pulse is in coincidence with the beam telescope signal BS to count electrons, and in anticoincidence for pions. The beam contaminations obtained are of the order of less than 10^{-6} π in the electron beams, and less than 10^{-7} e in the π beam.

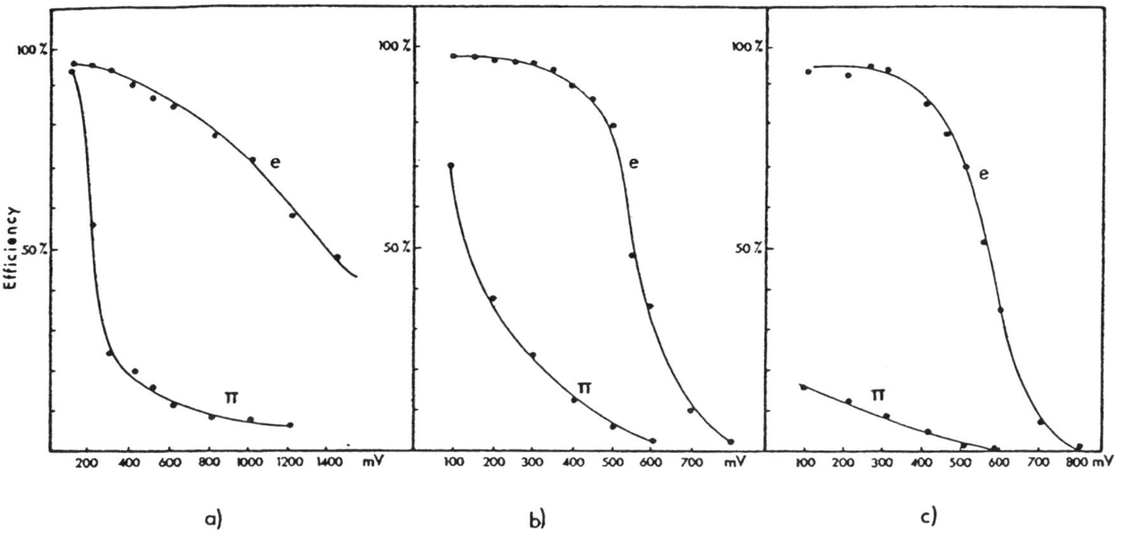

Fig. 8. Discrimination curves at 1100 MeV for electrons and pions: (a), (b) and (c) are as in fig. 7.

3.1.1. *Electronic calibration*

The optimum electronic conditions were determined at a central position of the detector and at the lowest momentum considered, 400 MeV, because at this momentum the efficiency for electron detection and the electron–pion discrimination are the lowest. Then the π rejection and e efficiency were measured at several momenta up to 1.1 GeV. Furthermore, the whole surface of the detector was scanned in order to know its response at any position, especially near the edges.

1) *Optimization at 400 MeV*. The efficiency of counter 1 was measured as a function of the discrimination level for pions and electrons. Fig. 7a shows that at 400 MeV a threshold of 300 mV is a good compromise: 18% efficiency for pions and 83% for electrons.

Similarly, a discrimination curve of pulse (Σ) was done for pions and electrons (fig. 7b). Notice that the electron–pion discrimination is poor at 400 MeV if only the total pulse height is used.

Finally, the pulse (Σ) discrimination curve was repeated with the counter 1 pulse, discriminated at 300 mV, in coincidence (fig. 7c).

These three sets of curves show that at 400 MeV a good compromise between low pion efficiency (i.e. good rejection) and high electron efficiency may be obtained for the following settings: (1) at 300 mV and (Σ) at 100 mV, which yield a 9% pion efficiency for a 82% electron efficiency.

2) *Measurements at higher energies*. Similar discrimination curves were done for electrons and pions at several energies. Fig. 8 shows the results at 1.1 GeV. With the discriminator settings optimized at 400 MeV, the efficiencies at 1.1 GeV are 93% for electrons and 16% for pions. Notice that better settings at 1.1 GeV would be (1) at 300 mV, (Σ) at 350 mV, yielding an efficiency of 90% for electrons and 6% for pions.

The discrimination level chosen for pulse (1) is 1.7 times the mean pulse height of minimum ionizing particles. That for pulse (Σ) corresponds to the mean pulse height for a 150 MeV electron. The energy dependence of electron and pion efficiencies, measured with these discriminator settings, is shown in fig. 20a.

3) *Scanning of the surface*. When the detector is used in an experiment, its structure (see fig. 2) and its distance from the target cause the effective solid angle to be significantly smaller than the solid angle for detection by the first counter only. In order to evaluate the efficiency for a particular experiment, a detailed study of the pulse-height distribution and of the efficiency for electrons and pions near the edges of the detector is needed. The variation of the electronic efficiency for electrons when the detector is rotated around its horizontal or vertical axis is shown in figs. 9a and 9b, respectively. In fig. 9c, the peak position of the (Σ) pulse-height distribution in the encoder is shown versus the momentum for different values of θ_H and θ_V. Notice that the peak position does not vary appreciably with angles, as long as the depth (d) of the Sandwich usable by the showers remains at its maximum; that is, as long

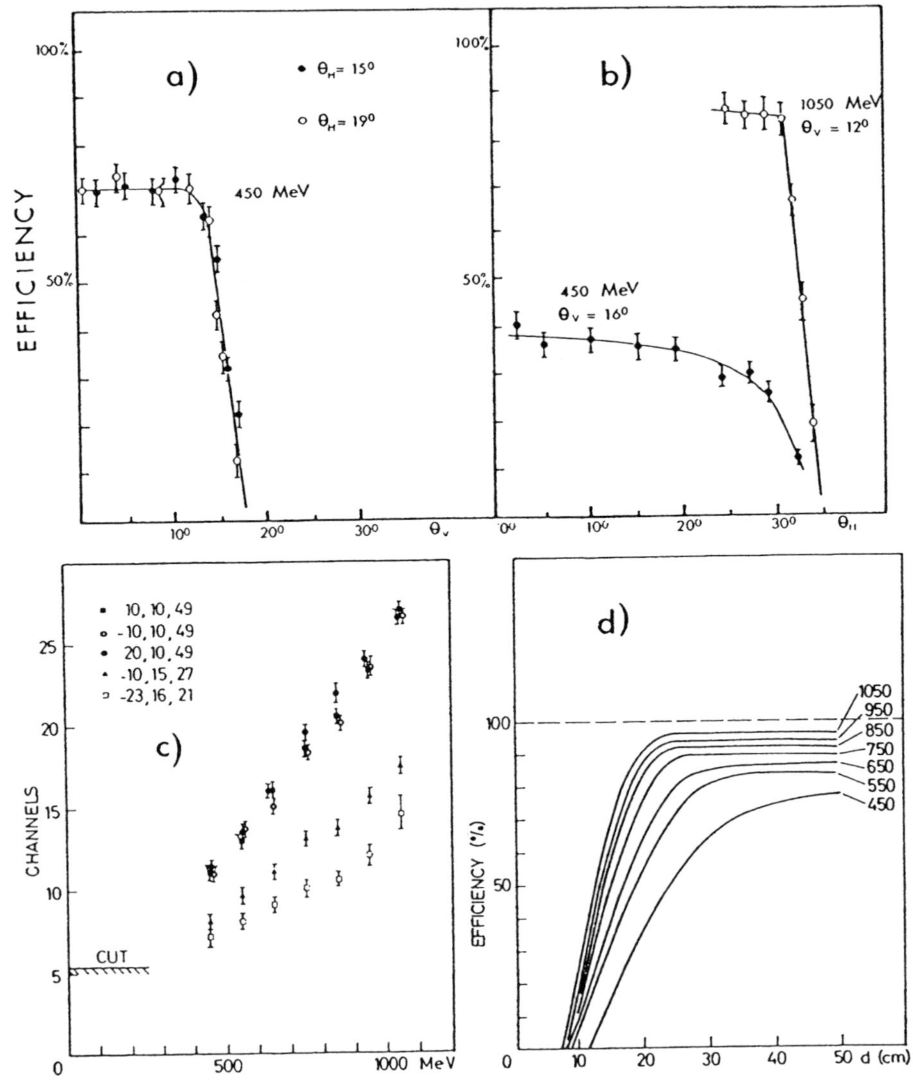

Fig. 9. (a) Electronic efficiency for electrons as a function of θ_V. (b) Electronic efficiency for electrons as a function of θ_H. (c) Peak of (Σ) pulse-height distribution versus momentum, for electrons sent into the Sandwich in various positions. The three numbers indicated for each series of points define θ_H, θ_V, in degrees; and the corresponding depth, in cm, usable for a shower. (d) Electronic efficiency for electrons at different energies as a function of the depth d.

as the shower does not go out of the sides of the detector. The depth was used to parametrize the electronic efficiency on the edges; the efficiency curves (fig. 9d) for momenta ranging from 450 to 1050 MeV/c are based on 170 calibration points.

3.1.2. *Alternative electronic triggers*

Four variants (fig. 10) of the circuit described in section 2.3 have been studied in attempts to improve the electron–pion discrimination. These studies did not bring about any significant improvement, but they were otherwise very instructive.

1) In addition to the pulse-height discrimination on pulses (1) and (Σ), more than one charged particle was required in each counter from 2 to 5. A threshold was set on the pulses and their coincidence signal (C), in coincidence with pulse (S)$=(\Sigma)\cdot(1)\cdot(M)\cdot(BS)$, gave a new trigger signal (E) (fig. 10a).

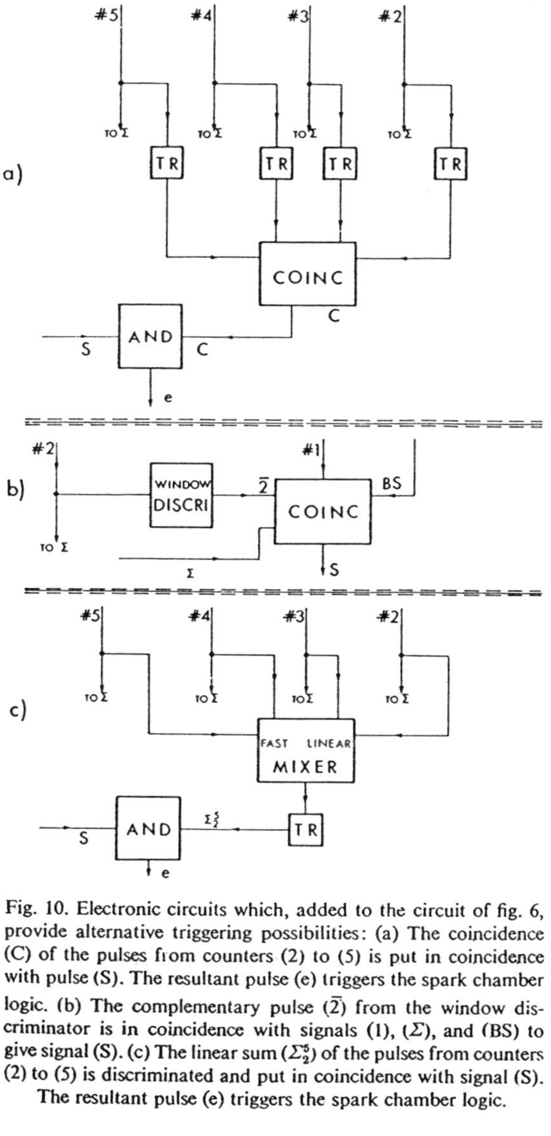

Fig. 10. Electronic circuits which, added to the circuit of fig. 6, provide alternative triggering possibilities: (a) The coincidence (C) of the pulses from counters (2) to (5) is put in coincidence with pulse (S). The resultant pulse (e) triggers the spark chamber logic. (b) The complementary pulse ($\overline{2}$) from the window discriminator is in coincidence with signals (1), (Σ), and (BS) to give signal (S). (c) The linear sum (Σ_2^5) of the pulses from counters (2) to (5) is discriminated and put in coincidence with signal (S). The resultant pulse (e) triggers the spark chamber logic.

Fig. 11 shows pion and electron efficiencies at 400 MeV as a function of the number of pulses added in the coincidence (C). The data in fig. 11a are for thresholds (1) and (Σ) arbitrarily set at 100 mV; fig. 11b corresponds to threshold (1) at 300 mV and threshold (Σ) at 100 mV. There is a sharp drop in electron efficiency when a pulse is required in more than three counters. Moreover, there is a decrease in electron efficiency as soon as one counter is put in coincidence (C). This is due to the possibility that an electromagnetic shower has only neutral components during the traversal of one counter. Therefore, this condition, although it does bring about a slight improvement in electron-pion discrimination at 400 MeV, causes a loss in electron efficiency and was therefore disregarded.

2) Particles producing, in counter 2, pulse heights in the region of the pion peak were vetoed in order to eliminate single track feedthrough due to specific ionization fluctuations and Landau tail. A window discriminator was used as shown in fig. 10b, and its output pulse ($\overline{2}$) was sent in coincidence with pulses (1), (Σ) and (BS) in the (S) coincidence unit.

Fig. 12a shows counter 2 pulse height spectrum obtained at 400 MeV by varying the threshold of the window discriminator, with a fixed window of 50 mV. In this measurement, the output of the discriminator was in coincidence with the monitor (BS). The spectrum shows that the window must be centred at 200 mV and be 150 mV wide in order to anticoincide most of the pions.

Fig. 12b shows pion and electron efficiencies at 400 MeV as a function of the width ΔE of the window centred at 200 mV. The coincidence with (1) and (Σ) was required. For $\Delta E = 150$ mV, the pion efficiency is

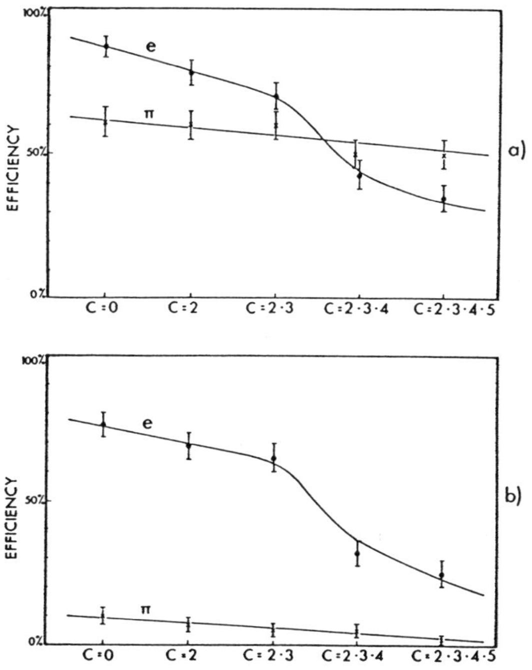

Fig. 11. Electronic efficiency as a function of the number of pulses from counters (2) to (5) in the coincidence (C): (a) threshold on (1) and (Σ) set at 100 mV; (b) threshold on (1) at 300 mV and (Σ) at 100 mV. The efficiency is given by $[(1)\cdot(\Sigma)\cdot(BS)\cdot(C)]/(BS)$.

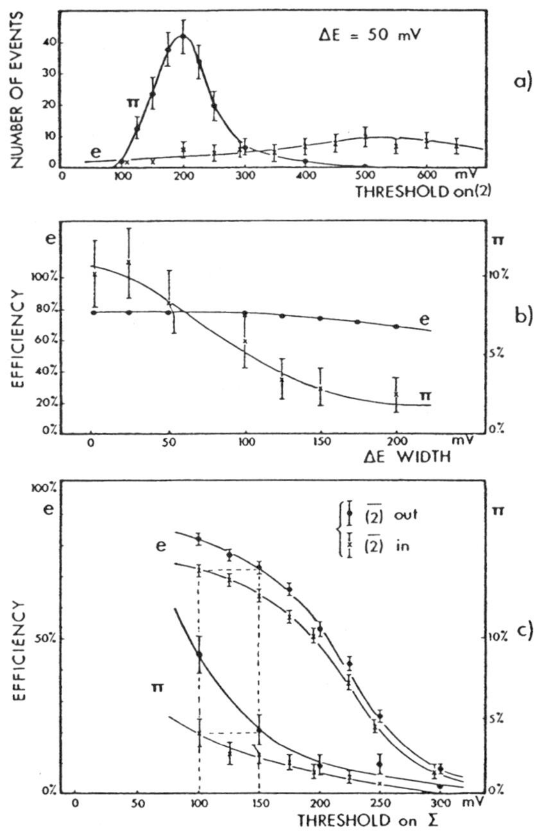

Fig. 12. Effect of the window discriminator on the pulse from counter (2): (a) Counter (2) pulse-height spectrum at 400 MeV/c incident beam momentum, obtained by varying the window discriminator threshold, with a fixed window of 50 mV. The discriminator output is in coincidence with (BS) for this measurement. (b) Efficiency versus width ΔE, with the window centred on 200 mV. The $(1)\cdot(\Sigma)$ coincidence is vetoed by the $(\bar{2})$ signal. (c) Efficiency versus threshold on (Σ). Signal (1) is in, and condition $(\bar{2})$ is in or out.

reduced by a factor of 3 and the electron efficiency drops by 5%. At 1.1 GeV, the effect of $(\bar{2})$ is a 40% reduction in pion efficiency and a 1.5% loss in electron efficiency. Fig. 12c shows the pion and electron efficiencies at 400 MeV observed by simply increasing the threshold on the (Σ) pulse.

3) A discrimination on the linear sum (Σ_2^5) was applied in addition to the discrimination on pulses (1) and (Σ), as shown in fig. 10c. An improvement in pion rejection at low momentum was expected, as most electron showers of 400 MeV stop in chambers 4 or 5. However, the efficiency measured as a function of threshold on pulse (Σ_2^5), in coincidence with (1) at 300 mV and (Σ) at 100 mV, shows a slow simultaneous decrease for e and π. Thus even at 400 MeV this condition does not improve the e–π discrimination.

4) Tests have been performed to find out whether it is advantageous to discriminate on total energy release by forming the linear pulse (Σ_1^9), or to keep this discrimination independent from the threshold on pulse (1) by using the linear pulse (Σ_2^9).

Discrimination curves for pulses (Σ_1^9) and (Σ_2^9) were done at 400 MeV, with the threshold on pulse (1) set at 400 mV. Fig. 13a shows that the reduction in pion efficiency of 3% at constant 80% electron efficiency, obtained by using (Σ_2^9) instead of (Σ_1^9), can also be obtained by setting (Σ_1^9) at 150 mV instead of (Σ_2^9) at 100 mV. The use of (Σ_1^9) is preferable because the sensitive area of the detector is increased by including

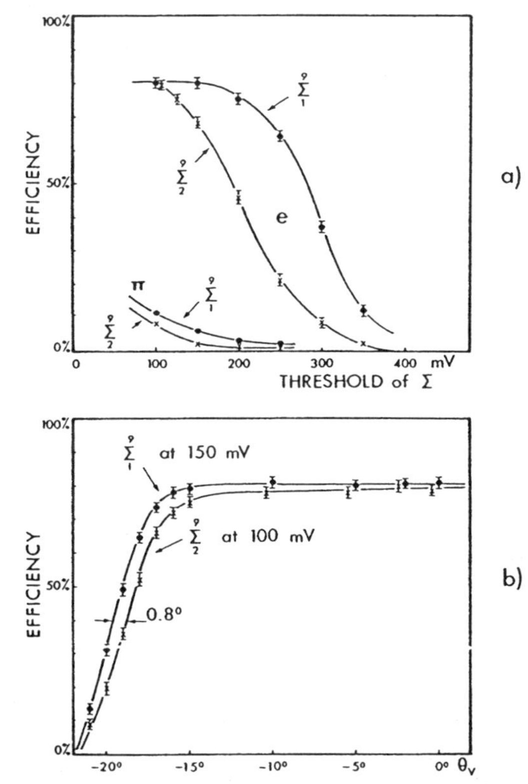

Fig. 13. Effect of counter (1) pulse in the linear sum (Σ); the efficiency is given by $[(1)\cdot(\Sigma)\cdot(\mathrm{BS})]/\mathrm{BS}$: (a) Efficiency at 400MeV versus threshold on (Σ_1^9) or (Σ_2^9). The threshold on pulse (1) is at 400 mV for this measurement. (b) Efficiency near the edge as a function of θ_V with $\theta_H = 10°$, for the two equivalent conditions, 150 mV on (Σ_1^9) or 100 mV on (Σ_2^9).

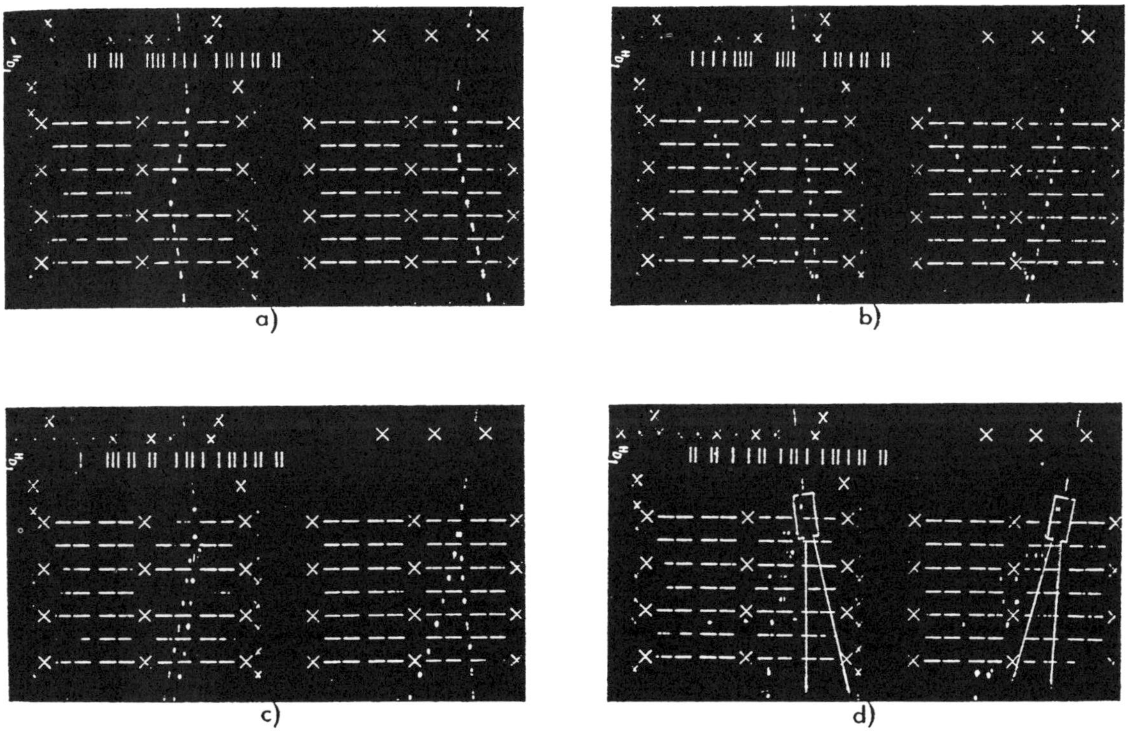

Fig. 14. Typical pion with our trigger conditions: (a) scattered track; (b) back-scattered pion; (c) π^-p scattering; (d) very asymmetric π^0 shower from π^- charge exchange. The counting-surface for spark counting is shown here.

counter (1) pulses in the linear sum; fig. 13b shows that, at 400 MeV, the drop in efficiency for electrons on the edge of the detector, shifts by an angle $\Delta\theta_V = 0.8°$.

3.1.3. *Picture calibration*

Using the triggering conditions defined in section 3.1.1, pictures of the spark chambers were taken at several momenta from 400 to 1100 MeV/c with highly purified ($<10^{-6}$ contamination from the other particle) pion and electron beams[11]) incident on various regions of the detector. Pictures of an electron and of four typical pion pattern are shown in fig. 4 and 14, respectively. A further rejection can be made on the pictures in two steps:

1) Many pions can be recognized and eliminated, such as straight or scattered pion tracks; nuclear-interacting pions; and very asymmetric π^0 showers from π^- charged exchange.

2) The remainder are essentially π^0 showers created by incident pions which are symmetric or nearly symmetric with respect to the incoming track in both direct and reflected views. To have a further rejection on this type of events, we use a method based on the fact that an electromagnetic shower, in our energy range yields a high density of sparks in a narrow forward cone, symmetric about the electron direction.

The method consists in counting, for electrons and pions, the number of sparks inside a surface defined by a cone having $\pm 6°$ opening angle in direct and reflected views; the bisector of the angle is aligned with the incoming track and its vertex is centred in the first slice of lead. To allow for the high multiplicity of the shower in the first chambers, resulting from the threshold on the counter (1) pulse, the "counting-surface" was defined by the opening angle $\pm 6°$ plus a rectangle covering the first two chambers, as shown in fig. 14d. The opening of the angle and the surface of the rectangle have been determined so as to obtain the best electron-pion discrimination over our energy range.

Fig. 15 shows scatter diagrams in which, for each calibration event, the number of sparks inside the counting-surface in the direct view is plotted versus the number of sparks inside the counting-surface in the

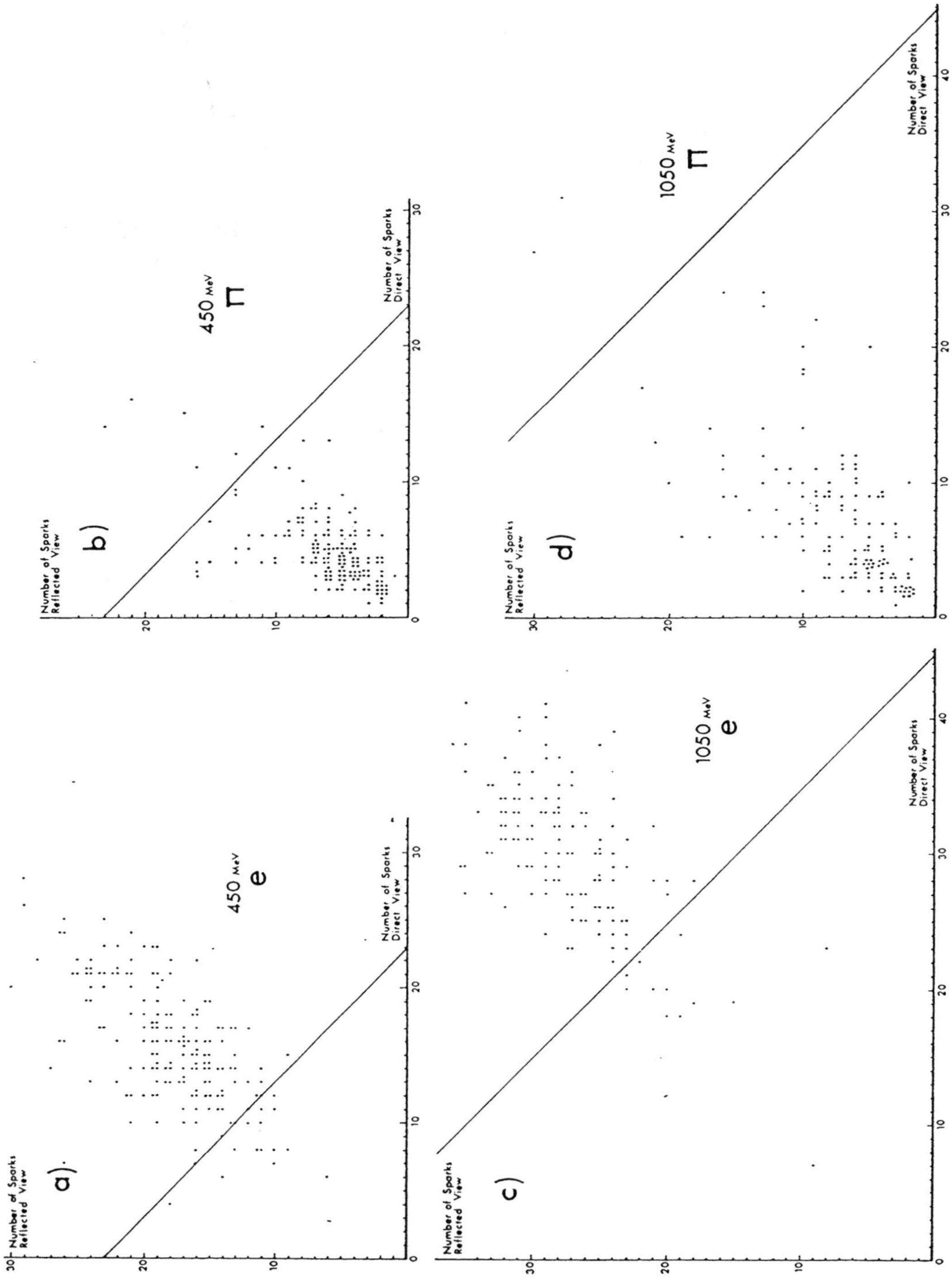

Fig. 15. Scatter diagrams of the number of sparks inside the counting-surface in the direct view versus the number of sparks in the reflected view: (a) electrons at 450 MeV; (b) pions at 450 MeV; (c) electrons at 1050 MeV; (d) pions at 1050 MeV. The straight lines correspond to $(90 \pm 2)\%$ picture efficiency for electrons.

reflected view. Figs. 15a and 15c represent the spark count distribution for electrons at 450 MeV and 1050 MeV in a central region of the detector. On the scatter diagram one can define a straight line corresponding to a threshold on the sum of the numbers of sparks inside the counting-surface in direct and reflected views, such that 90% of the electrons which have fulfilled the electronic triggering requirements produce enough sparks to be above the threshold. Figs. 15b and 15d represent the spark count distribution for pions of 450 MeV and 1050 MeV in the same region of the detector. If the total number of sparks produced inside the counting-surface by a pion is higher than the threshold, it simulates an electron; the number of pions above the threshold relative to the total number of pions which triggered the electronics gives the pion contamination in the picture analysis. This contamination increases from 0.6% at 1000 MeV to 1.7% at 450 MeV corresponding to a picture efficiency for electrons fixed at $(90 \pm 2)\%$, as shown in fig. 20b.

Pictures were taken with electrons and pions in many positions of the shower detector, in order to check homogeneity of the performances and thus determine the useful volume of Sandwich in which electron efficiency and pion rejection are acceptable.

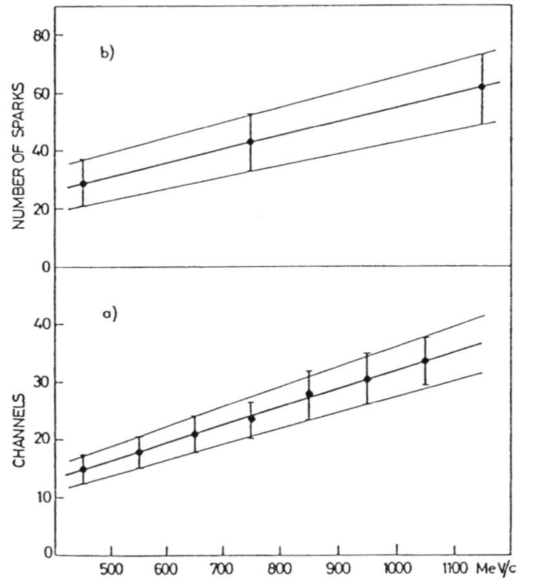

Fig. 16. (a) Mean and width of (Σ) pulse-height distribution versus the momentum of the electron beam. (b) Mean and width of the distribution of the total number of sparks inside the counting-surface versus the momentum of the electron beam.

3.2. Energy calibration

The energy released by a particle in the shower detector can be measured either by the linear sum of the pulse heights from the nine counters or by the total number of sparks produced in the spark chambers.

The sum (Σ) of all pulse heights is encoded and recorded for each event. If the position of the mean and the width of the pulse-height distribution is plotted as a function of the momentum of the electron beam, we observe a good linearity and an energy resolution of $\pm 14\%$ at 1000 MeV and $\pm 17\%$ at 500 MeV (fig. 16a).

On the other hand, the total number of sparks (i.e. direct plus reflected view) was counted, for each electron, inside the counting-surface used for pion rejection. The mean and width of the spark count distribution plotted against the momentum of the electron beam shows a good linearity up to 1200 MeV (fig. 16b). The energy resolution observed is about $\pm 37\%$ at 400 MeV and $\pm 25\%$ around 1000 MeV.

4. Calibrations as a γ-ray detector

To know the efficiency and the energy dependence of the response to γ-rays, an electron beam was generally used; in addition a γ-ray beam was set up in order to check these measurements and study the patterns of tracks produced in the kinematic chambers as a function of the thickness of the lead converter. To do this, an electronically selected electron beam was directed onto a 3 mm thick piece of lead, and the high-energy bremsstrahlung was then selected by vetoing all charged particles produced at an angle smaller than about 10°.

4.1. γ-ray detection efficiency

A maximum electronic efficiency for detection of γ-rays was desired. Therefore no condition was fixed for the counter (1) pulse, and the threshold on the linear sum (Σ) of the nine counters was set at 100 mV, corresponding to about 150 MeV energy release in the Sandwich. Fig. 17a shows the electronic efficiency as a function of momentum, measured with electrons but with the set-up and the electronic conditions used for γ-ray detection. Photographs were taken under these conditions at various momenta, and, from the spark-count distribution, the detection efficiency in the spark chambers was determined for various cuts on the total number of sparks. This efficiency, shown as a function of momentum in fig. 17b, has to be known because

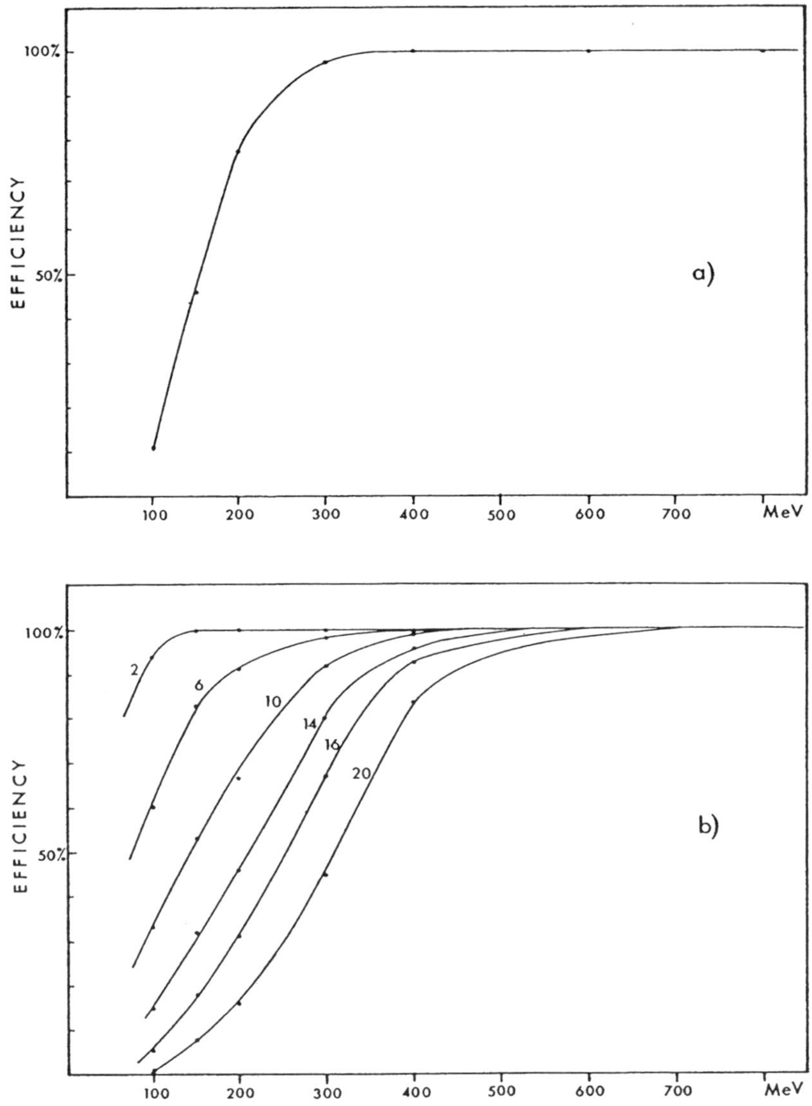

Fig. 17. (a) Electronic efficiency for γ-ray detection as a function of energy. (b) Efficiency of detection in the spark chambers as a function of energy, for various cuts which may be applied on the total number of sparks.

low-energy γ-rays showers (<400 MeV) need a minimum number of sparks to be characterized.

4.2. Measurable patterns

In principle, the direction of all γ-rays could be reconstructed using only the shower developed in the spark chambers of the Sandwich. When one single shower is produced, its direction is very near to that of the primary γ-ray, because no other branch has taken a significant amount of energy. However, the average precision of the measurement of the mean direction of a shower in our detector would be of the order of $\pm 5°$ and worse for low-energy showers (<600 MeV). Therefore, we first convert the γ-rays in a lead plate and guarantee the conversion by the scintillation counter M in coincidence in the electronic trigger. Thus the thin-plate chambers are used to show the direction of the shower at an early stage in its development. A negligible background of charged particles being expected in the case of γ-ray detection, a single track as well as multiple tracks in the kinematic chambers followed by a shower in the Sandwich can be considered as a converted γ-ray.

The conversion by lead plates of 0.3 cm, 0.5 cm, or 0.7 cm thickness was studied with the γ-ray beam. A plate of 0.52 cm (0.90 radiation length) was finally chosen as the best compromise between a good conversion efficiency ($\approx$60%) and a moderate multiplicity of secondary electrons at this early stage of the shower development. In fact our conditions correspond to a maximum probability for the production of pairs, which may be well measured in the kinematic chambers.

The γ-ray's direction is measurable with a precision of about $\pm 1°$ when, in both stereoviews, one or several tracks in the kinematic chambers face the shower produced in the Sandwich, i.e. when the direction of the most energetic branch of the γ-ray is clearly indicated. Measurable patterns are shown in fig. 18. Fig. 18a shows a narrow pair facing a shower in the Sandwich. This case occurs in about 13% of the converted events. Fig. 18b shows a track facing a shower with the possibility of any number of additional tracks in the kinematic chambers; in fact, we observed a maximum number of five tracks coming from the same point in the lead. This γ-ray pattern was present in about 37% of the converted γ-rays. Fig. 18c shows two tracks in the kinematic chambers, each facing a shower. This case occurs in about 18% of the converted γ-rays. When the difference in the development of the showers is large, the direction is clearly given by the most energetic branch and can be measured.

The measurable events of the three patterns a, b and c represent about 56% of the converted γ-rays. All the numbers quoted are based on the γ-ray beam calibrations at various energies and are compatible with the classification of patterns observed in experiments. The calibrations have shown little dependence of the measurability on primary γ-ray energy, between 400 and 1100 MeV.

Patterns other than a, b and c could be reconstructed, but with more precaution and a certain loss in accuracy. For instance, in the case of pattern d (fig. 18), observed

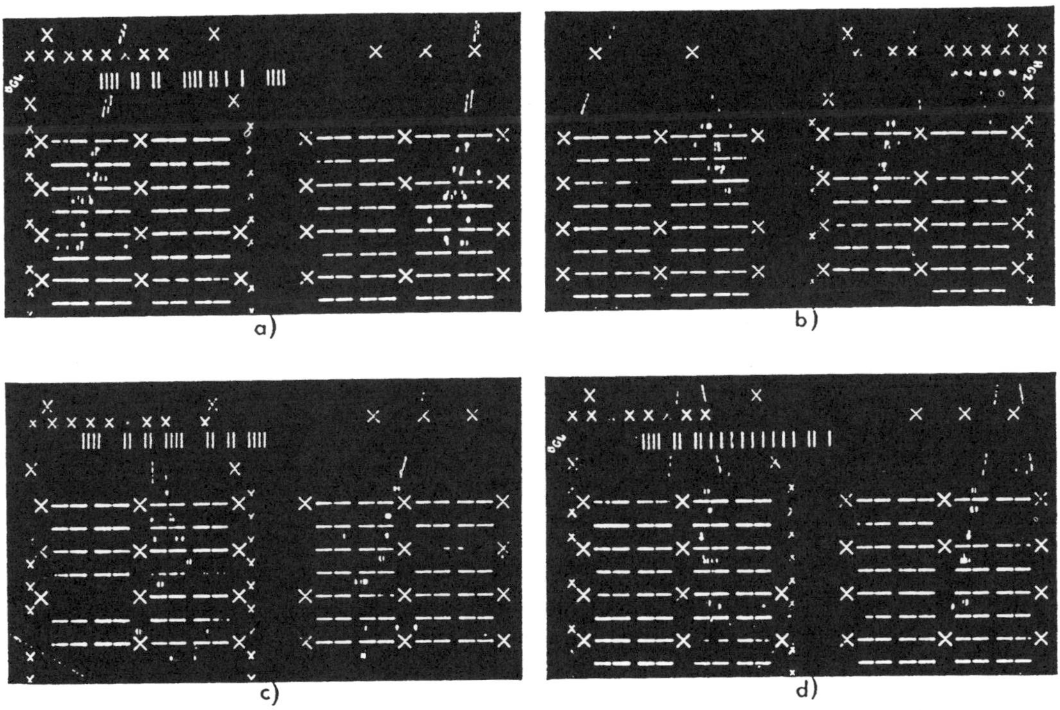

Fig. 18. Various patterns produced by γ-rays in the kinematic chambers: (a) Narrow pair facing a shower in the Sandwich. (b) Track facing a shower, with any number of additional tracks in the kinematic chambers. (c) Two tracks, each facing a shower. If the difference in development of the showers is large, the direction of the most energetic branch can be measured. (d) Showers in the Sandwich with one or more tracks in the kinematic chambers, not facing the shower.

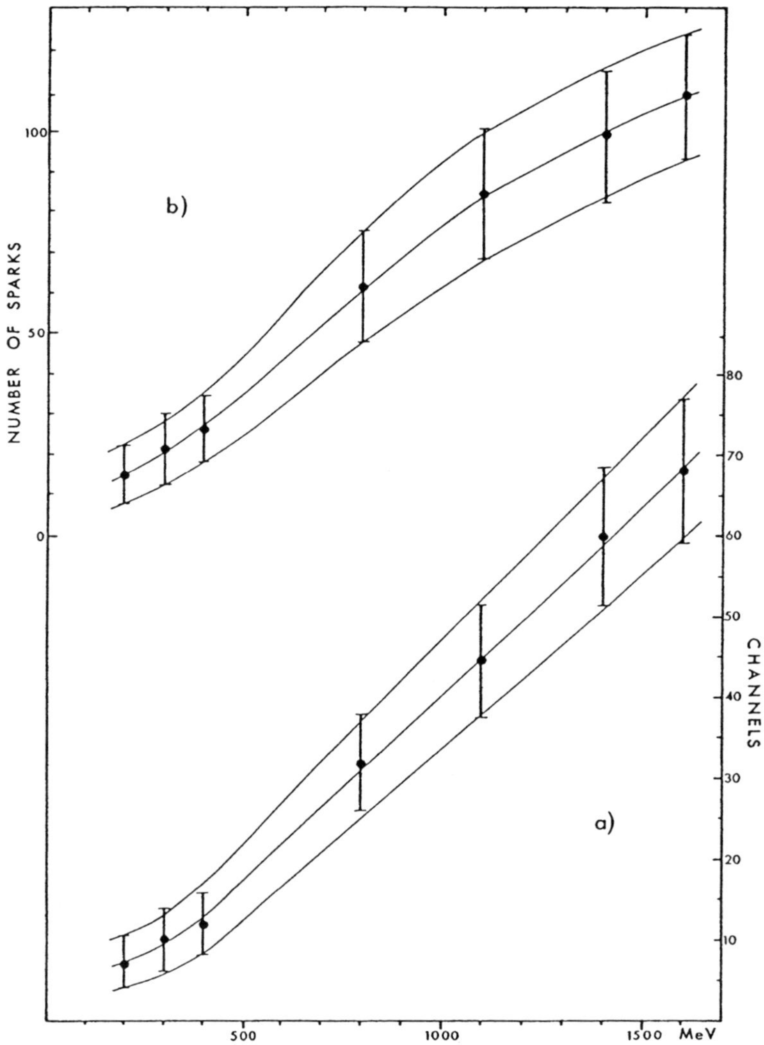

Fig. 19. (a) Mean and width of the pulse-height distribution versus the energy of the incoming γ-rays. (b) Mean and width of the total (direct plus reflected view) spark count distribution versus the energy of the incoming γ-rays. Here all the sparks of the shower are counted without surface-counting limits.

in ≈30% of the converted γ-rays, the conversion point in the lead can be reconstructed from the tracks. This point, together with the direction or the starting point of the shower in the Sandwich, can provide the desired accuracy, but only if the shower is sufficiently developed. On the other hand, in experiments where several particles are reconstructed, the conversion point in the lead, together with the vertex of the reaction studied, will give a good accuracy.

4.3. Energy calibration

The presence of the additional lead converter and the different electronic conditions causes the development of the γ-ray showers detected in the Sandwich to be slightly different in amplitude and geometry from the case of primary electrons. It was necessary to re-measure the response of the Sandwich in pulse height and number of sparks. Figs. 19a and 19b show the mean and the width of the pulse-height and spark-

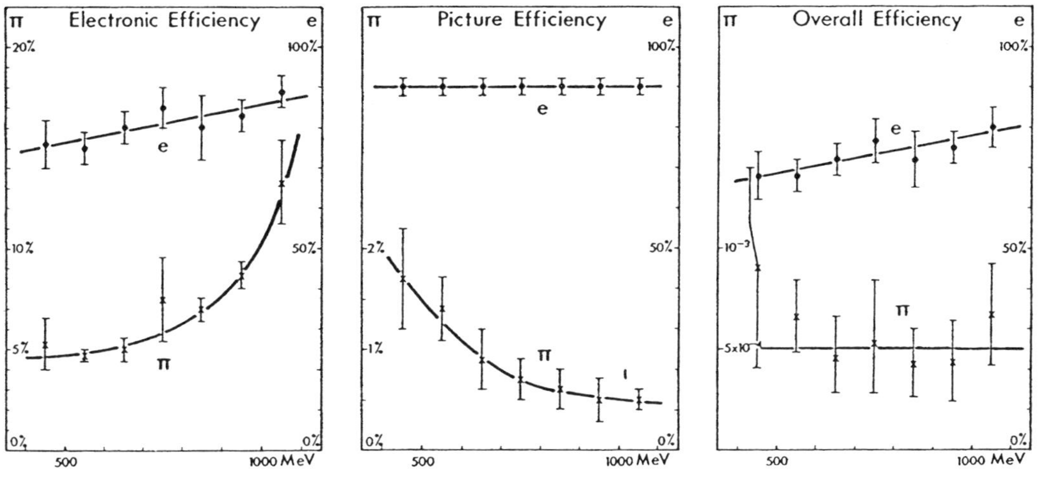

a) b) c)

Fig. 20. Efficiency for electrons and pions versus momentum obtained: (a) electronically; (b) from picture inspection, with respect to the number of triggered photographs; (c) by multiplying electronic and picture efficiencies.

count distributions, respectively, measured with the γ-ray beam at several energies. Note that the spark-count results cannot be compared in absolute value to the electron detector calibration (fig. 16b) because, for γ-rays, all sparks of the shower were counted without angular limits.

The energy resolutions observed from pulse-height measurements (±19% at 500 MeV and ±15% at 1000 MeV) are very near to the data found for the electron detector, whereas the resolutions from spark counting (±26% at 500 MeV and ±22% at 1000 MeV) are slightly better.

The calibration of the γ-ray detector with an electron beam, often repeated between and during experiments, gave results compatible with the above data.

5. Performances of the detector

5.1. Electron efficiency and pion rejection

The results of the electronic calibration (section 3.1.1) and of the picture calibration (section 3.1.3) versus momentum are shown in figs. 20a and 20b, respectively. The over-all efficiency for electrons and pions, shown in fig. 20c, is obtained by multiplying the electronic efficiency by the picture efficiency. We can see that the over-all efficiency for pions (5×10^{-4}) does not vary significantly with momentum, whereas electron efficiency increases by 12% between 450 MeV and 1050 MeV. The smooth behaviour of the instrument is proved by the fact that, as shown by a long series of calibrations, the over-all pion rejection power is not critically dependent on the choice of the discriminator settings.

The data shown in fig. 20 are detection efficiencies for pions and electrons, averaged over several regions of the detector and over different periods of time; the errors quoted contain the fluctuations with these two factors. When two such detectors were used in experiments[5,6]) where the average electron energy is 800 MeV, the following performance for the system of two detectors were obtained: 54% efficiency for electron pairs and 2.5×10^{-7} for pion pairs.

Table 1 shows the electron efficiencies and the pion rejections* at 400 and 1100 MeV, obtained in a central position of the detector in a period in which particular attention was given to spark chamber efficiency. This shows that, with extra precaution, performances for the system of two detectors down to 10^{-7} pion efficiency for 57% electron efficiency can be obtained in the energy range 400–1100 MeV.

5.2. Track reconstruction

The accuracy of reconstruction for electrons is only limited by the quality of the measurement (±0.3°) and by the multiple scattering in counters which may be placed on the electron trajectory. In experiments where

* We sometimes use rejection power instead of detection efficiency for π to emphasize that low detection efficiency means high rejection power.

TABLE 1
Electron efficiencies and pion rejections at 400 and 1100 MeV.

Momentum in MeV	Particle	Electronic efficiency	Picture efficiency	Over-all efficiency
400	e	$(77.5 \pm 2.2)\%$	$(89.0 \pm 2.2)\%$	$(69.0 \pm 2.6)\%$
	π	$(6.3 \pm 0.2)\%$	$(0.43 \pm 0.2)\%$	$(2.7 \pm 1.6) \times 10^{-4}$
1100	e	$(94.0 \pm 1.5)\%$	$(88.0 \pm 2.0)\%$	$(83.0 \pm 2.3)\%$
	π	$(17.6 \pm 0.6)\%$	$(0.16 \pm 0.16)\%$	$(2.8 \pm 2.8) \times 10^{-4}$

counter M (2.5 cm thick) was placed in front of the detector[5,6]), the error in reconstructing the direction of the electron was estimated to be $\pm 0.8°$. For γ-rays, with the method and the criteria of measurability defined in section 4.2 and used in experiments[7,8]), an accuracy of $\pm 1°$ was obtained. Fig. 21 shows a typical distribution of the minimum distances of two γ-rays directions and the corresponding distribution of the

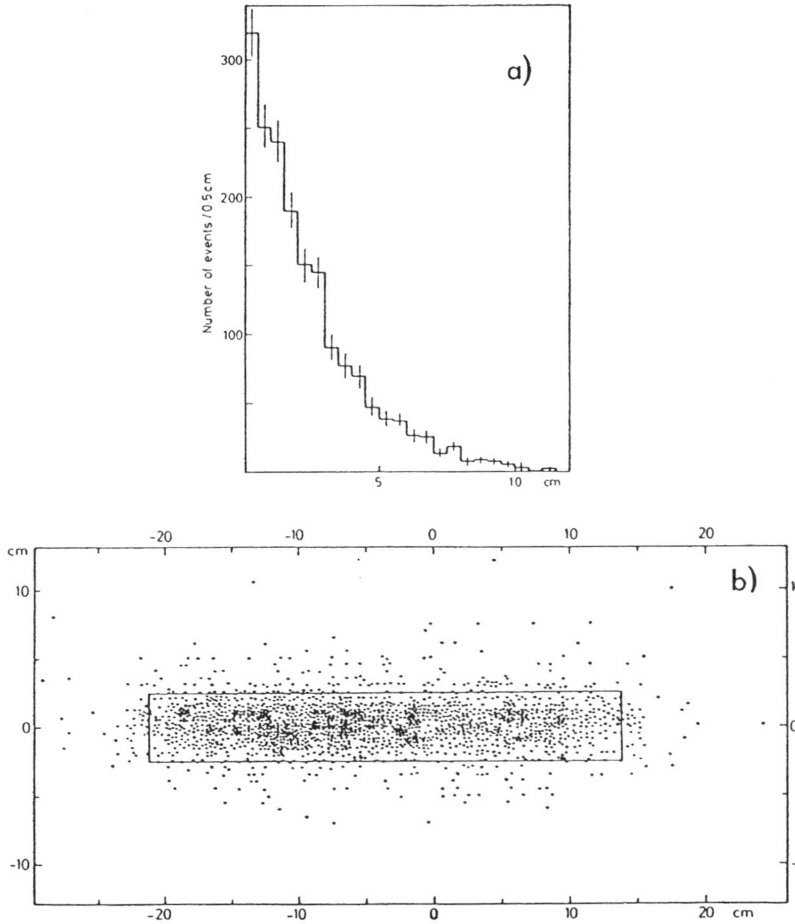

Fig. 21. (a) Distribution of the minimum distances of two reconstructed γ-rays for events of the type $\pi^- p \to n\gamma\gamma$. (b) The coordinates of the reconstructed vertices in the horizontal plane are shown together with the liquid hydrogen target (35 cm long, 5 cm diameter).

reconstructed vertices for events of the type $\pi^- p \to n\gamma\gamma$. The inner rectangle indicates the liquid hydrogen target.

5.3. Energy resolution

The measurement of the pulse-height and the counting of the sparks are two ways of sampling the total number of electrons in a shower produced in the Sandwich by an electron or a γ-ray. The uncertainties of these methods are rather large compared to the expected fluctuations in the number of electrons produced by a primary particle of given energy; thus the two methods may be considered as independent measurements of the primary energy. This statement was verified by calibrations at various momenta with electrons and γ-rays; the measured pulse-height was plotted against the spark count for each event and no correlation, at a given primary energy, was observed between the two measurements. The weighted mean of the energies determined by the two methods can then be used. Table 2 shows the energy resolutions obtained for electrons and γ-rays, by using either the pulse-height information, or the spark counting information or the weighted mean.

Table 2
Energy resolutions for electrons and pions by various methods.

		Pulse-height resolution (%)	Spark count resolution (%)	Weighted mean resolution (%)
e	500	±17	±37	±15
	1000	±14	±25	±12
	500	±19	±26	±15
γ	1000	±15	±22	±12
	1500	±13	±23	±11

Fig. 22 shows a 2γ invariant mass spectrum obtained measuring the opening angle and the energy of the 2γ detected. The width of the observed π^0 peak is an important check of the energy resolution proper to the instrument[8]).

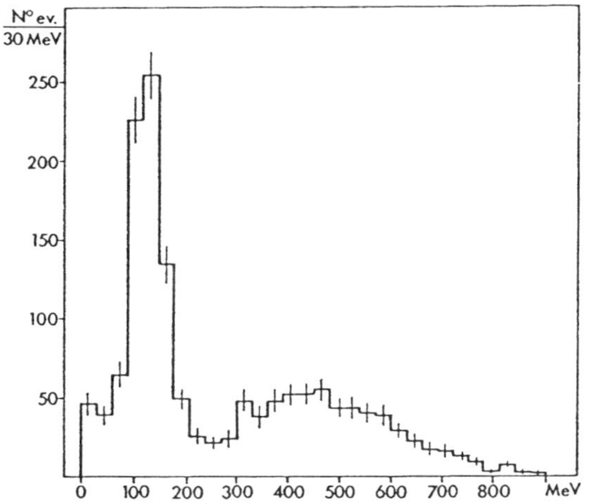

Fig. 22. A two-γ invariant mass spectrum. A π^0-peak is clearly visible around 130 MeV with a 66 MeV width.

6. Conclusion

The instrument described and the results obtained show that this detector is a powerful electron and γ-detector to be operated in presence of high π-background, a typical and very serious problem in the study of electromagnetic processes associated with hadronic reactions.

The solution of this problem, i.e. to achieve a high rejection power against pions, keeping a good electromagnetic detection efficiency has allowed the study of the electromagnetic decay modes of mesons to be pursued in a systematic way[4]) with a simple and powerful instrument.

We are very grateful to Prof. W. Paul and Prof. P. Preiswerk for the interest shown and for the support given to our work as Directors of Physics Dep. I and of the NP Division respectively.

We would like to thank Messrs. G. Muratori and F. Doughty for the construction of the optical system, Mr. Ch. Bregy for the realization of the cylindrical lenses, Mr. L. Thornhill for manufacturing the plastic scintillator counters, Messrs. F. Blythe and J. Dixon for the design of the mechanics, and the CERN PS surveyor staff, in particular Messrs. L. Perret, J. Leault and W. Coosemans, for their assistance in the determination of the reference systems.

We wish also to thank our technicians G. P. Ferri, K. Ley, O. Polgrossi and E. Stocco for their contribution to the construction and the running of this instrument.

References

[1]) T. Massam, Th. Muller, M. A. Schneegans and A. Zichichi, Nuovo Cimento 39 (1965) 464.

[2]) D. Bollini, A. Buhler-Broglin, P. Dalpiaz, T. Massam, F. Navach, F. L. Navarria, M. A. Schneegans and A. Zichichi, Rev. Phys. Appl. 4 (1969) 108.

[3]) D. Bollini, A. Buhler-Broglin, P. Dalpiaz, T. Massam, F. Navach, F. L. Navarria, M. A. Schneegans, F. Zetti and A. Zichichi, Nuovo Cimento 61 A (1969) 125.

[4]) For a review see A. Zichichi, Ann. Phys. (USA) 66 (1971) 405.

[5]) D. Bollini, A. Buhler-Broglin, P. Dalpiaz, T. Massam, F. Navach, F. L. Navarria, M. A. Schneegans and A. Zichichi, Nuovo Cimento 56 A (1968) 1173.

6) D. Bollini, A. Buhler-Broglin, P. Dalpiaz, T. Massam, F. Navach, F. L. Navarria, M. A. Schneegans and A. Zichichi, Nuovo Cimento **57** A (1968) 404.

7) D. Bollini, A. Buhler-Broglin, P. Dalpiaz, T. Massam, F. Navach, F. L. Navarria, M. A. Schneegans and A. Zichichi, Nuovo Cimento **58** A (1968) 289; M. Basile, D. Bollini, P. Dalpiaz, P. L. Frabetti, T. Massam, F. Navach, F. L. Navarria, M. A. Schneegans and A. Zichichi, Nucl. Phys. **33** B (1971) 29.

8) M. Basile, P. Dalpiaz, P. L. Frabetti, T. Massam, F. L. Navarria, M. A. Schneegans and A. Zichichi, Further evidence for the decay mode $\phi \to \eta\gamma$ and the measurement of the branching ratio $\Gamma(\phi \to \eta\gamma)/\Gamma(\phi \to \text{total})$, submitted to Nucl. Phys. B.

9) M. Basile, P. Dalpiaz, P. L. Frabetti, T. Massam, F. Navach, F. L. Navarria, M. A. Schneegans and A. Zichichi, Lettere Nuovo Cimento **4** (1970) 838.

10) T. F. Anderson, P. M. Blackall, J. Daub, A. E. Head, M. B. Metcalf and V. J. Weights, The measurement of spark chamber film at CERN using Luciole, 1967 Intern Conf. *Programming for flying-spot devices* (Munich, 1967) p. 87.

11) G. Fortunato, Un nuovo rivelatore di elettroni ad alto potere di rigezione contro pioni, Thesis (Universita di Bologna, 1966) unpublished.

Section 2

THE TOF DEVICE WITH ± 70 psec RESOLUTION AND THE NBC TECHNOLOGY

The possibility to reach ± 70 psec time resolution allowed to open a new series of experimental investigations: i) the separation of antideuterons from other lighter negative particles via TOF measurements; ii) the position resolution in plastic scintillator counters, for fast kinematic reconstruction of charged particles; iii) the position resolution for neutrons in thick scintillating counters thus allowing neutron-missing-mass spectroscopy.

This section contains the reproduction of some original papers, where TOF technology was essential.

Section 2

(i)

THE SEPARATION OF ANTIDEUTERONS FROM OTHER LIGHTER NEGATIVE PARTICLES

G. BRAUTTI, *et al*
16 Agosto 1965
Il Nuovo Cimento
Serie X, Vol. 38, pag. 1861-1874

CERN
SERVICE D'INFORMATIONS
SCIENTIFIQUE

A High-Intensity, Partially Separated, Beam of Antiprotons and K-Mesons.

G. BRAUTTI (*), G. FIDECARO, T. MASSAM, M. MORPURGO, TH. MULLER (**),
G. PETRUCCI, E. ROCCO (*), P. SCHIAVON (*), M. SCHNEEGANS (**) and A. ZICHICHI
CERN - Geneva

(ricevuto l'11 Marzo 1965)

Summary. — A beam of small production angle and large acceptance has been designed at the CERN Proton Synchrotron in order to yield partially separated, high-intensity beams of antiprotons and K-mesons. The new features of the beam are discussed together with the experimental results.

---

1. – Introduction.

The use of separated beams has been limited mostly to bubble chamber experiments. Counter experiments can make use of other « separating » techniques and are generally performed in unseparated beams.

However, if very-high-intensity beams are to be developed, separation becomes necessary also for counter experiments. In this case the use of the other techniques to reject unwanted particles makes poor separation acceptable. This fact can be exploited to build more intense beams.

We present hereafter the basic design consideration for « high-intensity separated beams for counter experiments », the description of a particular beam made at the CERN Proton Synchrotron, and the results obtained (***).

(*) Istituto di Fisica dell'Università di Trieste, Istituto Nazionale di Fisica Nucleare, Sezione di Trieste.

(**) Institut de Recherches Nucléaires, Université de Strasbourg, Strasbourg.

(***) A description of this beam and the first results have been presented at the *International Conference on Elementary Particles at Dubna* (August 1964).

The most powerful negative beam of the world.

T. MASSAM, *et al.*
1º Settémbre 1965
Il Nuovo Cimento
Serie X, Vol. 39, pag. 10-14

Experimental Observation of Antideuteron Production.

T. MASSAM, TH. MULLER (*), B. RIGHINI, M. SCHNEEGANS (*) and A. ZICHICHI
CERN - Geneva

(ricevuto il 13 Marzo 1965)

Summary. — The results of an experiment which show the existence of antideuterons in the production process proton-beryllium are reported.

We report here the results of an experiment on the production of antideuterons in proton-beryllium collisions. The beam used for the investigation was the high-intensity, partially-separated negative beam of the CERN Proton-Synchrotron. The beam layout is shown in Fig. 1 and its caracteristics are summarized in Table I. The details of the beam are described elsewhere ([1]), so we will mention only the points relevant to the detection of antideuterons

TABLE I. – *Basic parameters of the beam.*

Production angle	111 mrad
Target	Be; ∅ 1×20 mm; at 100 mrad
Horizontal angular acceptance	± 32 mrad
Vertical angular acceptance	± 6.2 mrad
Maximum momentum band	± 2%
Total length	61 m
Beam size in the experimental area	horizontal 3 cm; vertical 2.2 cm

(*) Institut des Recherches Nucleaires, Strasbourg.

([1]) G. BRAUTTI, G. FIDECARO, T. MASSAM, M. MORPURGO, TH. MULLER, G. PETRUCCI, E. ROCCO, P. SCHIAVON, M. SCHNEEGANS and A. ZICHICHI: *Nuovo Cimento*, 38, 1861 (1965). See also report presented by A. ZICHICHI at the *International Conference on High-Energy Physics* (Dubna, August 1964): «*A high-intensity, partially-separated, beam of antiprotons and* K*-mesons*».

The first experimental observation of Antideuteron production. The paper was submitted to "Il Nuovo Cimento" on March 13, 1965 and it was published on September 1st, 1965.

Section 2
(ii)

THE POSITION RESOLUTION IN PLASTIC SCINTILLATOR COUNTERS, FOR FAST KINEMATIC RECONSTRUCTION OF CHARGED PARTICLES

D. Bollini, P. Dalpiaz, P.L. Frabetti, T. Massam, F. Navach,
F.L. Navarria, M.A. Schneegans and A. Zichichi

ACCURATE POSITION DETERMINATION OF PARTICLE TRAJECTORIES USING PLASTIC SCINTILLATOR COUNTERS

From
Nuclear Instruments and Methods 81, *56 (1970)*

1970

Reprinted from:
NUCLEAR INSTRUMENTS AND METHODS 81 (1970) 56–60; © NORTH-HOLLAND PUBLISHING CO.
Printed in the Netherlands

ACCURATE POSITION DETERMINATION OF PARTICLE TRAJECTORIES USING PLASTIC SCINTILLATOR COUNTERS

D. BOLLINI, P. DALPIAZ, P. L. FRABETTI, T. MASSAM, F. NAVACH, F. L. NAVARRIA, M. A. SCHNEEGANS and A. ZICHICHI

CERN, Geneva, Switzerland
Istituto di Fisica dell'Università di Bologna, Italy
Istituto Nationale di Fisica Nucleare, Sezione di Bologna, Italy
Centre de Recherches Nucléaires, Strasbourg, France

Received 21 November 1969

Using the CDF technique for locating charged particles in plastic scintillator counters, it has been found that the accuracy in determining the position of an incident particle in the counter is essentially a function of the distance between the two photomultipliers.
Moreover it has been shown that a position resolution of $\Delta x = \pm 0.15$ cm and a time resolution of $\Delta t = \pm 0.07$ nsec can be achieved in a simple way.
Finally it has been shown that it is possible to measure simultaneously the coordinates of an incident particle in two orthogonal directions. This would allow to measure particle trajectories in high rate beams, when spark chambers cannot be used.

1. Introduction

The purpose of this paper is to report the results of a study of the position determination of a charged particle incident on a plastic scintillator counter and to describe the set-up that was needed for this experiment.

The position was measured using the CDF technique[1]), and the experiment included the study of two types of plastic scintillators and two types of photomultipliers.

The results achieved represent a substantial improvement on previous measurements[1,2]).

2. The set-up

The measurements were performed using the 540 MeV CERN Synchro-cyclotron scattered out proton beam, at a typical incident rate of 10^2 protons/sec · cm^2.

The experimental set-up (fig. 1) was kept as simple as possible. The beam was defined by two counters, X and Y, each 2 mm wide and 1.5 cm thick, placed 30 cm apart along the beam.

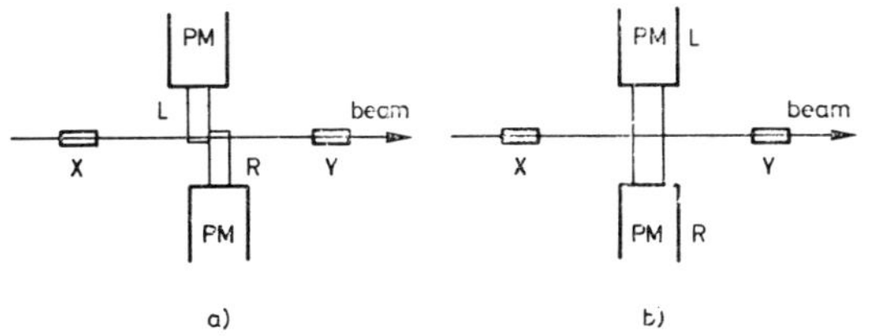

Fig. 1. a) Experimental arrangement for the study of the absolute time resolution. b) Experimental arrangement for the study of the position resolution.

The apparatus for the study of the time resolution consisted of two scintillators R and L viewed each one by a photomultiplier without a light guide. The overlap in space between R and L was 2 mm, equal to the transverse dimensions of the beam-defining counters (fig. 1a).

The accuracy in locating incident particles was studied using a plastic scintillator block viewed by two photomultipliers (fig. 1b).

The types of plastic scintillator investigated were NE 102A and NE 111, in small blocks of various dimensions coupled to XP 1021 or 56 AVP photomultipliers. The optical contact was made with silicon grease (Dow Corning 20-057) or with solid siliconoloid discs (4 cm × 0.4 cm thick). In some cases, in order to get the maximum photostatistics, the scintillator was viewed directly by the photomultipliers without any light guides. The dimensions of all scintillators are summarized in table 1.

3. Electronics

The electronic diagram (fig. 2a) is identical for both the time and the position resolution measurements. Each signal from the photomultipliers R and L enters one into a low level discriminator (CERN NP-2620, called TR in fig. 2a*), where they are shaped 15 nsec wide. The pulse height varied between 0.7 and 1.5 V, and either >50 mV or >350 mV discriminator inputs were tried. The shaped pulses, one delayed by 5 nsec

* The ZC circuits indicated in fig. 2a were used for some measurements, as mentioned later in the text.

TABLE 1

Position resolution.

Test	Scintillator	Dimensions* (cm^3)	Distance between the PM'S (cm)	PM	Electronics†	n^+	Resolution fwhh (cm)
1	NE 102A	5×3×(2)	5	XP 1021	CS (50)	4.8	0.6
2		5×2×(3)	5	XP 1021	CS (350)	—	0.4
3	NE 102A	20×3×(2)	32	56 AVP	CS (50)	2.6	1.3
4		20×3×(2)	32	XP 1021	CS (50)	2.6	1.1
5	NE 102A	10×10×(2)	56	56 AVP	CS (50)	2.5	1.8
6	NE 111	8×2×(3)	8	XP 1021	CS (350)	6.9	0.35
7		8×2×(3)	8	XP 1021	CS (50)	6.2	0.35
8		8×3×(2)	8	XP 1021	CS (350)	5.3	0.45
9		8×3×(2)	8	XP 1021	ZC	3.1	0.7

* The dimension given in parentheses in this column is the counter thickness along the proton beam direction.
† ZC means Zero Crossing circuits. CS are the CERN-shapers, and the threshold in mV is given in parentheses.
$^+$ n is the apparent refractive index of the scintillator as determined from the slopes of the position and time calibration.

relative to the other, are gated by the X and Y signals in a fivefold coincidence unit (CERN NP-6234), the linear output of which goes to a stretcher amplifier (CERN NP-4142) and then to a TMC multichannel analyser. The timing of the coincidence unit is shown in fig. 2b.

The overlap area between the R and L pulses gives the zero reference point for the time and the position calibration. The time calibration is obtained by varying the delay of R and keeping L fixed, while the position calibration is obtained by displacing the XY telescope and keeping the delays of R and L fixed. A photograph of the output of the stretcher amplifier is shown in fig. 3: three well resolved bands can be seen, corresponding to a variation of 0.5 nsec in the R delay.

During some measurements, more sophisticated discriminators were used for R and L signals, namely Zero Crossing (ZC) circuits[3]), which trigger for a constant fraction of the input pulse height. These specially designed circuits improved the time resolution, and were particularly useful for understanding the effect of the light attenuation in the scintillator.

During all the measurements the pure electronic resolution was checked with a pulse generator, sending

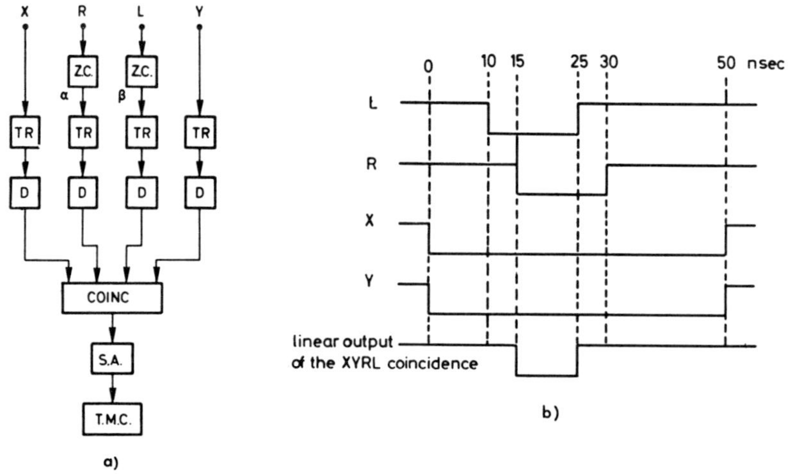

Fig. 2. a) Diagram of the electronic logic: ZC are zero crossing circuits[3]) [used only for some measurements], TR are discriminators, D are delay boxes, COINC is a coincidence unit, S.A. is a stretcher amplifier, TMC is a multichannel analyser. b) Timing of the XYRL coincidence.

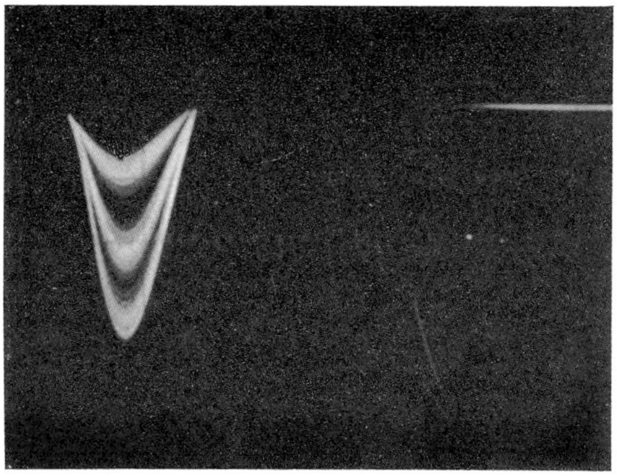

Fig. 3. Output of the shaping amplifier: 50 mV/cm, 1 μsec/cm. The delay of R is varied by 0.5 nsec each step.

the same pulse into the R and L discriminators: the time jitter introduced by the electronics at points α and β of fig. 2a was always between 70 and 90 psec fwhh.

4. Results

4.1. Time resolution

Fig. 4 shows the time resolution obtained using two NE 111 ($5 \times 3 \times 2$ cm^3) counters coupled to XP 1021 photomultipliers, as shown in fig. 1a. The electronics used for this measurement includes ZC circuits before L and R triggers, as indicated in fig. 2a. The resolution obtained with this method is 0.14 nsec fwhh, and is constant over the range scanned. The linearity is satisfactory, the differential non linearity (at most 10%) being probably introduced by the delay box used. This measurement was repeated with the same counters and the same photomultipliers, but sending the anode pulses directly into the 50 mV input of the CERN shapers (TR in fig. 2a). The time resolution obtained by this method is 1.2 times wider than using ZC circuits.

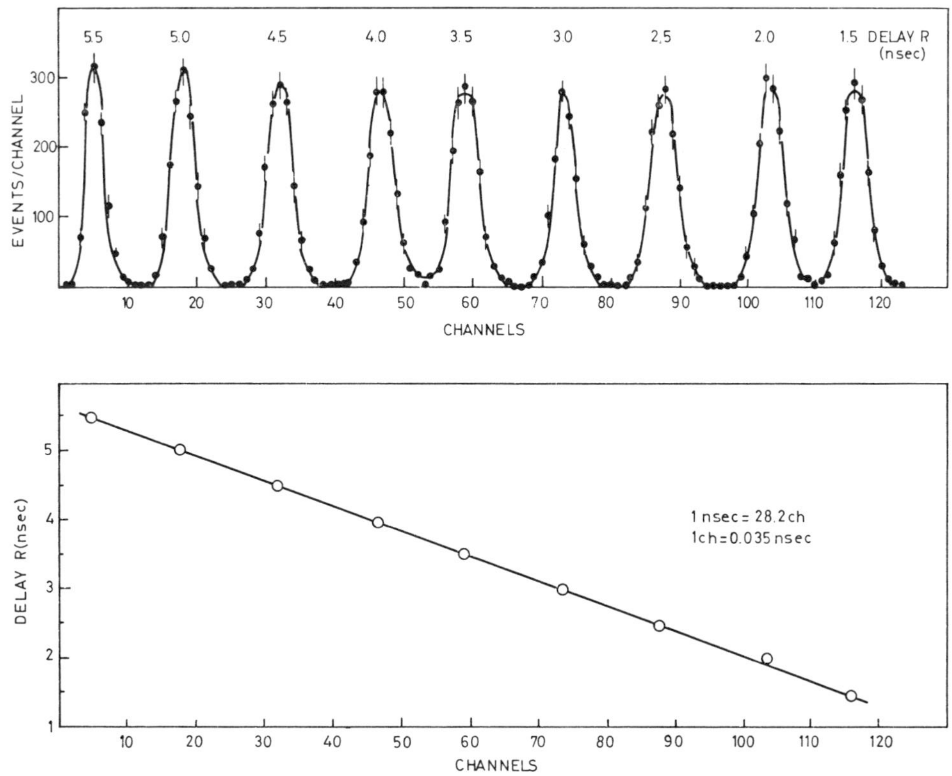

Fig. 4. Time resolution using zero crossing circuits with two NE 111 ($8 \times 3 \times 2$ cm^3) counters viewed by XP 1021 photomultipliers. The overlap in space between the two counters was 2 mm. The time resolution which can be deduced from this graph is 0.14 nsec fwhh. The linearity of the response is satisfactory.

4.2. Resolution in locating particle trajectories

Fig. 5 shows the position resolution obtained using one NE 111 ($8 \times 2 \times 3$ cm^3) counter coupled, without light guides, to XP 1021 photomultipliers, as shown in fig. 1b. The measurement was performed by displacing the XY telescope. The observed resolution is 0.35 cm fwhh and this stays constant along the counter from -3.5 to $+3.5$ cm. (Outside these limits the spectra become wider and a strong non linearity shows up.) This was the best result obtained. When corrected for the width of the XY telescope (0.2 cm) this gives a position resolution $\Delta x = 0.3$ cm fwhh. The counter thickness along the proton beam direction was 3 cm.

5. Systematic study of the positional resolution

A series of tests were made in order to study the variation of the position resolution under different experimental conditions. Table 1 is a summary of the results obtained. The first observation is that the resolution does not depend on the fixed threshold of the CERN shapers (tests 6 and 7). This will allow us to compare data even if they are taken with different thresholds.

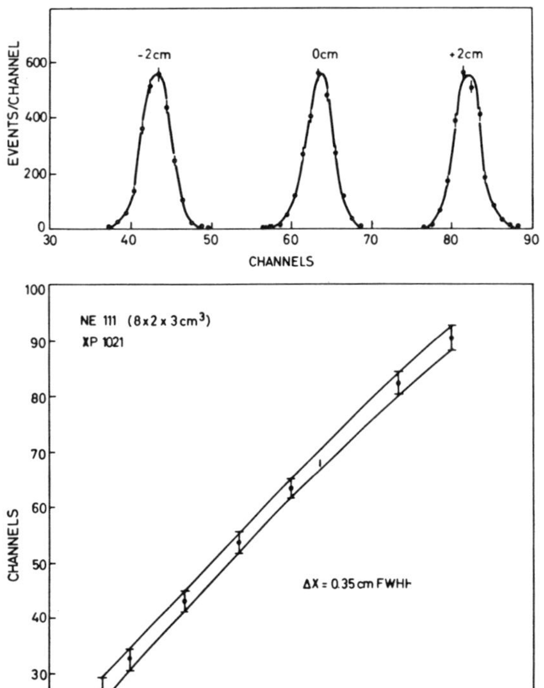

Fig. 5. Position resolution obtained using as counter a plastic scintillator NE 111 ($8 \times 3 \times 2$ cm^3, 3 cm thick along the beam direction) and a XP 1021 photomultiplier followed by CERN shapers (no ZC circuits were used). The counter was viewed directly by the PM's without light guides. Typical overlap spectra are shown in the upper insert. The resolution obtained from this measurement is 0.35 cm fwhh.

By comparing test 1 with test 4, and test 3 with test 5, one can derive the following important conclusion: the accuracy depends strongly on the distance between the photomultipliers. This effect is due to the absorption of scintillation light in the plastic counter, thus reducing the number of photoelectrons produced in the PM photocathode. Consequently the resolution is worsened. The same happens when the thickness of scintillator along the beam (i.e. the light produced) is decreased. This is derived from the comparison of test 1 with test 2, and of test 6 with test 8.

The quantity n reported in table 1 is the apparent index of refraction, which is defined from the apparent velocity of light in the plastic scintillator counter, measured from the time and position calibrations.

From tests 8 and 9 one can see that in spite of the better time resolution obtainable with ZC circuits, the position resolution is better with the CERN shapers. The reason is the high attenuation of the NE 111 plastic scintillator, which produces an apparent index of refraction that is higher for fixed threshold discriminators than for ZC circuits, thus compensating the loss in intrinsic resolution. For a detailed discussion of the optical properties of plastic scintillator see ref. [4]).

The type of scintillator also plays an important role. In fact the NE 111 counters (see tests 6 and 8) have a resolution better than that of the NE 102A counters (see tests 2 and 1 respectively), in spite of the fact that the distance between PM is greater. Notice that the improvement is less important for the counter with higher photostatistics (i.e. with the thickness extending further along the beam). Finally, the XP 1021 photomultipliers show a better resolution than do the 56 AVP, as one can see by looking at tests 3 and 4.

6. Conclusions

The measurements reported in this paper show that the accuracy achieved in the position resolution depends on the energy lost by the particle in the scintillator. Since 540 MeV protons produce about 30% more light than a minimum ionizing particle, one could expect a resolution that is ≈ 18 % worse for a higher energy beam.

The position resolution reported in section 4.2 has been given without taking into account multiple scattering in counter X, because this effect is small. In fact, for 540 MeV protons and for the present geometrical configuration, the contribution to the uncertainty in position localization, due to multiple scattering, is ≈ 0.1 cm fwhh.

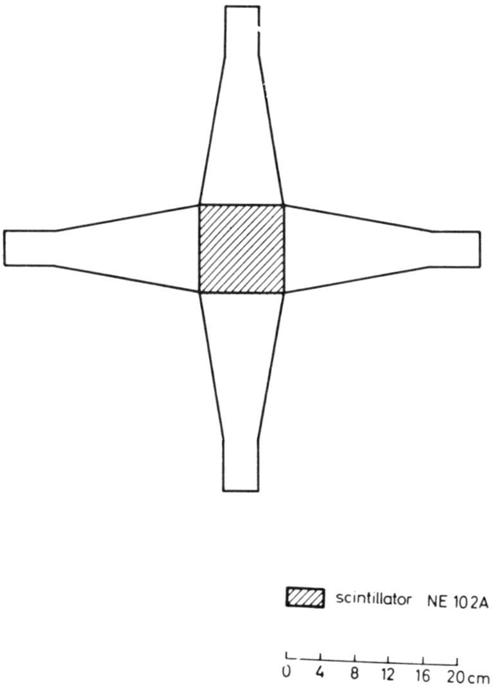

Fig. 6. Plastic scintillator counter coupled to four light guides.

As mentioned above, a rough parametrization of the position resolution can be done as a function of the distance between the two photomultipliers. The other relevent parameters, such as electronics, material and geometry of the scintillator and of the optical coupling, yield variations in the observed resolutions of, at the most, 30 %.

A particular example of the independence of the positition resolution from the geometrical shape of the counter is given by test 5, which was performed with a $10 \times 10 \times 2$ cm^3 block of plastic scintillator with four conical plexiglass light guides, 23 cm long, coupled as shown in fig. 6. This test also shows that it is possible to measure simultaneously the coordinates of an incident particle in two orthogonal directions with good accuracy and good linearity. This would allow to measure particle trajectories in beams with high rates, when spark chambers cannot be used.

Finally, it should be notice that the tests reported in this paper were repeated fifteen days later and gave essentially the same results. Moreover, a stability better than 25 psec has been observed over a period of twelve hours, compatible with statistical fluctuations and binning effects. This confirms the reliability and the stability of the time difference method for the localization of particle trajectories, a method which has already been used in large scale counter experiments[2]).

References

1) G. Charpak, L. Dick and L. Feuvrais, Nucl. Instr. and Meth. **15** (1962) 325.

2) P. Dalpiaz, G. Fortunato, T. Massam, F. Zetti and A. Zichichi, Proc. Intern. Symp. *Electron and positron storage rings* (Saclay, Sept. 1966) p. V-6-1; D. Bollini, A. Buhler-Broglin, P. Dalpiaz, T. Massam, F. Navach, F. L. Navarria, M. A. Schneegans, F. Zetti and A. Zichichi, Nuovo Cimento **61** A (1969) 125.

3) D. A. Gedke and W. J. McDonald, Nucl. Instr. and Meth. **55** (1967) 380; Nucl. Instr. and Meth. **58** (1968) 253; D. A. Gedke and W. J. McDonald, University of Alberta Preprint (1969).

4) V. Alles-Borelli, D. Bollini, P. Dalpiaz, E. Fiorentino, T. Massam, L. Monari, F. L. Navarria, F. Palmonari, F. Zetti and A. Zichichi, Measurement of wave and group velocity in plastic scintillator, Nuovo Cimento, to be published.

Section 2
(ii) cont'd

THE HIGH PRECISION TOF TECHNOLOGY IN SPACE

The TOF technology developed at CERN in the sixties is an important component of the experiment to search for Antimatter in Space. The precision of ± 70 psec in time accuracy achieved at CERN is now made "space" resistant.

F. Palmonari

ANTIMATTER IN SPACE
THE AMS APPARATUS

From
"The Discovery of Nuclear Antimatter", L. Maiani and R.A. Ricci (eds), Vol. 53

Reproduction of pages 99, 108, 111, 112, 113, 114, 115

SIF
BOLOGNA
1996

Conference Proceedings Vol. 53
«The Discovery of Nuclear Antimatter»
L. Maiani and R.A. Ricci (Eds.)
SIF, Bologna, 1996

ANTIMATTER IN SPACE

THE AMS APPARATUS

F. Palmonari

Department of Physics
INFN and University of Bologna

AMS [1] means Alpha Magnetic Spectrometer for Extraterrestrial Study of Antimatter, Matter and Missing Matter on the International Space Station Alpha (ISSA). The construction of the apparatus joins in an international collaboration people and institutions from China, Finland, Germany, Italy, Russia, Switzerland and United States.

The origin and the scientific aims of the experiment have been already beautifully exposed in the previous talk by Samuel C.C. Ting, spokesman of the collaboration.

Therefore I will briefly overview the AMS apparatus, as it will fly on a 10-day "precursor" Shuttle mission in June 1998. It consists of three main components: a permanent magnet, a Silicon microstrip tracker and scintillation counters for anticoincidence and for the time of flight.

I will describe in some detail the time-of-flight (TOF) system, because the Bologna group is responsible for building it.

Then, given the argument, I have the unique opportunity to remind the experimental work done in Bologna, under the leading inspiration of Antonino Zichichi, in scintillation counter techniques, since almost thirty years.

To conclude, I will show some preliminary results, obtained in a cosmic ray test, of a prototype TOF scintillation counter.

OVERVIEW OF THE AMS APPARATUS

In Fig. 1, a global view of the AMS apparatus, following the label numbers (bold in the text), one can see that all components are assembled on an Aluminium barrel structure (2), 114 cm in diameter and 80 cm in height, supporting the **permanent magnet**. The magnet is made of Nd-Fe-B blocks inserted in between two cylindrical walls as shown in the cut-out (9). They magnet aperture, and the two planes are oriented along orthogonal axes. On each side short trapezoidal plexiglass light-guides collect the scintillation light on the 1.8 x 1.8 cm^2 square photocathode of three R5900 type photomultipliers. Figure 9b is a lateral view of the two TOF planes attached to the rigid Aluminium honeycomb support. In this figure a cut-out view of the magnet and of the anticoincidence scintillation counters, covering the inner magnet wall, is shown.

The time-of-flight system has three tasks:

1) Give the first-level trigger to the AMS experiment, by selecting single particles traversing the spectrometer.

2) Distinguish up/down moving particles and measure their velocity. The four planes of counters provide two independent measurements.

3) Determine the absolute charge Z of cosmic nuclei by four measurements of their ionization energy loss.

Figure 10 is a schematic diagram of the electronics: photomultiplier (PM) signals from each counter end are summed and sent to discriminators to define three different trigger levels: $Z = 1$, $Z > 1$ and $Z > 2$. At the same time the low threshold discriminator signals are sent to four pipeline time-to-digital converters. Two are for the velocity measurement with two time scales (0.1 and 1 ns least-count respectively), and two for the particle ionization measurement (low Z and high Z scales), by the time-over-threshold method.

Figure 11 is a picture of the new photomultiplier that will be used in the TOF system. Recently developed by Hamamatsu [3], it is very compact (2.5 x 2.5 x 3 cm^3), light (25 g), and with 10 stages of a new design "metal channel dynode" it reaches a very high gain of 10^6 at only 800 V. The above characteristics, the fast risetime of 1.4 ns, an excellent linearity of response up to 50 mA and finally an already tested ruggedness to vibrations and accelerations, make this PM ideal for space applications.

The expected performances of the scintillation counters are, concerning the charge Z measurement, a separation of nuclei good up to Carbon and fair up to Iron. For the time-of-flight measurement, an overall time resolution of 100 ps is expected. This would ensure a particle separation as shown in Fig. 12, i.e. the possibility to distinguish antiprotons from the overwhelming background of electrons and pions up to about 2 GeV/c momentum.

THE TOF TECHNIQUE FOR LARGE SCINTILLATION COUNTERS

The principle used to measure the time of traversal of a charged particle in a long plastic scintillation counter is sketched in Fig. 13. The light produced by the

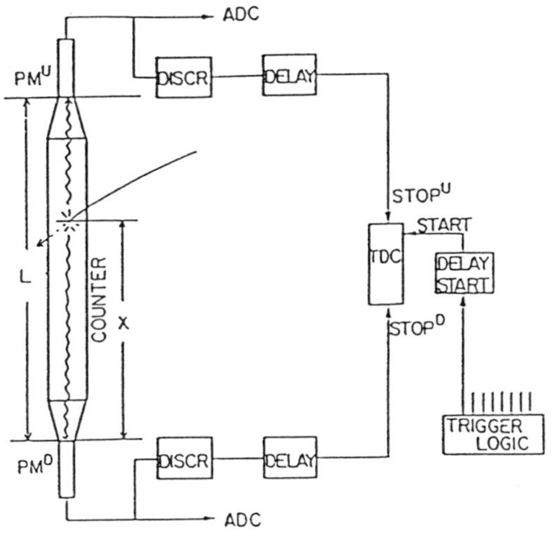

let L = distance between PMD and PMU
X = distance of the particle hit from PMD
V = 'effective' speed of collected light (about 15 cm/ns)
T0 = particle's hit time relative to the trigger time

$$T1 = TstopD - Tstart = T0 + X/V + const$$

$$T2 = TstopU - Tstart = T0 + (L-X)/V + const$$

- hit position:

$$X = V \times (T1-T2)/2 + const$$

- hit time:

$$T0 = (T1+T2)/2 + const$$

Fig. 13: Time measurement principle in scintillation counters.

particle crossing the counter at distance x from PM^D, travels in both directions along the counter at velocity v (L is the total length of the counter), is collected by PM^D and PM^U and transformed into electric pulses by sending the anode signals to threshold discriminators. Two time-to-digital converter (TDC) channels measure their time difference T_1 and T_2 with respect to a common reference signal (as given for example by the trigger signal). Let us call T_0 the particle hit time with respect to the same reference signal, the two measurements will be:

$$T_1 = T_0 + x/v + C_1,$$
$$T_2 = T_0 + (L-x)/v + C_2,$$

where C_1 and C_2 are constants depending on the individual PM transit times and on the signal cable lengths. The two TDC measurements can be combined to give two independent relevant quantities:

i) The time difference, giving the hit position along the counter:

$$x = v \cdot (T_1 - T_2)/2 + \text{const.}$$

ii) The time sum, giving T_0 independently of the hit position x:

$$T_0 = (T_1+T_2)/2 + \text{const.}$$

Here the constants depend on the electronics set-up, and do not change during the experiment.

The above principles had been developed as early as in '65 by Zichichi's group (I joined the group later for the ADONE experiment in Frascati), in a large acceptance neutron spectrometer [4] to study at the CERN PS the production of neutral vector mesons in the process pion + proton → neutron + electron pairs.

Figure 14 is an impressive picture of one neutron spectrometer arm, a pile of twelve 100 cm long plastic scintillator bars, with 18 x 18 cm^2 square cross-section, built in Bologna, seen by 5 in. Philips 56AVP photomultipliers at both ends. Thanks to the time-of-flight measuring technique described above, the spectrometer measured simultaneously the emission angle of the neutron and its velocity. The reference time was given by the relativistic electron-positron pair giving the trigger in the electromagnetic shower counters. As one can see in Fig. 15a and b, the achieved space and time resolutions were respectively 1.5 cm and 0.3 ns (one standard deviation).

Four years later, in '69 at ADONE, the electron-positron collider in Frascati, our group applied the same TOF technique [5]. Figure 16 is a perspective view of

Fig. 14: Neutron spectrometer arm of Zichichi's group vector meson experiment at CERN.

(a)

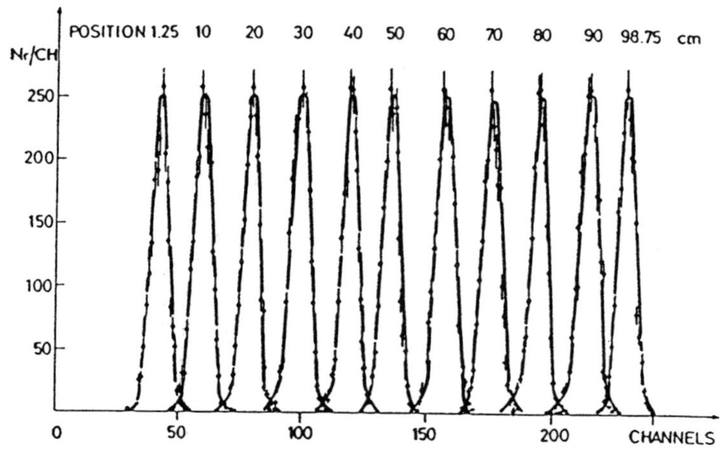

(b)

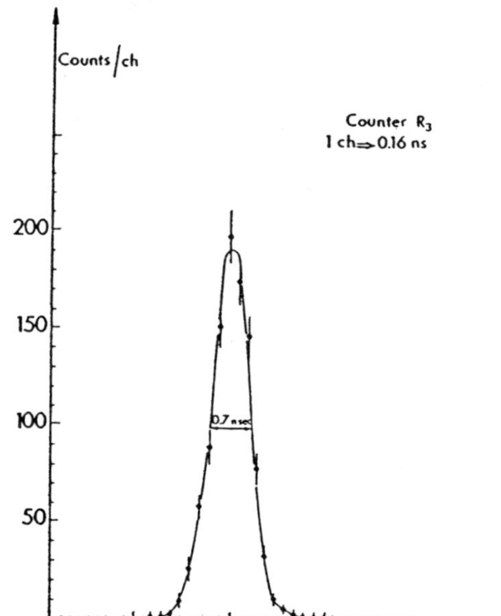

Fig. 15: Performances of the scintillation counters of the neutron spectrometer: (a) space resolution and (b) time resolution.

the ADONE apparatus. The two hodoscopes of six 120 cm long plastic scintillator counters with a thickness of 12 cm, at opposite sides of the interaction region, are clearly visible.

In this experiment the time of flight was used already at the trigger level. By a fast comparison of the mean time of the hodoscopes, particle pairs coming from the interaction could be distinguished from the background cosmic ray muons traversing the apparatus from both sides. In Fig. 17 the TOF peak produced by good events (a) is compared with the two cosmic ray background peaks (b). Their separation was ±3.5 ns and the time resolution better than 0.5 ns (one standard deviation). Thanks to the rejection power of the time of flight, the muon-pair annihilation cross-section at all ADONE energies could be measured with 0.3% accuracy.

The Zichichi group, with its great expertise in TOF technology, has the best record with the ± 75 psec accuracy achieved in these studies. Let me quote a statement by A. Zichichi published in Ref. 6[(*)]: "moreover it has been shown that a time resolution of $\Delta t = \pm 0.07$ nsec can be achieved in a simple way". This time resolution was and still remains — after so many years — a world record in TOF technology.

Finally I would like to mention the CERN ISR experiment which has shown the first evidence for the production of heavy baryons in proton-proton collisions. Figure 18 is a picture of the apparatus in the interaction region R4 of the ISR, equipped with the Split-Field Magnet spectrometer; the large area TOF scintillator counters [7] are clearly visible in the foreground.

The ISR being a continuous collider, no reference interaction time was available. The meantime technique was therefore applied in the following way. First all tracks hitting the TOF counters were reconstructed by the SFM wire chambers, then the track length and curvature was used to best fit the relative times to a common interaction time, by giving a proton, kaon, pion or electron weight to each track.

Figure 19 shows the interaction time resolution obtained (a) with the trigger electron track, and (b) using all tracks hitting the TOF counters. The separation between protons, kaons and pions in ISR multiparticle events (shown in Fig. 20) was an useful tool to reduce the combinatorial background in two- and three-particle invariant mass spectra.

(*) Reference 6 is the paper reproduced in this section: D. Bollini et al., *NIM* 81, (1970) 56.

Section 2
(iii)

THE POSITION RESOLUTION FOR NEUTRONS IN THICK SCINTILLATING COUNTERS ALLOWING NEUTRON-MISSING-MASS SPECTROSCOPY

D. Bollini, A. Buhler-Broglin, P. Dalpiaz, T. Massam, F. Navach,
F.L. Navarria, M.A. Schneegans, F. Zetti and A. Zichichi

A NEW LARGE-ACCEPTANCE AND HIGH-EFFICIENCY NEUTRON DETECTOR FOR MISSING-MASS STUDIES

From
Il Nuovo Cimento <u>*61A*</u>*, 125 (1969)*

1969

D. Bollini, *et al.*
21 Maggio 1969
Il Nuovo Cimento
Serie X, Vol. 61 A, pag. 125-172

A New Large-Acceptance and High-Efficiency Neutron Detector for Missing-Mass Studies.

D. Bollini, A. Buhler-Broglin, P. Dalpiaz, T. Massam, F. Navach, F. L. Navarria, M. A. Schneegans, F. Zetti and A. Zichichi

CERN - Geneva
Istituto di Fisica dell'Università - Bologna
Istituto Nazionale di Fisica Nucleare - Sezione di Bologna
Centre de Recherches Nucléaires - Strasbourg

(ricevuto il 20 Dicembre 1968)

Summary. — A large-acceptance and high-efficiency neutron detector is described. The sensitive surface and volume of the detector are 2.16 m^2 and 0.78 m^3, respectively. The detector consists of twenty-four elements of plastic scintillator, each having dimensions $(100\times18\times18)$ cm^3. The large volume of scintillator, in the particular geometrical arrangement chosen, allows a mean detection efficiency of about 25% in the range $(70\div390)$ MeV neutron kinetic energy for a laboratory solid angle of 0.14 sr at 4 m radial distance. With the techniques adopted, calibrations with charged particles can be easily performed in a few hours using a low beam intensity. An interesting feature of this instrument is the accuracy achieved in locating incident particles, which is ±1.4 cm for charged particles, and ±2.5 cm for neutrons. The accuracies achieved for the time-of-flight measurement are ±0.35 ns for charged particles and ±0.7 ns for neutrons. With these resolutions in the neutron time of flight and angle, the uncertainty in the missing mass is ±4 MeV for η, ±10 MeV for ω, and ±15 MeV for φ mesons.

1. – Introduction.

This paper describes a large detector which allows accurate simultaneous measurements of a neutron's time of flight and of the co-ordinates of its interaction point. Methods have been developed to include this information in the electronic logic, thus allowing a fast electronic selection of interesting events.

The detector consists of twelve plastic scintillation counters, each of dimensions $(100 \times 18 \times 18)\ \text{cm}^3$ and viewed from each end by one photomultiplier. These twelve counters, with a total weight of 0.45 ton of scintillator, are arranged in two stacks so as to cover a detection area of $1.08\ \text{m}^2$ with a depth of 36 cm.

Details of the construction, calibration procedure, performances and other details concerning the use of the instrument under various experimental conditions are also reported. Several experiments (1-5) have been completed using either one or two such detectors in a missing-mass spectrometer system. The detectors, though designed in particular for the investigation of neutral resonances in the reaction

$$\pi^- + p \rightarrow n + B^0 , \tag{1}$$

may be used to study many other processes and their modular construction makes them as flexible as possible for adaptation to different geometrical arrangements.

2. – Design considerations.

Since the detector was designed for a particular purpose, for use in a missing-mass spectrometer, this has had an influence on the final choice of the design. For this reason and also in order to help the reader to understand more easily some of the calibrations which were made, it will be described in this context.

2·1. *The missing-mass method.* – The simultaneous measurement of the neutron's velocity and its production angle θ_n relative to the beam direction allows the complete reconstruction of reaction (1): the mass, velocity and direction of the B^0 are all determined. In particular, the expression for the mass is

$$m_{B^0}^2 = m_\pi^2 + m_p^2 + m_n^2 + 2E_\pi m_p - 2E_\pi E_n - 2E_n m_p + 2p_\pi p_n \cos\theta_n , \tag{2}$$

(1) A. Buhler-Broglin, P. Dalpiaz, T. Massam, F. L. Navarria, M. A. Schneegans, F. Zetti and A. Zichichi: *Nuovo Cimento*, **49** A, 183 (1967).

(2) D. Bollini, A. Buhler-Broglin, P. Dalpiaz, T. Massam, F. Navach, F. L. Navarria, M. A. Schneegans and A. Zichichi: *Nuovo Cimento*, **56** A, 531 (1968).

(3) D. Bollini, A. Buhler-Broglin, P. Dalpiaz, T. Massam, F. Navach, F. L. Navarria, M. A. Schneegans and A. Zichichi: *Nuovo Cimento*, **56** A, 1173 (1968).

(4) D. Bollini, A. Buhler-Broglin, P. Dalpiaz, T. Massam, F. Navach, F. L. Navarria, M. A. Schneegans and A. Zichichi: *Nuovo Cimento*, **57** A, 404 (1968).

(5) D. Bollini, A. Buhler-Broglin, P. Dalpiaz, T. Massam, F. Navach, F. L. Navarria, M. A. Schneegans and A. Zichichi: *Nuovo Cimento*, **58** A, 289 (1968).

where m, E and p denote respectively the mass, total energy and momentum of the particles in reaction (1), all measured in the laboratory system. Figure 1 shows the missing-mass curves, that is the relation between the neutron kine-

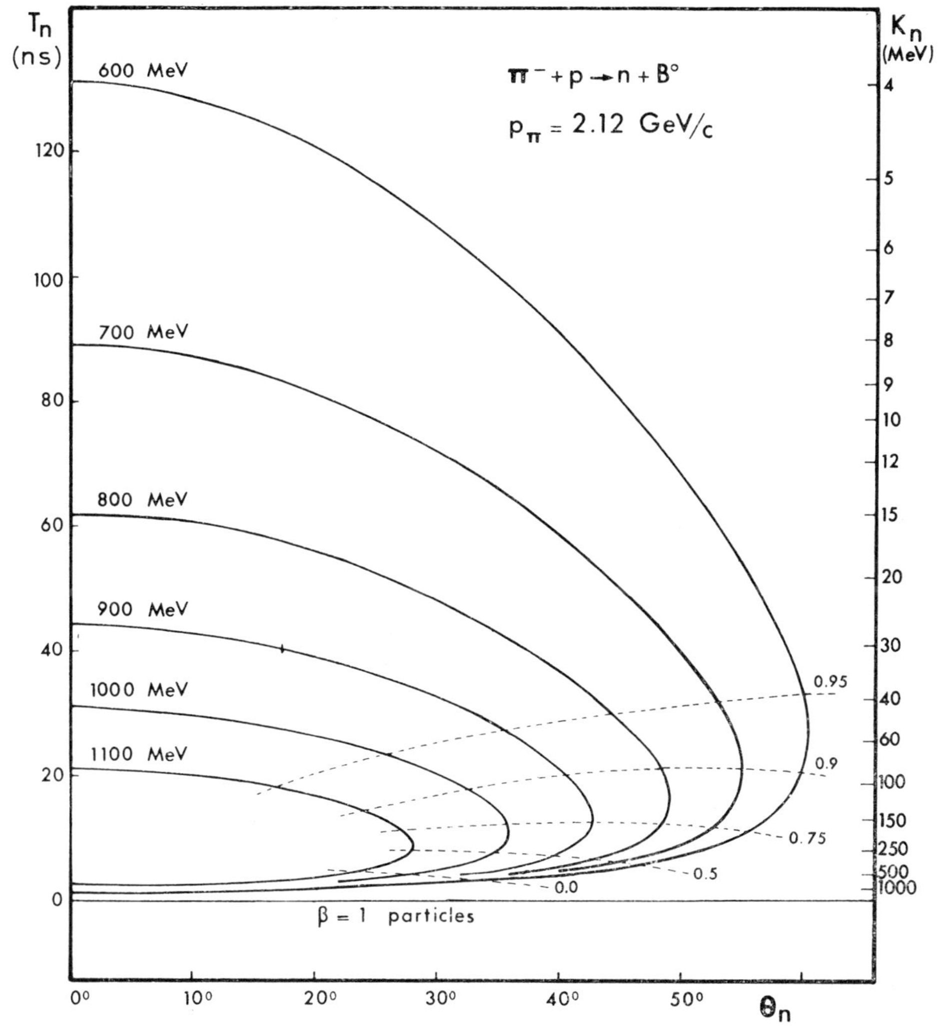

Fig. 1. – The neutron time of flight T_n is plotted *vs.* the neutron emission angle θ_n in the laboratory system. The ordinate on the right refers to the neutron kinetic energy. The kinematic curves are labelled by the corresponding neutron missing masses. The dashed lines indicate constant values of $\cos\theta_n^*$.

tic energy K_n and the production angle θ_n plotted for constant values of m_{B^0} in eq. (2). The time of flight T_n of the neutron over a 4 m path length is marked on the ordinate on the left.

In general, a good compromise between centre-of-mass acceptance and wide separation of the isomass curves is obtained by working near the Jacobian peak where the laboratory angle of the neutron is a maximum. On the other hand, in the case of peripheral processes, the number of events with small angles and large times of flight is very large. The detector should thus be capable of operating in both these regions.

2'2. *Possible designs.* – For a wide range of values of the missing mass, the neutron kinetic energy in the region of the Jacobian peak ranges from a few tens of MeV to a few hundred MeV. The simplest way of detecting neutrons over all this energy range is to detect their interaction in a scintillation counter. Three possible counter designs have been considered and the choice was based on obtaining maximum accuracy in the two regions of the missing-mass curves mentioned above.

i) Large plates of scintillator placed normal to the incident neutron direction and viewed from all four sides by photomultipliers.

This has the advantages that the position may be well defined along the neutron flight path and that the number of photomultipliers is small. The measurement of position in the transverse direction would be made by using the time differences between the signals from the photomultipliers. However, this technique requires good light collection and high photoelectron statistics, which are not easy to attain with this geometry.

A test was performed with a $(40\times40\times5)\ cm^3$ scintillator coupled with four plexiglass light guides of the same dimensions, each guide being viewed by a 56 AVP photomultiplier. A position resolution of ±5.5 cm was obtained with charged particles.

ii) A hodoscope system of rectangular blocks with the light collected from behind the block.

This system has the advantage that the incident neutron travels in the same direction as the light from the interaction point towards the phototube. These velocities are equal for a time of flight equal to 20 ns, just where the curves are crowding close together. Thus, just in this critical region, errors due to the uncertainty in the depth in the counter at which the neutron interacts cancel to first order. The disadvantage is that the transverse position is defined by the dimensions of the counter and in order to have a good mass resolution and a large solid angle either a prohibitive number of elements and photomultipliers would be required, or the counter could only be used in the region away from the Jacobian peak.

iii) Long blocks of scintillator viewed by one photomultiplier from each end and using the time-difference method for position determination [6] in

[6] G. Charpak, L. Dick and L. Feuvrais: *Nucl. Instr. Meth.*, **15**, 323 (1962).

one direction and the physical size of the counter in the other two orthogonal directions.

The light collection is good and therefore the position along the counter at which a charged particle interacts is well defined (± 1.4 cm), which is better than in either of the other two solutions. The number of photomultipliers is intermediate between i) and ii). This is the solution which was chosen.

2'3. *Characteristics of the selected design.* – There are several aspects to be taken into account:

i) Resolution. The error introduced by the uncertainty in the transverse position is minimized by placing the counter so that the polar angle of the neutron is measured along the counter by the accurate time difference method as indicated in Fig. 2. Then, uncertainty in one transverse direction corresponds to uncertainty in the azimuthal angle and has only a second-order effect on the polar angle and on the mass measurement. The other transverse dimension defines the distance which the neutron travels from the target and so is important for the resolution for the time-of-flight determination. Again, this is only a second-order effect if one is at the Jacobian peak, but the instrument is more flexible and has a larger acceptance if also events not in this region may be accepted. For these reasons the transverse dimensions were chosen to be (18×18) cm^2. As a result of this finite depth, the contributions to the uncertainty in the mass determination from uncertainties in the length of the neutron flight path and from uncertainties in the time measurement

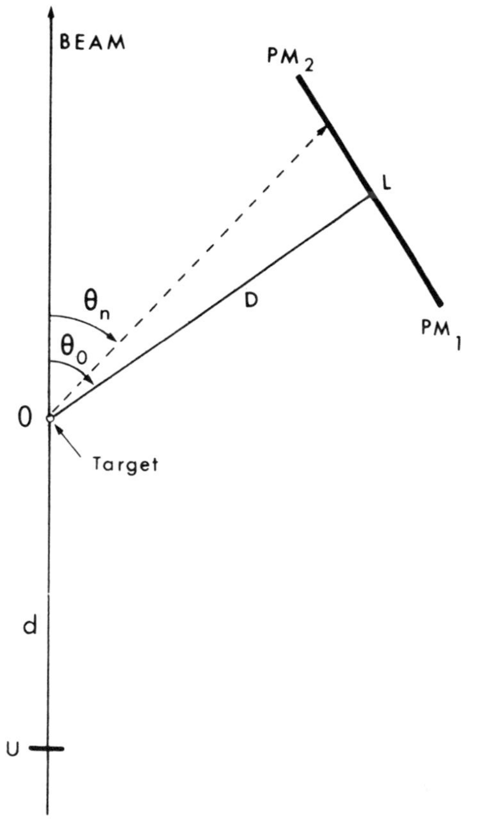

Fig. 2. – Scheme of principle for missing-mass study: d is the distance between counter U and the centre of the target; θ_n is the neutron emission angle; PM_1 and PM_2 are two photomultipliers viewing the ends of a neutron counter of length L, which is set at distance D from the centre of the target.

were equal near the Jacobian peak. Comparable errors from the uncertainty in the vertical position of the interaction occur only at neutron production angles less than 2°. These dimensions were also convenient for coupling the counter to an XP 1040 photomultiplier.

ii) Uniformity and linearity. Uniformity of response, namely constant efficiency along the counter, and good linearity of the indicated position and time of flight as a function of the true values are achieved by using a scintillator with a low optical attenuation and by the correct design of the light guides.

Some preliminary tests were made to study the best geometrical features of a counter for good light collection and uniformity in response. These characteristics are strongly determined by the length and shape of the light guides: they should be sufficiently long that the photocathode is uniformly illuminated independent of the interaction point so as to always average the transit time spread of the photomultiplier in the same manner. They should be rectangular rather than conical so that the total quantity of light reaching the photocathode is independent of the transverse position of the interaction point. This is important both for transit time spread effects and also because it simplifies the calculation of the neutron efficiency. Plexiglass light guides with dimensions $(18\times18\times15)\ cm^3$ were found to meet these requirements.

iii) Solid angle and efficiency. The product of solid angle and efficiency should be as large as possible, but is restricted partly by cost and partly by the requirement of compatibility with other parts of the experimental set-up. Notice that the final solution which was chosen provides a modular system which can be extended or reduced with a minimum of change in the electronic logic and has a good flexibility for arrangement in different geometrical configurations. For instance, in order to increase the detection efficiency, without loss in resolution, two stacks of elements were used one behind the other (giving a total thickness of 36 cm).

iv) Fast electronic selection of events. In certain experiments it may be necessary to select a given missing-mass region (or impose kinematic constraints), so a fast, low pile-up position and time-of-flight selection technique has been developed. It will be described in Subsect. **6·3**.

v) Material. The preference for plastic rather than liquid scintillator was partly a matter of taste; it was mechanically simpler, the problems of handling half a ton of liquid in a mobile and compact detector are not trivial and further we were able to manufacture the plastic scintillator ourselves.

vi) Mechanical supports. The two stacks of 24 elements should be easily and rapidly displaced in the vertical and horizontal directions for calibrations in a direct beam.

vii) Background. Once the dimensions and material of the counter are decided, the only way of acting on the background is by varying the energy threshold of the detector. This will be discussed in Sect. **9**.

3. – Principle of the method.

The methods to be used for measuring the neutron time of flight and its interaction point in the scintillator are described below. It is also pointed out how these measurements are applied in missing-mass studies.

3·1. *Determination of the neutron interaction point.* – The light, produced by charged products of the neutron interaction in the counter, reaches the two opposite PM's at times t_1 and t_2. The measurement of the relative delay $t_1 - t_2 = \tau_\theta$ enables one to know the position P_θ of the neutron along the line joining the two PM's; τ_θ is determined by measuring the area of the overlap of the two pulses superposed in an AND circuit. In order to avoid negative-time ambiguity arising from the symmetry of the counter with respect to its centre, a fixed delay Δt, larger than the time needed by the light to cross half a counter, is added to one of the pulses. Figure 3 exemplifies the relative positions of the two pulses corresponding to a particle crossing the counter in its centre. The overlap pulse P_θ is integrated, then analysed on a multi-channel analyser (MCA) or/and digitized and recorded on magnetic tape.

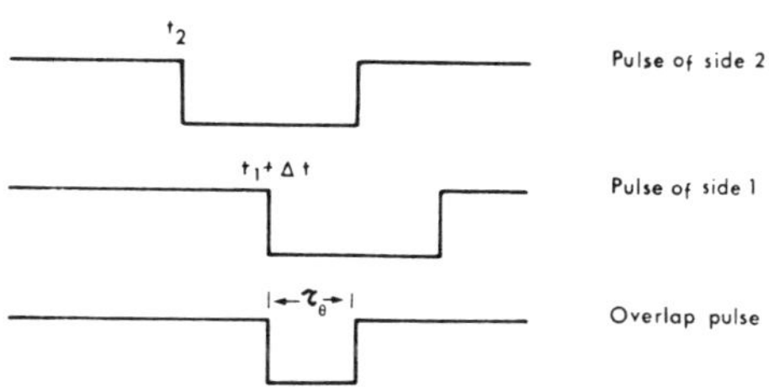

Fig. 3. – Relative timing and overlap of PM_1 and PM_2 pulses.

This technique is especially suitable for measuring the position of the neutron interaction point because in the measurement of τ_θ, timing errors in the electronics, produced by variation in the amount of light, cancel to first order, provided that the difference in the optical attenuation in the two sides of the counter is small. For this reason, particular care has been used in order to obtain a large attenuation length for our scintillators (see Subsect. **4**·1 and **7**·1).

3·2. *Measurement of the neutron time of flight.* – The time differences T_1 and T_2 between a signal from counter U and each of the two PM's of a neutron counter element are given by (see Fig. 2)

$$(3)\qquad \left\{\begin{aligned} T_1 &= \frac{d}{c} + T_n + \frac{L/2 - D\,\mathrm{tg}\,(\theta_n - \theta_0)}{v}\,,\\ T_2 &= \frac{d}{c} + T_n + \frac{L/2 + D\,\mathrm{tg}\,(\theta_n - \theta_0)}{v}\,,\end{aligned}\right.$$

where T_n is the neutron time of flight between the target (assumed to be point-like) and the interaction point in the scintillator; v and c are the velocities of light in the scintillator and in the vacuum respectively. The constants

$$(4)\qquad T_1^{(c)} = T_2^{(c)} = \frac{d}{c} + \frac{D}{c} + \frac{L}{2v}$$

may be determined by placing the counter in a direct beam ($\theta_n = \theta_0 = 0°$) of relativistic particles. The knowledge of these constants allows the adjustment of the relative delays between the signals from U and the neutron counters, in order to work directly with the quantities

$$(5)\qquad t_1 = T_1 - T_1^{(c)}\,,\qquad t_2 = T_2 - T_2^{(c)}\,.$$

By adding and taking into account (3) and (4), one obtains an expression for T_n which is independent of v and θ_n:

$$(6)\qquad T_n = \frac{t_1 + t_2}{2} + \frac{D}{c}\,.$$

The kinematic curves for reaction (1) in the (T_n-θ_n)-plane are shown in Fig. 1.

The technique used to measure t_1 and t_2 is the same as that used to measure P_θ. The precision on T_n obviously depends on that of t_1 and t_2. These are determined, amongst other factors, by the counter thickness along the neutron direction, which has been taken to match the expected overall time resolution, as mentioned in Sect. **2**.

3·3. *The skew kinematics.* – T_n is the sum of two quantities. To use such information in a trigger implies integration and consequently an integrating time constant with the risk of serious pile-up effects. On the other hand, the signals t_1 and t_2 have remarkable advantages. Not only they can be directly used in a fast electronical trigger, but they produce nicely skew kinematic

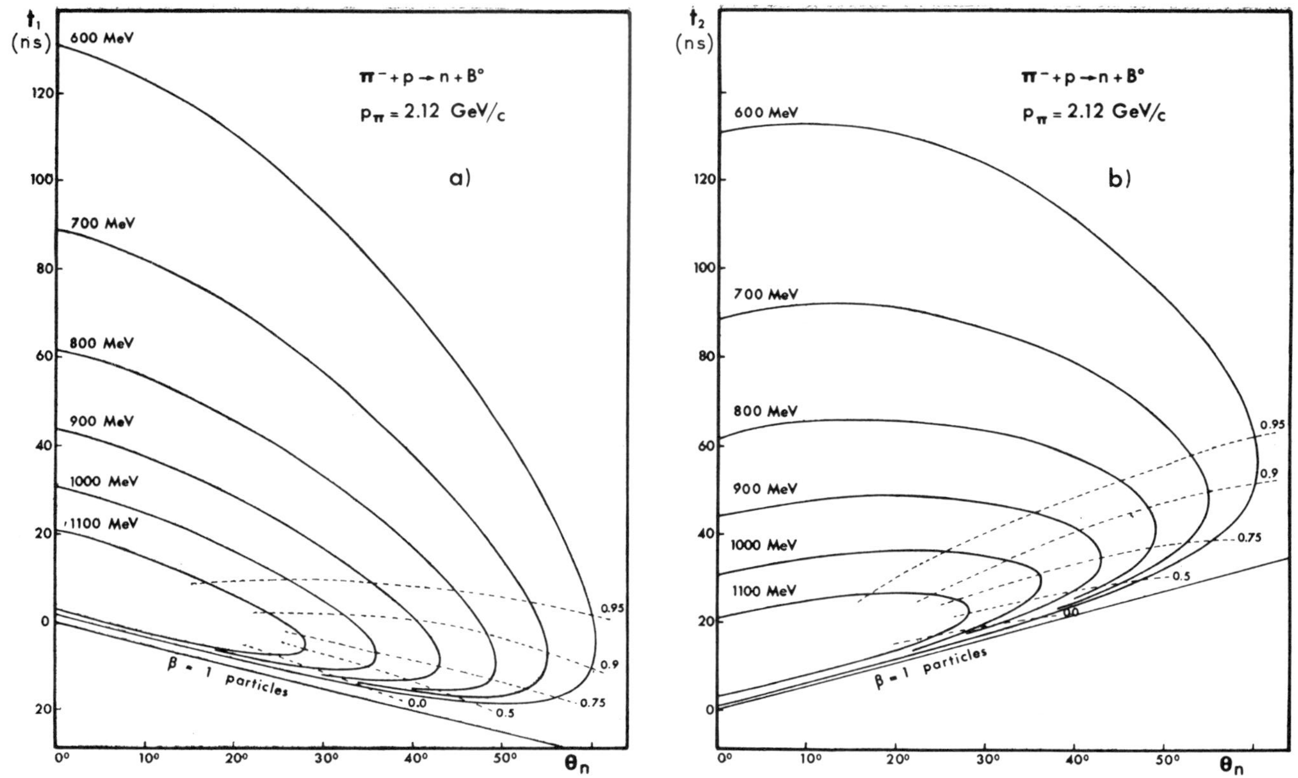

Fig. 4. – The quantities t_1 and t_2, which are strictly related to the neutron time of flight, are plotted *vs.* θ_n in *a*) and *b*) respectively. Notice the deformation with respect to the curves of Fig. 1, due to the dependence of t_1 and t_2 from θ_n.

curves because t_1 and t_2 depend not only on T_n, but also on θ_n and v (*), as can be seen from (3) and (5). Figures 4 *a*) and 4 *b*) show the missing-mass curves plotted using the values of t_1 and t_2 obtained by setting $\theta_0 = 0°$ (**).

The particular shape of the curves is very convenient for direct observation on the MCA of peaks in the θ_n, t_1 or t_2 distributions. In fact, in the $(\theta_n\text{-}t_2)$ or $(\theta_n\text{-}t_1)$ planes, the parts of the kinematic curves nearly perpendicular to the axes correspond to larger intervals of $\cos\theta_n^*$ than in the $(\theta_n\text{-}T_n)$-plane, for those regions where the projections of the curves are well resolved for different mass values.

4. – Mechanical construction.

4·1. *Construction of a single counter.* – Figure 5 shows the structure of one of the twelve identical counters which all make up the detector. The two plexiglass light guides of $(18\times18\times15)\,\text{cm}^3$ are continued by two cylinders of the same material, each 11.5 cm in length, having the same diameter as the photocathode. The purpose of such cylinders is to permit the extension of the μ-metal shielding beyond the end of the photocathode, and thus to obtain a more efficient magnetic shielding of the phototube. The external part of the shield, *i.e.* the soft-iron cylinder, was plugged in a plastic frame of $(18\times18\times2.5)\,\text{cm}^3$ glued onto the light guide. The assembly is completed by a thin seal, shaped to facilitate making a light-tight covering, and by two insulating rings. The soft-iron cylinders support the scintillator only by means of these connections in order to avoid the presence of a metallic structure in the solid angle covered by the neutron detector, and to avoid contact with the reflecting surfaces of the guide.

The scintillator used for the neutron counters is organic, with a tetraphenylbutadiene wave shifter. It was made in Bologna, at the « Augusto Righi » Institute of Physics, and has the following properties:

(*) A measurement of the velocity of light in the scintillator together with a systematic study of the basic properties of light propagation in plastic scintillator is reported elsewhere (7).

(7) V. Alles-Borelli, D. Bollini, P. Dalpiaz, E. Fiorentino, T. Massam, L. Monari, F. Palmonari, F. Zetti and A. Zichichi: *Measurement of wave and group velocity in plastic scintillator*, to be published in *Nuovo Cimento*.

(**) The kinematic curves reported in the published papers (1-4), refer to the quantities t_1 and t_2 which differ by the constant d/c from those defined in (5): *i.e.* the zero reference point is taken at the centre of the target rather than at the centre of the neutron detector ($\theta_n = \theta_0 = 0°$).

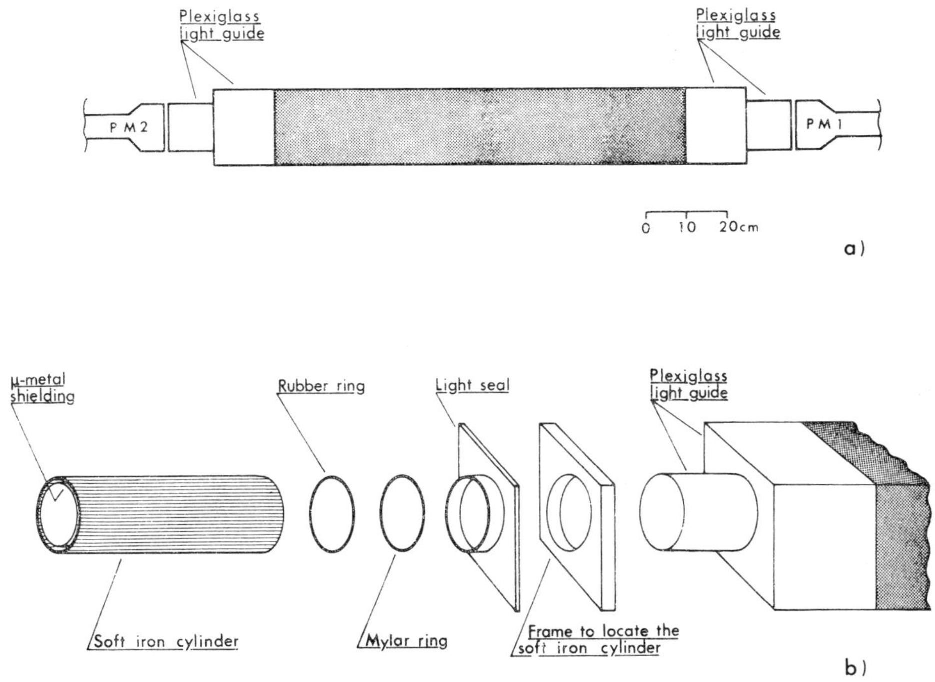

Fig. 5. – *a*) Assemblage of a neutron counter. *b*) The connections between a light guide and the PM base are shown. ▒ Scintillator.

a) attenuation length (*): $\lambda = (385 \pm 50)$ cm ,

b) refractive index (7): $n = 1.56 \pm 0.03$,

c) density: $\varrho = (1.17 \pm 0.01)$ g/cm^3.

In order to test the mechanical strength of a counter, the glued parts were submitted to loads of 500 kg. They did not break. Furthermore, the deformation of the scintillator bar under the effort of its own weight was measured in its centre, using a comparometer. No variations greater than the instrument precision, *i.e.* 10 μm, were observed during a period of two weeks.

4'2. *Mounting an array of counters.* – As mentioned before, a neutron detector is constructed of twelve counters assembled in two stacks of six: a forward set (counters 1 to 6) and a backward set (counters 7 to 12). The two photomultipliers of one counter are identified by whether they are at the end corresponding to large (1) or small (2) neutron emission angles.

(*) The attenuation length λ was maximized by studying the production process of the scintillator. For the measurement of λ, see Subsect. **7**'1.

TABLE I.

Forward stack		Backward stack	
High angle	Low angle	High angle	Low angle
$R_{1/6}$	$R_{2/6}$	$R_{1/12}$	$R_{2/12}$
$R_{1/5}$	$R_{2/5}$	$R_{1/11}$	$R_{2/11}$
$R_{1/4}$	$R_{2/4}$	$R_{1/10}$	$R_{2/10}$
$R_{1/3}$	$R_{2/3}$	$R_{1/9}$	$R_{2/9}$
$R_{1/2}$	$R_{2/2}$	$R_{1/8}$	$R_{2/8}$
$R_{1/1}$	$R_{2/1}$	$R_{1/7}$	$R_{2/7}$

Table I shows the labelling of all the photomultipliers of the detector placed on the right (R) of the incident beam.

The scintillators are supported by means of the soft-iron cylindrical shielding of the photomultipliers, which are fixed in iron supports A, as shown in Fig. 6 a); these iron supports clamp the soft-iron cylinders, and the photo-

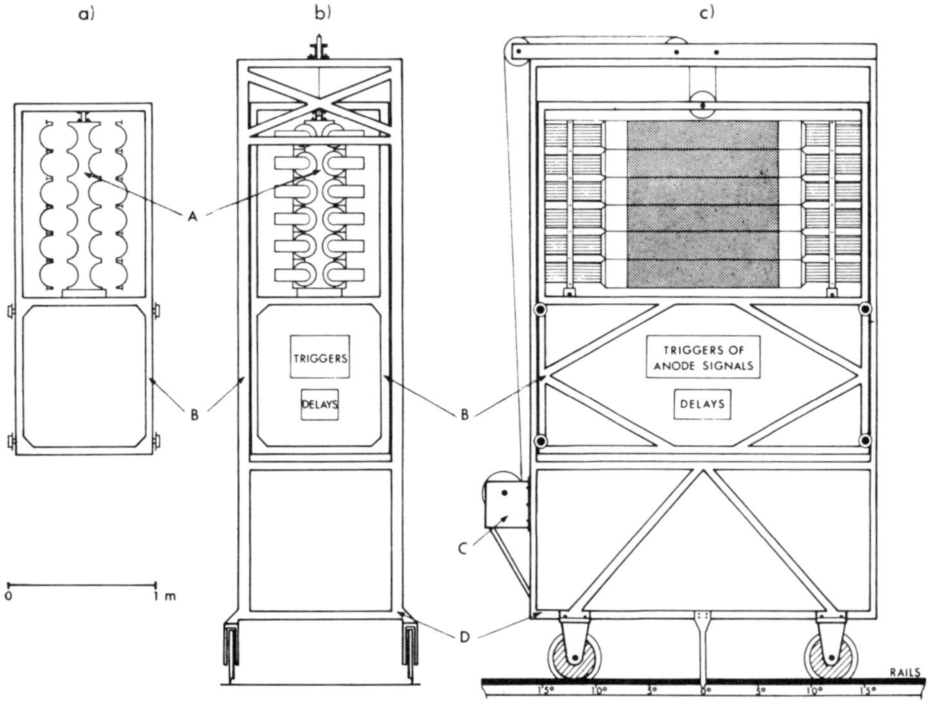

Fig. 6. – *a*) Side view of the metallic structure supporting the bases of the twelve counters. *b*) Side, and *c*) back view of the detector, showing the metallic cage mounted on wheels, in which the above structure supporting the counters can slide vertically.

Fig. 7. – Photo of a neutron detector seen from the back.

multipliers are pressed against the light guides by springs mounted in the soft-iron cylinders. The supports A are mounted in a metallic frame B with dimensions $(2.30\times2.20\times0.95)$ m³ shown in Fig. 6 *a*). The winch C allows this structure to be moved up to 1 m vertically inside the 4 m high, 2.30 m wide, 1.10 m deep metallic cage D which is mounted on wheels (Fig. 6 *b*) and Fig. 6 *c*)). The wheels and the winch allow any point of the detector to be placed in the direct beam. Figure 7 shows a photo of the neutron detector, as seen from the back.

A large anticoincidence counter G_R, fixed on the neutron detector frame, guarantees with high rejection efficiency $(<4\cdot10^{-4})$ that charged particles coming from the target area cannot trigger the electronic logic. G_R is composed of three plastic scintillator elements of $(117\times49\times2.5)$ cm³, each element being viewed through its smallest faces by two 56 AVP photomultipliers.

5. – The experimental arrangement in a missing-mass study.

In this Section the set-up (Fig. 8), used for missing-mass studies, will be briefly described.

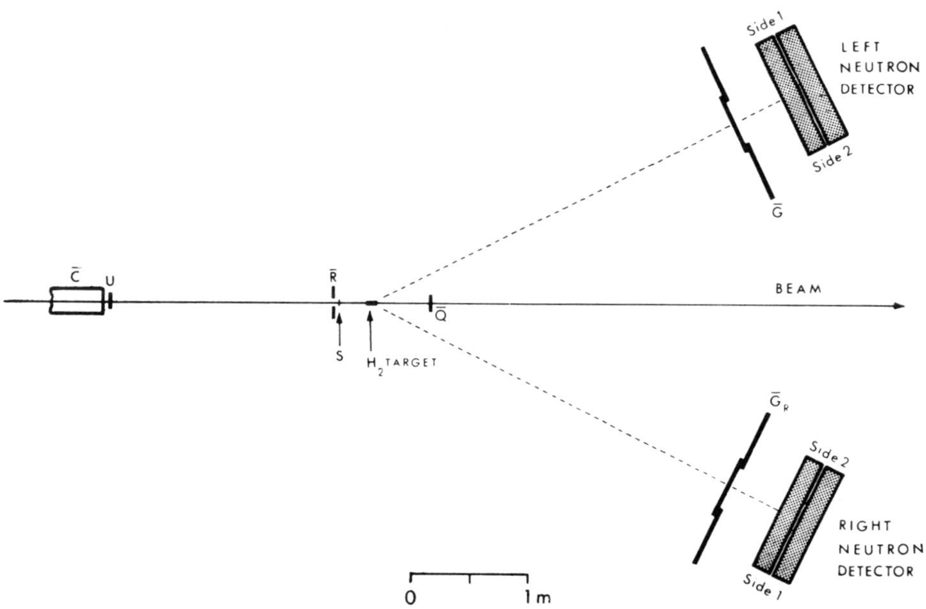

Fig. 8. – Set-up for missing-mass studies. An ethylene Čerenkov counter C and four scintillation counters U, S, R, Q, define pions interacting in the target by the coincidence $US\bar{R}\bar{Q}\bar{C}$. Notice that the 10 cm long liquid-hydrogen target can be considered to be pointlike. $\bar{G}_R$ and $\bar{G}_L$ are veto-counters which prevent charged particles from triggering the neutron detector logic.

A π^- beam (*) traverses an ethylene Čerenkov counter which rejects electrons and then crosses counter U which yields the zero time signal (for dimensions of all counters, see Table II). This counter is split into three hori-

TABLE II.

Counter	Dimensions (cm)		Distance from the target (cm)
U { U_1	$10\times\ 2\times\ 1$		-190
U { U_2	$10\times\ 2\times\ 1$		
U { U_3	$10\times\ 2\times\ 1$		
$\bar{R}$	$20\times\ 20\times\ 1.5$ with a central hole $\varnothing$ 4		$-\ 36$
S	$\varnothing\ 4\times 0.1$		$-\ 30$
$\bar{Q}$	$5\times\ 5\times\ 0.5$		$+\ 35$
H_2 target	$L=10$, $\varnothing=\ 4$	$L=40$, $\varnothing=\ 5$	—
G_R { G_R^1	$49\times 117\times\ 2.5$		$+350$
G_R { G_R^2	$49\times 117\times\ 2.5$		
G_R { G_R^3	$49\times 117\times\ 2.5$		
A_3	$10\times\ 5\times\ 3$		$+300$
A_4	$18\times\ 10\times\ 1$		$+800$

zontal scintillator elements in order to handle high rates, without deterioration of the time accuracy. Counter S, together with counter R, defines a 4 cm diameter beam, well centered in the target. S is a thin (0.1 cm) scintillator designed to have a small interaction rate compared to the target, and R is a counter with a hole ($\varnothing = 4$ cm) in its centre, which, used in anticoincidence,

(*) This beam ($m4_e$) lies in the South Hall of the CERN 28 GeV proton synchrotron and is similar in lay-out to the high-intensity antiproton beam $m4_a$ (8).

(8) G. BRAUTTI, G. FIDECARO, T. MASSAM, M. MORPURGO, TH. MULLER, G. PETRUCCI, E. ROCCO, P. SCHIAVON, M. A. SCHNEEGANS and A. ZICHICHI: *Nuovo Cimento*, **38**, 1861 (1965).

removes beam halo, protects against interactions in counter U and rejects pions bombarding the heavy parts of the target. Counter U was placed far from counter S to reduce the background from U. Neutrons from interactions in counter U could only be recorded if the interaction also produced charged particles in the small solid angle of $5\cdot10^{-4}$ sr subtended by counter S.

Two types of cylindrical liquid-hydrogen target were used: one 10 cm long and 4 cm in diameter and the other 40 cm long and 5 cm in diameter. In the latter case, a reconstruction of the event is needed in order to keep the good missing-mass resolution. Counter Q vetoes pions which do not interact in the target. The signals of the other veto counters, surrounding the target, may be mixed in the same electronic channels as counter Q.

Counters A_3 and A_4 are positioned in the beam line respectively in front and behind the neutron detector, when the latter is set in the direct beam for calibrations.

The beam is dumped in a cavity of 1 m^2, 2 m deep, built in the concrete shielding wall at about 8 m behind the target. This is necessary in order to eliminate back-scattering from the wall which, because of a very small inefficiency ($< 4\cdot10^{-4}$) in the veto counters after the target, could produce a spurious peak at large neutron times of flight.

Two identical neutron detectors, of the type previously described, are placed on the right (R) and on the left (L) of the beam line. They can be displaced around the target on semi-circular rails at a radial distance of 4 m. In this case the polar angle covered by each neutron detector is 14° in the range $0°\div90°$ with an azimuthal acceptance ranging from almost 360° at small polar angles to 14° at 90° polar angle.

6. – Electronics.

6'1. *A single counter logic.* – The diagram of the electronic logic used to obtain the quantity P_θ, t_1 and t_2 (*) from one neutron counter is shown in Fig. 9.

In order to obtain two pulses N_1 and N_2 carrying the time information with a limited dependence on pulse height and at the same time be able to use a reasonably high counting rate, the « high-low » discrimination technique was used. The anode signal of PM_1 (PM_2) is sent to a trigger, the threshold of which is set at a level about ten times lower than the average pulse height [9]; the

(*) For the definition of these symbols, see Sect. 3.

(9) E. Gatti and V. Svelto: *Nucl. Instr. Meth.*, **43**, 248 (1966); W. J. McDonald and D. A. Gedcke: *Internal Report of the Nuclear Research Centre*, University of Alberta, Edmonton, Canada (April 10, 1967).

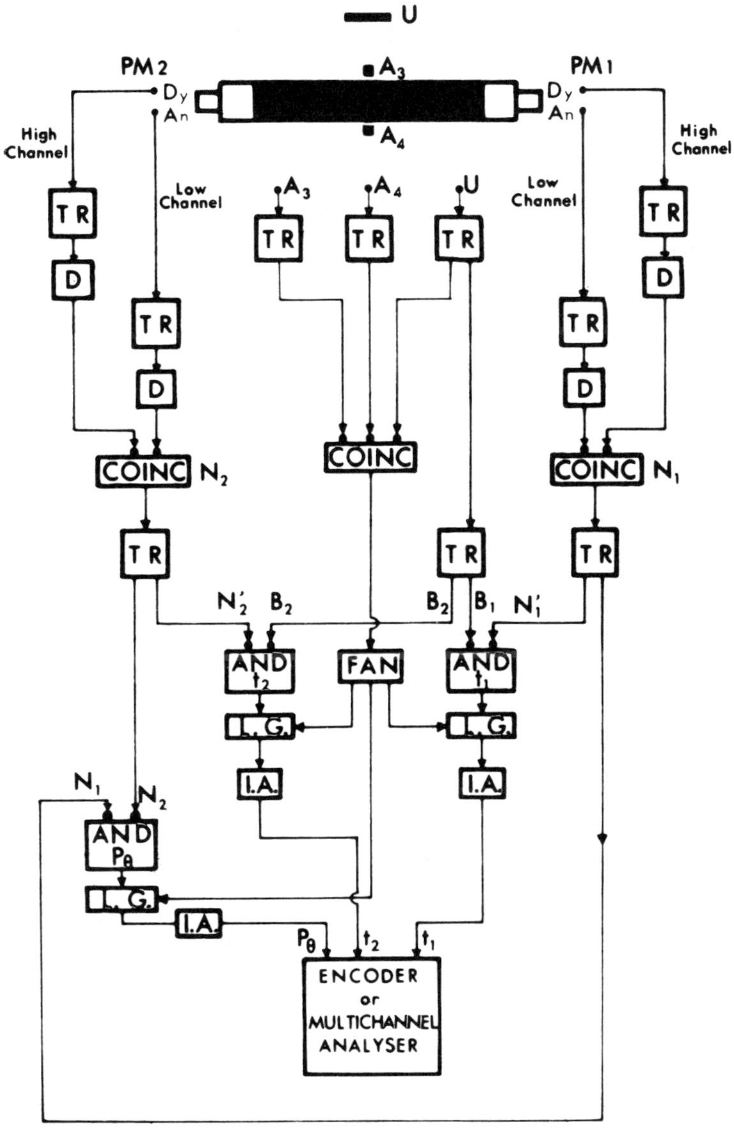

Fig. 9. – Block diagram of the electronics associated with one neutron counter. TR is a trigger, D is a delay line, COINC is a coincidence unit, L.G. is a linear gate and I.A. is an integrator amplifier. In the coincidence units, as well as in the AND circuits, the symbols • or o have been adopted for coincidence and anticoincidence inputs respectively.

signal from the last dynode of PM_1 (PM_2) is sent to a high-threshold trigger, set at a level depending on the physical problem to be investigated. The « high » pulse gates the « low » pulse yielding the signal N_1 (N_2), which is reshaped to be 50 ns long. This length is determined by a clip-cable and so is very stable, which is essential for the technique described below.

The overlap of pulses N_1 and N_2 in an AND circuit yields the signal P_θ. N_1 is delayed with respect to N_2 by 10 ns in order to avoid negative-time ambiguity; in this way the leading edge of signal P_θ is always synchronized with the pulse N_1, but its length, τ_θ, depends on the interaction point and is greater for interaction points nearer to PM_1.

The pulse from beam counter U, carrying the zero-time information, is shaped in a trigger to give identical signals B_1 and B_2. Two signals N_1' and N_2', identical to N_1 and N_2, are available. The overlap of B_1 and N_1' produces t_1 and the overlap of B_2 and N_2' produces t_2. The timing is such that the leading edges of t_1 and t_2 are synchronized with N_1' and N_2' respectively, and a larger overlap corresponds to a shorter time of flight.

The signals P_θ, t_1 and t_2 pass through a linear gate, L.G., opened by a final coincidence which, both for tests and calibrations with charged particles, consisted of pulses from counters U, A_3 and A_4.

The gated signals P_θ, t_1 and t_2 are sent to MCA's or to encoders through an Integrator Amplifier (I.A.).

6·2. *Extension to an array of counters.* – The electronic logic for the neutron detector (Fig. 10), is a superposition of twelve elementary circuits identical to that described above.

The anode pulses from the six PM's at one end of a set of counters, for instance $R_{1/1}$ to $R_{1/6}$, carefully equalized in height using a relativistic muon beam, are separately discriminated at a low level of 0.1 V. The timing of the six discriminator outputs is equalized to within ± 0.1 ns using continuously variable delay lines (see Fig. 11) (*) and then the signals are mixed in an OR circuit. The electronics of the « low discrimination » channel is placed near the PM bases in order to avoid long cables which would produce deterioration of the rise time: a vital feature of the signals which carry time information.

The dynode pulses, equalized and timed, enter a linear mixer, whose output pulse is submitted to a high-level discrimination. This is the « high discrimination » channel. « High » and « low » channels enter the $R_{1/F}$ coincidence unit in such a way that the « high » pulse gates the « low » pulse, the rising edge of

(*) Each line consists of coaxial pipes sliding into each other. The radial dimensions are such that the mean characteristic impedance is 50 Ω. Reflections are less than 7%. The movement of the pipes is remotely controlled, and microswitches indicate the range limits ($0 \div 1.5$) ns.

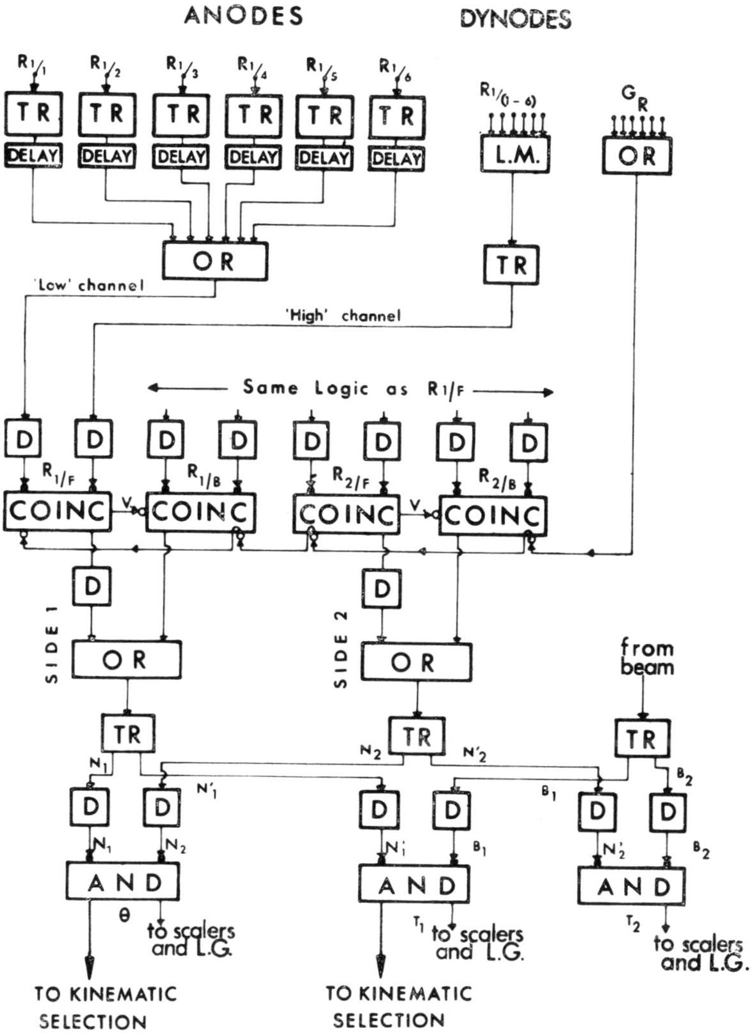

Fig. 10. – Block diagram of the electronics for a neutron detector built of twelve counters. L.M. is a linear mixer, DELAY is the variable delay line shown in Fig. 11. Other symbols are those of Fig. 9.

which carries the time information. In the same way, the three other groups of PM's yield the timing signals $R_{2/F}$, $R_{1/B}$ and $R_{2/B}$.

These four coincidences are all vetoed whenever a charged particle is detected by counter G_R, *i.e.* when any one of the six photomultipliers which view the three segments of the counter give a signal. Furthermore, coincidences $R_{1/B}$ and $R_{2/B}$ are vetoed by the pulses $R_{1/F}$ and $R_{2/F}$ respectively, in order to take the timing only from the forward set of counters, if the neutron converts in the front layer. This is because a signal could be produced in the back layer by a long recoil proton, thus spoiling the measurement of the neutron interaction point in the front layer.

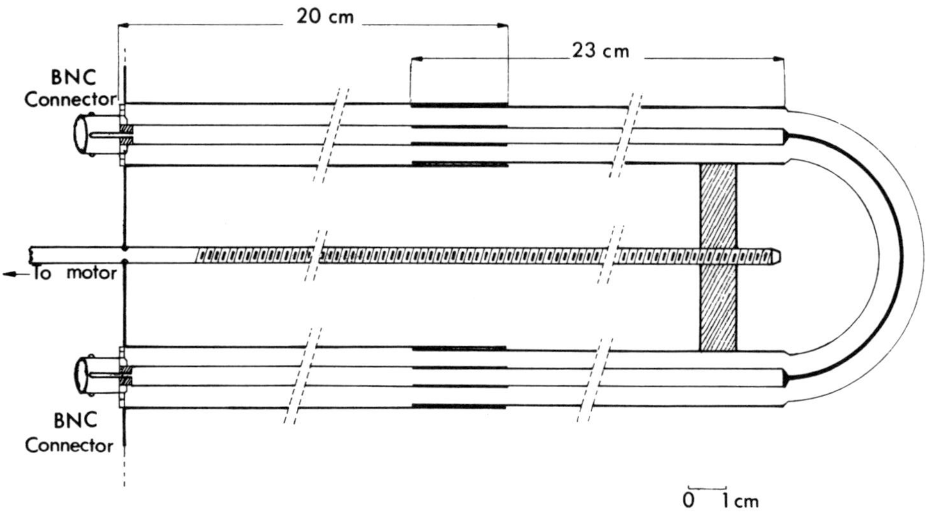

Fig. 11. – Variable delay line used for the fine timing equalization of all counters.

Pulses $R_{1/B}$ and $R_{1/F}$ are mixed in an OR circuit, the output of which is called N_1, and analogously $R_{2/F}$ and $R_{2/B}$ give N_2. Pulses N_1 and N_2, corresponding to each side of the neutron detector, can be considered as coming from one single scintillation counter. They provide the pulses P_θ, t_1 and t_2 in the way described in Subsect. **6**'1 for one element of the detector. These signals, gated by the FINAL coincidence which will be described in Subsect. **6**'3, are integrated and amplified and then sent to encoders and MCA's.

A set of pattern units indicates which neutron counter is hit, in order to know the azimuthal angle and the distance of the counter from the target. The knowledge of the neutron pattern also allows one to eliminate multiple events.

6'3. *Auxiliary electronics for missing-mass studies.* – The neutron-detector electronics has been described in the previous paragraph for one neutron detector only. When we use two detectors, « Right » and « Left », signals $L_{1/B}$ and $L_{1/F}$ are mixed with $R_{1/B}$ and $R_{1/F}$ in the OR circuit producing SIDE 1 pulse (see Fig. 10), and $L_{2/B}$ and $L_{2/F}$ are mixed with $R_{2/B}$ and $R_{2/F}$ in the OR circuit producing SIDE 2 pulse. Figure 12 shows the diagram of the overall logic.

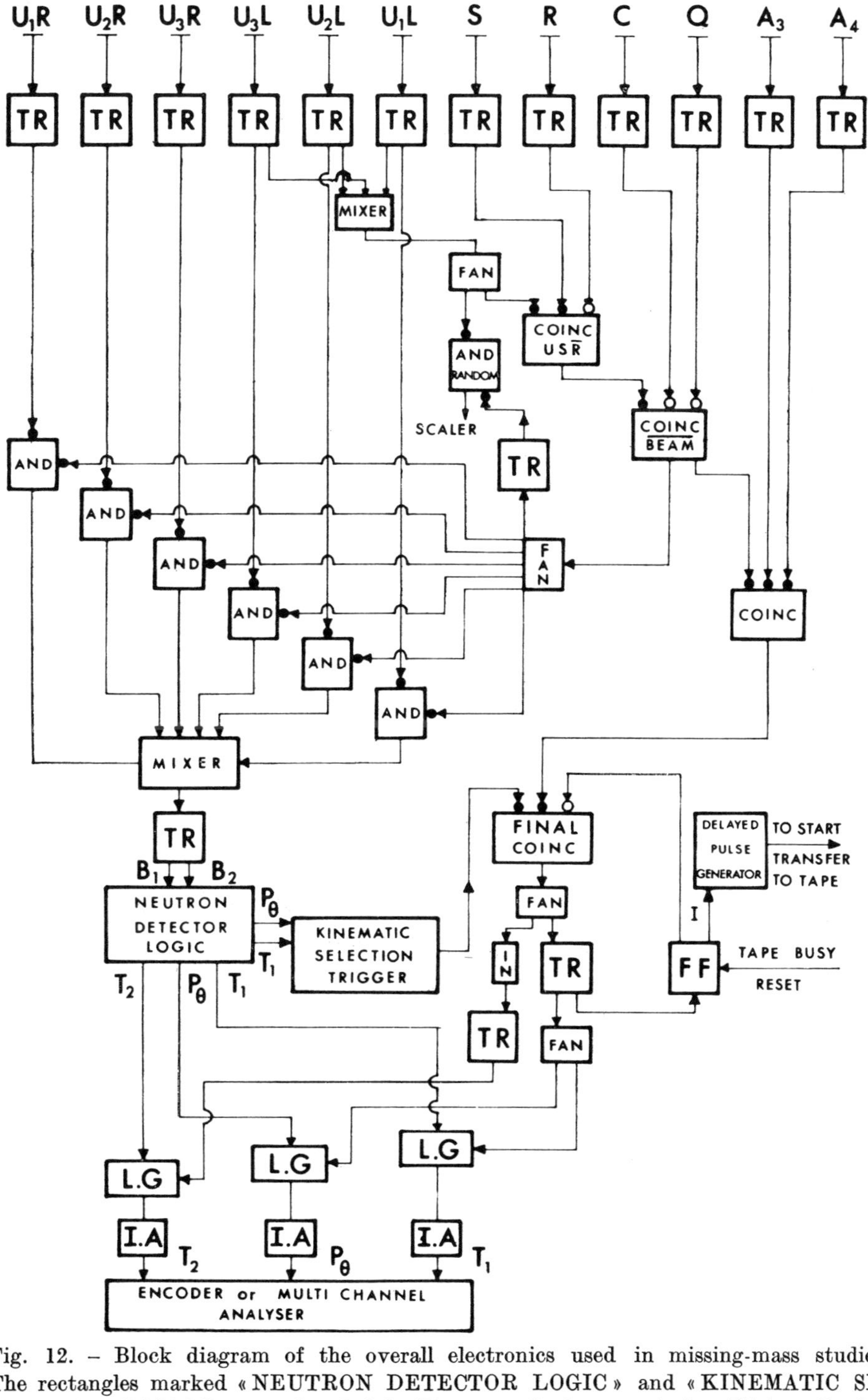

Fig. 12. – Block diagram of the overall electronics used in missing-mass studies. The rectangles marked « NEUTRON DETECTOR LOGIC » and « KINEMATIC SELECTION TRIGGER » are detailed in Fig. 10 and 13 respectively. FF is a flip-flop circuit, IN is an inverter, other symbols are those of Fig. 9.

6˙3.1. Beam telescope. Counter U consists of three segments of plastic scintillator, each viewed by two 56 AVP photomultipliers labelled U_R or U_L according to their position relative to the incident beam.

The six pulses of counter U, all equalized at (1.5 ± 0.15) V, are discriminated at a low level of 0.1 V (in this way, the leading edges of the six trigger output pulses carry the smallest time uncertainty which we were able to achieve: $\pm$ 0.35 ns).

The three pulses $U_{L/1}$, $U_{L/2}$, $U_{L/3}$, are mixed to give pulse U, which is put in coincidence with pulses S and is anticoincided by pulse R. $US\bar{R}$ indicates the number of pions incident on the target and was used as the monitor for all experiments. In a second coincidence unit, pulse $US\bar{R}$ is vetoed by counter Q pulse and by the Čerenkov pulse C; the output pulse of the coincidence $US\bar{R}\bar{Q}\bar{C}$, called BEAM, indicates the number of pions which interacted in the target. This low rate signal, BEAM, is now used to gate all six signals from U.

The gated signals are timed to $\pm$ 0.1 ns and mixed to produce the pulse U_T, which is reshaped by a trigger giving B_1 and B_2, the zero-time references for neutron time-of-flight measurements.

Another of the BEAM signals is reshaped to 100 ns width, and then put in coincidence with U. The number of output pulses of this coincidence, called RANDOM, measures how many times a second U pulse occurs during the 100 ns time interval following a BEAM signal.

6˙3.2. Fast kinematic selection (MASS BOX). The application of kinematic restrictions to the final electronic trigger is a powerful method of reducing background triggers. Figure 13 shows the electronic block-diagram of the kinematic-selection system. A fast signal M is produced only when the emission angle measured by the signal P_θ, and the time of flight of a neutron measured by the signal t_1, satisfy certain conditions.

The two pulses P_θ and t_1 are treated electronically in identical ways. Pulse $P_\theta(t_1)$ is fanned out into eight channels, and each channel i (j) is clipped by an open-ended cable of length $\tau_\theta^i/2$ $(\tau_t^j/2)$. Figure 14 shows the three cases which can occur depending on the length τ_θ of the signal P_θ. If the clipped pulse is applied to a discriminator which is set at a threshold higher than the single-pulse level, the circuit will operate only if the pulse is longer than τ_θ^i (τ_t^j). This is equivalent to selecting neutron emission angles (times of flight) greater (smaller) (*) than θ_n^i (t_1^j).

If an event occurs between the angle (time of flight) $\theta_n^i(t_1^j)$ and $\theta_n^{i+1}(t_1^{j+1})$, the triggers corresponding to $\theta_n^i, \theta_n^{i-1}, \theta_n^{i-2}, \ldots$ $(t_1^j, t_1^{j-1}, t_1^{j-2} \ldots)$ will operate. A sys-

(*) See Subsect. **6**˙1.

tem of coincidence units, indicated by $(i, \overline{i+1})$, is arranged so that only a signal corresponding to $\theta_n^i \leqslant \theta_n < \theta_n^{i+1}(t_1^j > t_1 \geqslant t_1^{j+1})$ passes and the others are vetoed. A careful adjustment of timing is done before all the coincidence units.

Since the leading edges of both θ_n and t_1 were synchronized with N_1, the two signals corresponding to $\theta_n^i \leqslant \theta_n < \theta_n^{i+1}$ and to $t_1^j > t_1 \geqslant t_1^{j+1}$ may be brought

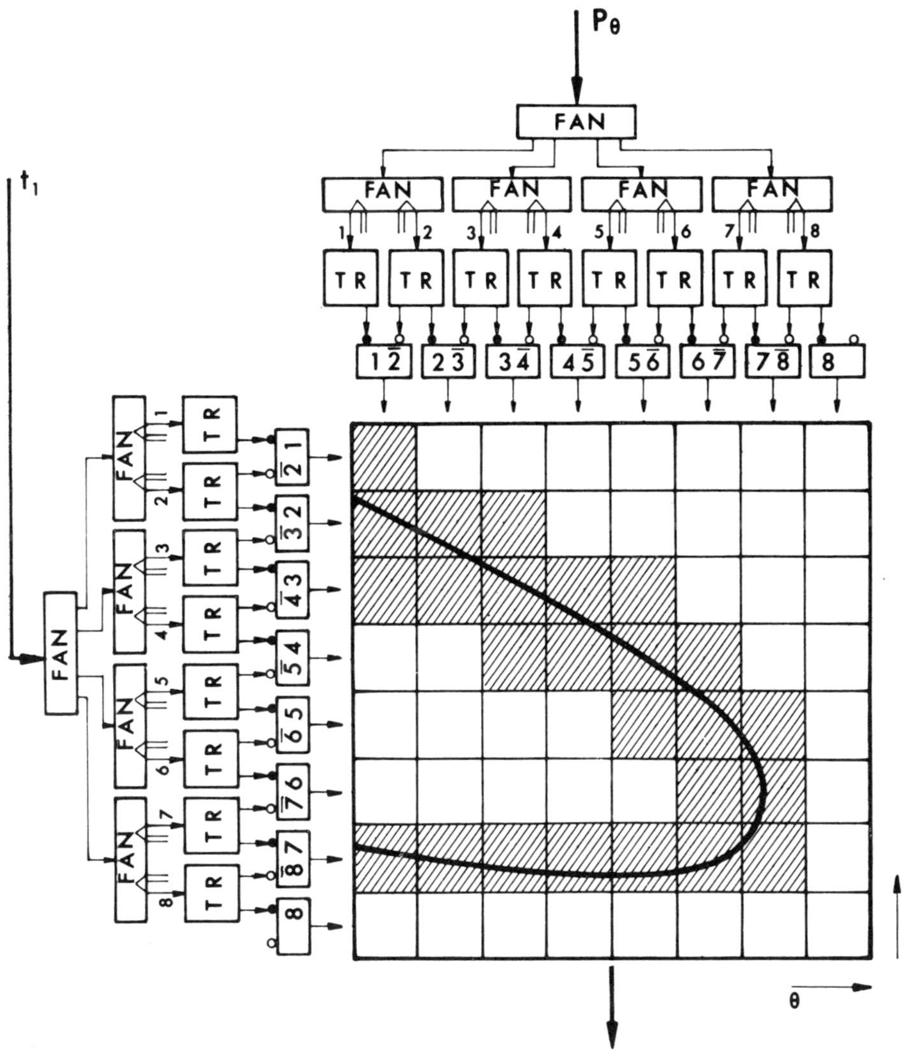

Fig. 13. – Schema of the blending of P_θ and t_1 information in the fast matrix coincidences. The hatched squares represent regions of the $(\theta_n$-$t_1)$-plane, selected by means of active elements plugged in the corresponding position of the matrix. ▨ coincidence on, □ coincidence off.

into sharp coincidence (5 ns resolution) in a fast matrix coincidence system in which an active element, plugged into the i (j) position, produces the fast output pulse M, also synchronized with N_1. Because this short resolving time may be used, randoms are kept at a low level, and negative-time ambiguity in t_1 is avoided.

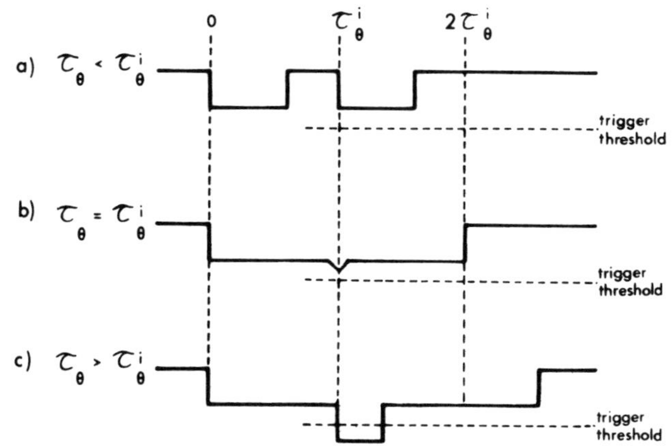

Fig. 14. – Clipped pulses P_θ for different physical situations: *a*) $\theta_n < \theta_i$, *b*) $\theta_n = \theta_i$, *c*) $\theta_n > \theta_i$. The horizontal dashed lines refer to the threshold of the triggers placed before the coincidence-anticoincidence units (see Fig. 13).

Figure 13 also shows a possible application of the kinematic-selection system: active elements are plugged in certain positions corresponding to a desired missing-mass curve in the $(\theta_n\text{-}t_1)$-plane; this allows, for instance, the triggering of the data-recording system only when a given resonance is produced in association with a neutron in a two-body process.

Another possible way of using the system is to explore large missing-mass regions in the $(\theta_n\text{-}t_1)$-plane, excluding only those regions with high background.

The same electronic system can also be used for the selection of collinear events in colliding-beam experiments ([10]).

6'3.3. The FINAL coincidence. A derivation of BEAM and the kinematic-selection pulse M (and A_3 and A_4 signals during calibrations with charged particles) are sent into the FINAL coincidence unit. Further coincidence conditions may be introduced at this point, for example the selection of particular final-state configurations. The relative timing of these pulses is

---

([10]) P. DALPIAZ, G. FORTUNATO, T. MASSAM, F. ZETTI and A. ZICHICHI: *Proceedings of the International Symposium on Electron and Positron Storage Rings, Saclay, 26-30 September 1966*, p. V-6-1.

such that the leading edge of the signal FINAL is synchronized with N_1, while the falling edge is synchronized with BEAM.

Signal FINAL fulfils three important purposes:

it gates P_θ, t_1, t_2 and the neutron hodoscope signals;

it starts the transfer onto magnetic tape;

it provides a dead time during encoding and transfer to tape.

In order to minimize the effects of jitter between the gate signals and P_θ, t_1 and t_2 signals, the gate for P_θ and t_1 is derived from the leading edge of FINAL, but that for t_2 is derived from the falling edge of the FINAL pulse by differentiating, inverting and reshaping the FINAL output.

Another derivation of FINAL pulse enters a flip-flop circuit (FF), which produces pulses I, one of which blocks the FINAL coincidence. The second I pulse enters a delayed pulse generator, which, 200 μs later, sends a « print command » to the scalers and a « start transfer » to the magnetic-tape read-out.

This delay ensures that the scalers have finished their carry before reading takes place. At the end of the transfer, a pulse is produced, which resets the (FF) circuit, opens the FINAL coincidence and allows the experiment to continue.

7. – Calibration with charged particles.

The calibration of the neutron detector is essential in order to know the correspondence between the quantities θ_n, t_1 and t_2 and the respective encoder or MCA channels; to check pulse stability; and to record the relative delays of the counters.

Since it was impossible to get a narrow, monochromatic neutron beam in the experimental area where the set-up is positioned for the main experiment, most of the calibrations were performed with charged particles. Nevertheless, the response of the detector to neutrons has been checked in special runs and will be reported in Sect. **8**.

7·1. *Performances.* – In this Subsection, we will describe the measurements performed in each of the twenty-four neutron counter elements using the electronic logic described in Subsect. **6**·1.

7·1.1. Pulse height resolution. The pulse height resolution was measured using μ^- of momentum $\geqslant 1$ GeV/c; therefore the energy lost in the plastic scintillator was ~ 40 MeV. The telescope A_3A_4 was set at the middle of the scintillator element to be calibrated, and the anode pulses of the latter, gated by A_3A_4 coincidence, were sent to the Integrator-Amplifier and were then analysed on a MCA. (For this measurement, both counters A_3 and A_4 had dimensions

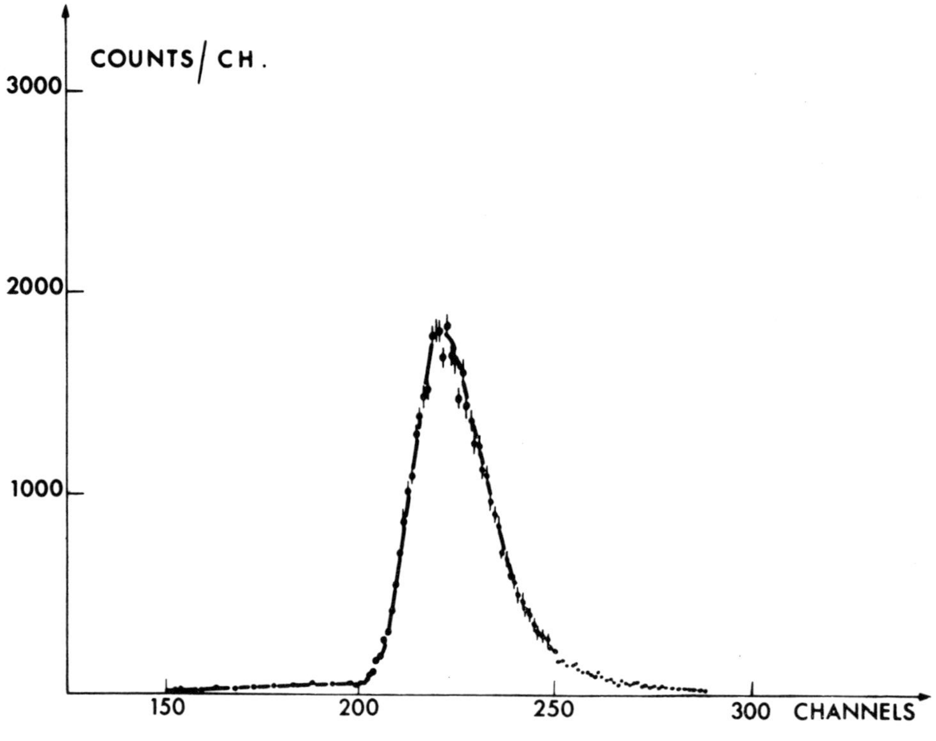

Fig. 15. – Energy resolution of the neutron counter for muons. The pulses of the two PM's have been added and then analysed on MCA. The zero-point energy corresponds to channel No. 1, and is not reported in the Figure.

$(2.5\times2.5\times18)\,\text{cm}^3$.) The pulse height resolution was found to be $\pm5.75\%$ when only one PM was used and $\pm4\%$ when both PM pulses were added together. An example of the distribution of the pulse amplitudes obtained adding the two PM signals of one scintillator element is shown in Fig. 15.

7·1.2. Attenuation length. With the same electronic arrangement as in 7·1.1, a measurement of the scintillator's attenuation length has been performed using cosmic-ray muons. In Fig. 16 the peak position of the pulse-height distribution for PM_1 is plotted *vs.* position of the A_3A_4 telescope along the scintillator. This curve yields an attenuation length: $\lambda=(385\pm50)$ cm.

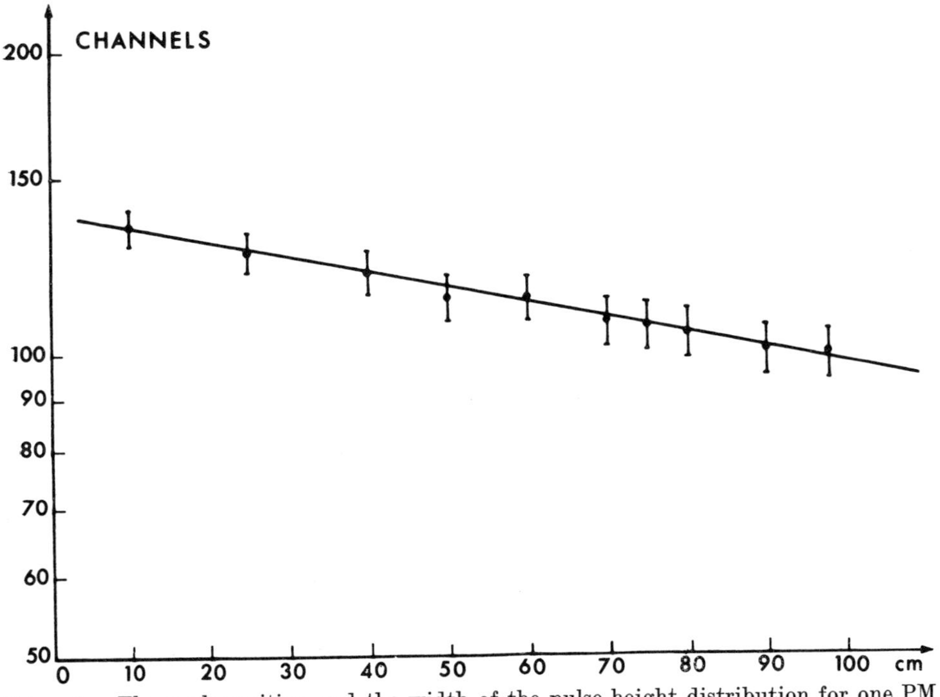

Fig. 16. – The peak position and the width of the pulse-height distribution for one PM is plotted *vs.* the position of the A_3-A_4 telescope along the scintillator. This graph allows the determination of the scintillator attenuation length.

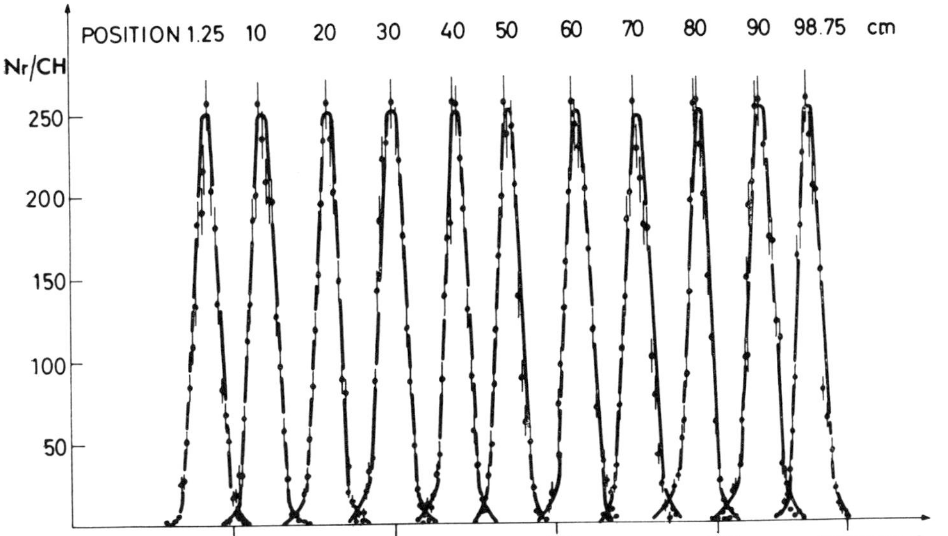

Fig. 17. – Each of the peaks in the P_θ spectrum corresponds to a given position of the A_3-A_4 telescope along the neutron counter. This spatial resolution has been measured with a muon beam. Notice that it is constant along the whole counter.

7·1.3. Spatial resolution. The position of the muon interaction point along the scintillator was selected by the telescope A_3A_4, and the electronics used was the one described in Fig. 9 (excluding counter U). Figure 17 shows

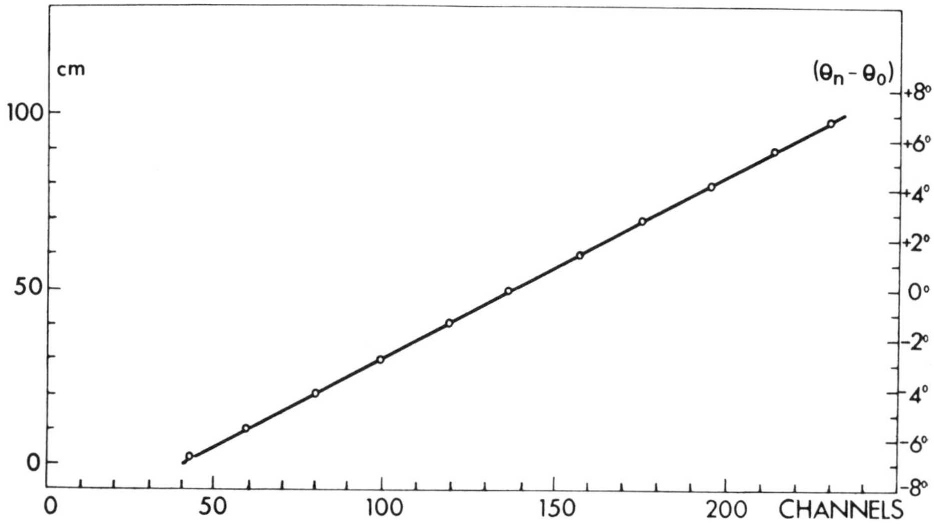

Fig. 18. – Degree and cm to channel calibration for P_θ. Notice the linearity of the response.

the position distribution for various telescope settings. Correcting for finite A_3 and A_4 width (2.5 cm), a resolution for charged particles of ± 1.4 cm is obtained. Figure 18 represents a typical example of a check measurement, which shows the linearity of the instrument.

7·1.4. Uniformity of response. This test was done with cosmic rays, since the distribution of incoming particles is then uniform all over the counter. The P_θ spectrum on MCA presents a flat part, if the response is uniform. Fig. 19 shows a typical cosmic-ray spectrum; the flat part differs from a straight line by less than 5%. The length of the flat part corresponds to the length of the scintillator, which is a rough check of the channel to position calibration.

The slopes of the two steep edges of the spectrum depend on the spatial resolution. The observed slopes agree with the resolution measured using counters A_3 and A_4.

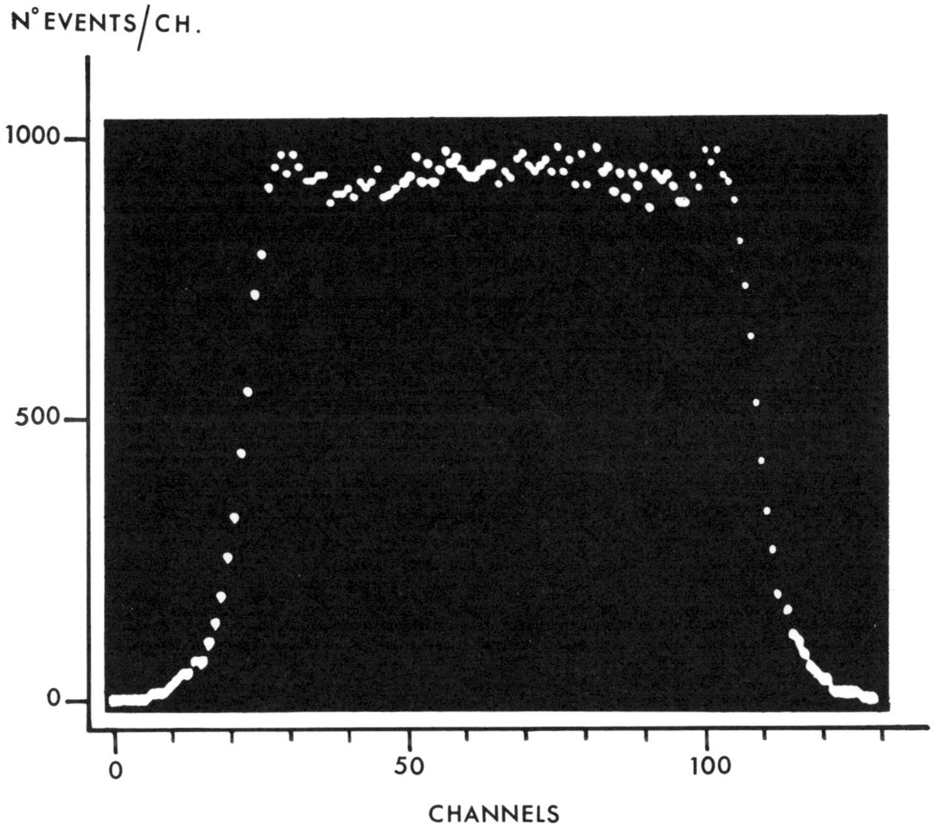

Fig. 19. – Test performed with cosmic rays: the incident-particle flux being uniform all over the counter, a flat part is expected corresponding to the length of the scintillator. The uniformity of the response can be evaluated to be more than 95%.

7'1.5. Time-of-flight resolution. Using a relativistic pion beam with the electronic arrangement of Fig. 9, the resolution in t_1 and t_2 was measured and found for both of ± 0.35 ns. Figure 20 shows the spread of t_1 directly observed on the MCA.

7'1.6. Pulse stability. To have a neutron detector reliable over long periods of time, a high stability of the PM pulses of all neutron counters is required.

Stability in transit time and gain of the PM depend very closely on internal factors and on the stability of the high-voltage supplies.

The effect of the intrinsic instability on all the neutron counters was checked, and it was found that the PM's become more stable with time. Figure 21 *a*)

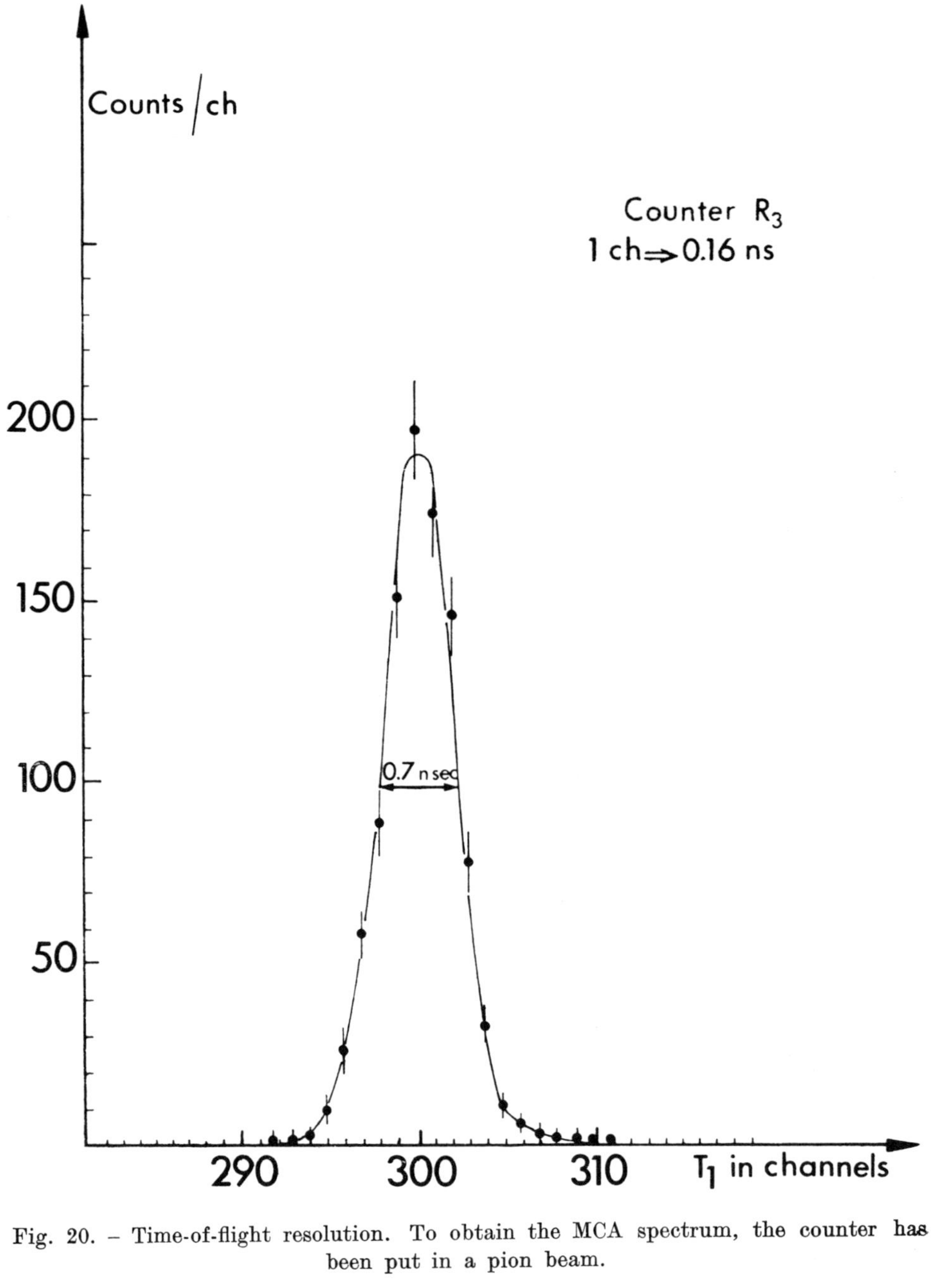

Fig. 20. – Time-of-flight resolution. To obtain the MCA spectrum, the counter has been put in a pion beam.

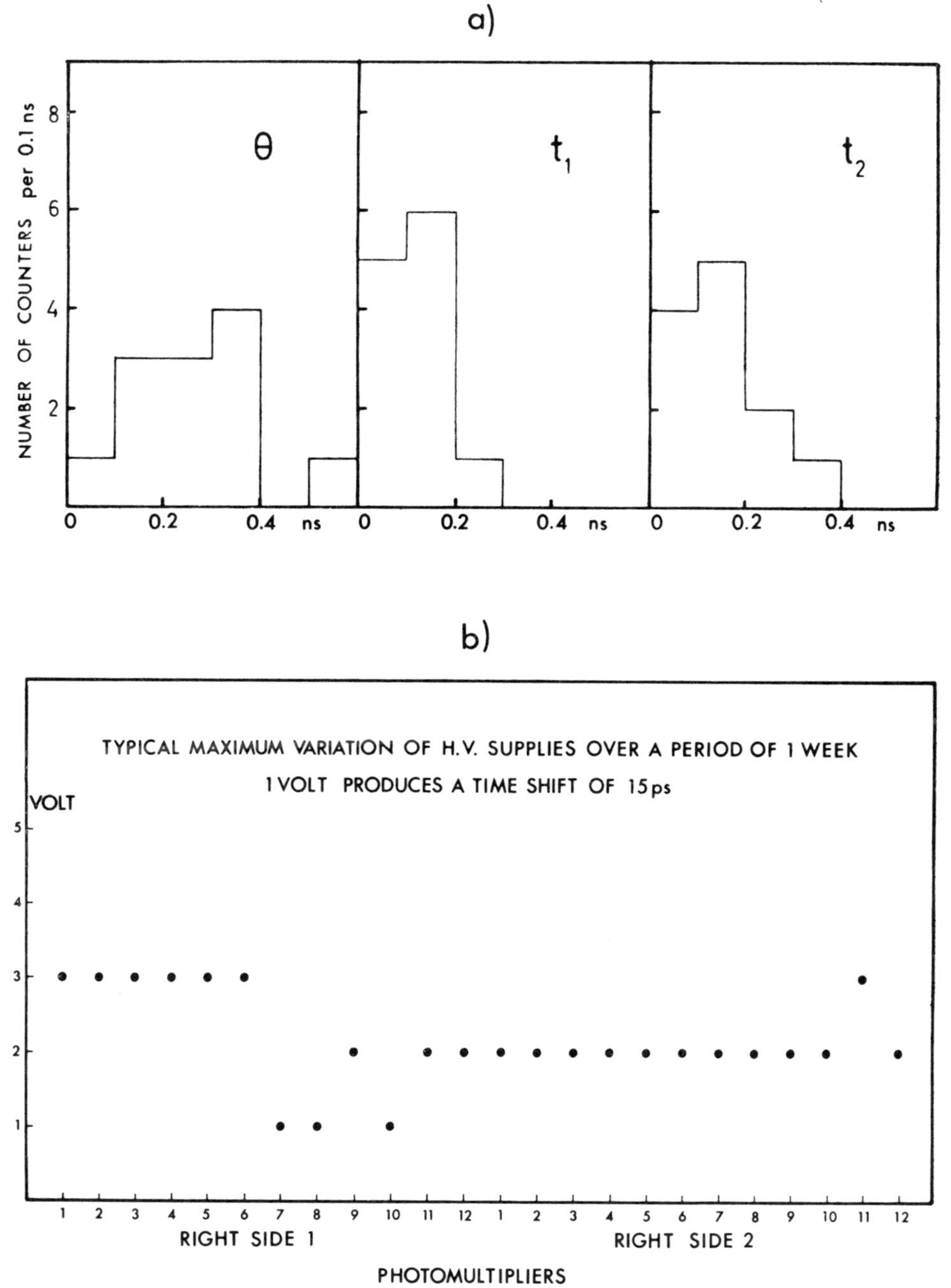

Fig. 21. – *a*) Neutron counter stability over a week: the time variation for θ, t_1 and t_2 signals is plotted for each counter. *b*) Typical maximum variation of the H.V. supplies over a period of one week, for all the PM's of one neutron detector. Notice that in fact a variation of 1 V produces a 15 ps shift in the time definition.

shows the stability of P_θ, t_1 and t_2 over a period of one week for the twelve counters of the detector « Right ».

For the high voltage, we used highly stabilized power supplies, which were periodically checked with a digitized voltmeter and adjusted to ± 0.5 V whenever they drifted by more than 3 V. Figure 21 *b*) shows the H.V. stability of the PM's.

7·2. *Pulse-height equalization.* – The dynode pulse heights of all PM's were equalized to (1.25 ± 0.1) V. Each scintillator in turn was centred in the direct beam and the pulse heights were measured using polaroid pictures of the pulses displayed on a fast oscilloscope, which was triggered by a beam telescope signal. Periodic checks were performed using the same technique.

7·3. *Calibration in position and time of flight.* – Each counter was put successively in the direct beam which was defined electronically by BEAM$\cdot A_3 \cdot A_4$. The signals P_θ, t_1 and t_2 were analysed on three MCA's, in order to check the resolution in position and time of flight and to know the relative timing of the electronic channel corresponding to each counter. The signals were also encoded and recorded on magnetic tape. Calibrated delay lines were used to vary the delay of N_1 and N_2 relative to B_1 and B_2, in order to calibrate the analysers

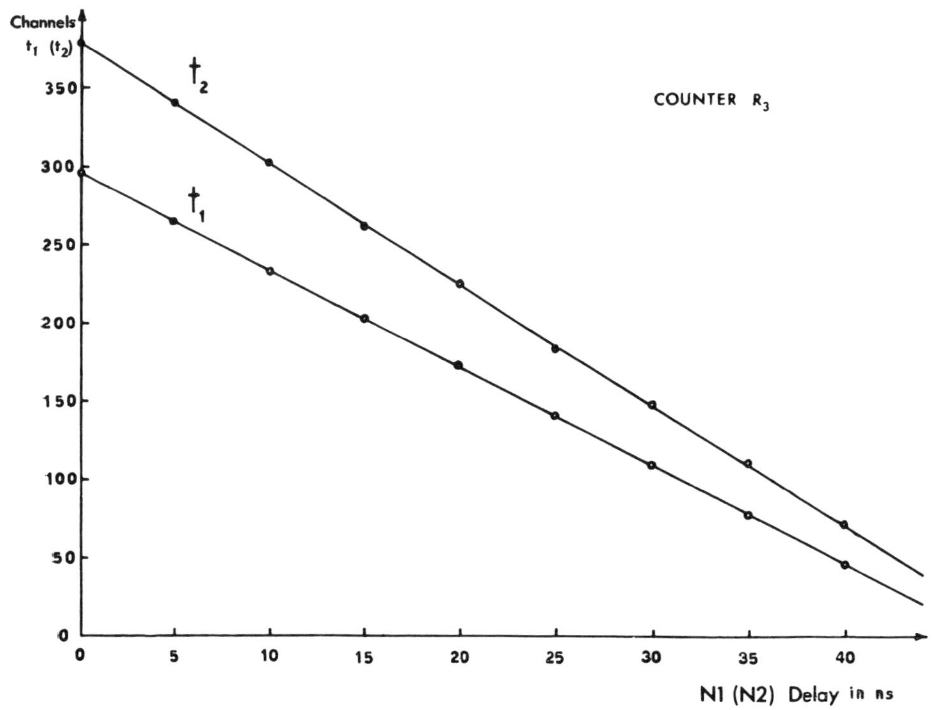

Fig. 22. – Calibrations of t_1 and t_2 to the respective encoder channels.

and encoders used for t_1 and t_2. Figure 22 shows the calibration and linearity of the t_1 and t_2 encoders. To calibrate the analyser and the encoder used for the signal P_θ, the detector was moved on its rails and positioned at known angles from $-7°$ to $+7°$ (Fig. 18). The linearity of this curve over the full width of the counter shows that edge effects are small.

7'4. *Timing.* – In principle it is sufficient to know the relative delays of the electronic channels corresponding to each counter to compute the missing mass. However, in order to have clean electronic selection of time and angle, and to be able to observe the mixed spectra of twenty-four counters directly on the MCA, it is necessary to time the neutron-counter electronic channels very carefully.

This timing was done relative to counter U, whose six anode pulses had first been timed amongst themselves to ± 0.1 ns. Each neutron counter

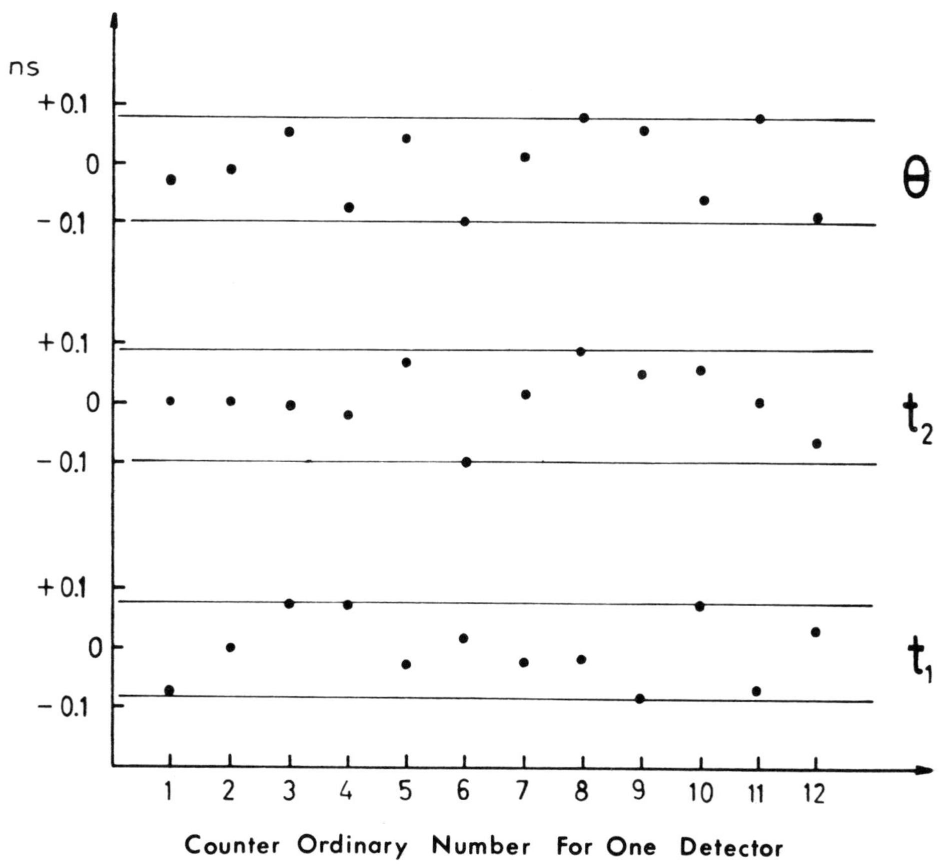

Fig. 23. – Timing amongst the counters: the strips indicate the precision achieved for P_θ, t_1, t_2.

element in turn was centred in the direct beam and spectra of P_θ, t_1 and t_2 were taken on the MCA and on the encoder. The delays of the anode pulses were adjusted until the peak positions for all counters coincided; the dynode delays were consequently adjusted to maintain the correct time relation between the « high » and « low » signals. Rough adjustments were done with cables, and the fine time equalization of anode channels was performed before the mixer, using the variable delay line shown on Fig. 11.

Figure 23 shows the time equalization achieved. All counters of both detectors were equalized in time for P_θ, t_1 and t_2 to ± 0.1 ns.

8. – Performance in use as a neutron detector.

The performance of the neutron detector was studied using neutrons associated with the well-known resonances η and ω in reaction (1).

8'1. *The position resolution.* – In order to calibrate the resolution in neutron position of the detector, the reaction

$$\pi^- + p \to n + \eta \quad (\eta \to \text{neutrals})$$

was studied at 0.81 GeV/c incoming pion momentum. The centre of one neutron detector was set at 28° on the right-hand side of the beam line and three anticoincidence counters surrounding the target were added to the set-up shown in Fig. 8, in order to select the neutral decay modes of the η-meson. The electronics used was explained in Subsect. **6'**3.

The missing-mass distribution in the region *a* of the $(t_1\text{-}\theta_n)$-plane (Fig. 24), where the isomass curves are nearly perpendicular to the θ_n-axis, was measured. This allowed a good calibration of the resolution in neutron position. To select this region, the time acceptance of the electronics was reduced to 5 ns with the help of the kinematic-selection trigger (see Subsect. **6'**3.2). In this way, neutrons with energies ranging from 70 to 90 MeV were selected.

Figure 25 *a*) shows the neutron missing-mass distribution obtained. The smooth background outside the η peak allows a fit with a polynomial function. Figure 25 *b*) shows the η peak after subtraction of the background.

The observed mass resolution obtained from a least-square fit of a Gaussian to the missing-mass distribution is (8.4 ± 0.4) MeV FWHH. This resolution depends on several external contributions which can be calculated (see Table III), as well as on the precision in measuring the neutron interaction point. Combining the quantities given in Table III, comparing the result with the observed resolution, and taking into account that 1 cm change in the neutron

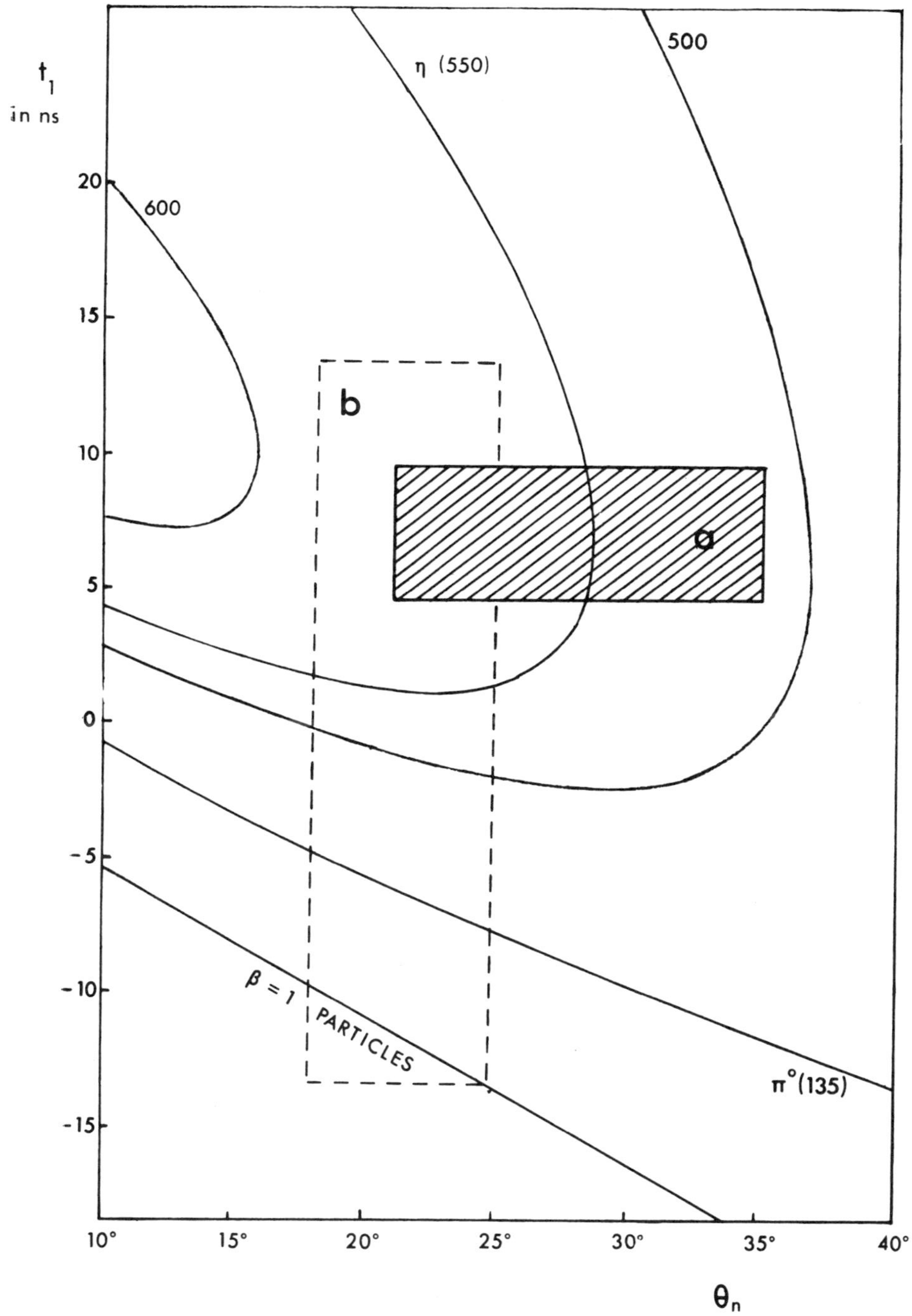

Fig. 24. – Kinematic curves of the reaction $\pi^- + p \rightarrow B^0 + n$ at 0.81 GeV/c. The areas *a* and *b* indicate the regions used for the measurement of the spatial and of the time resolution respectively.

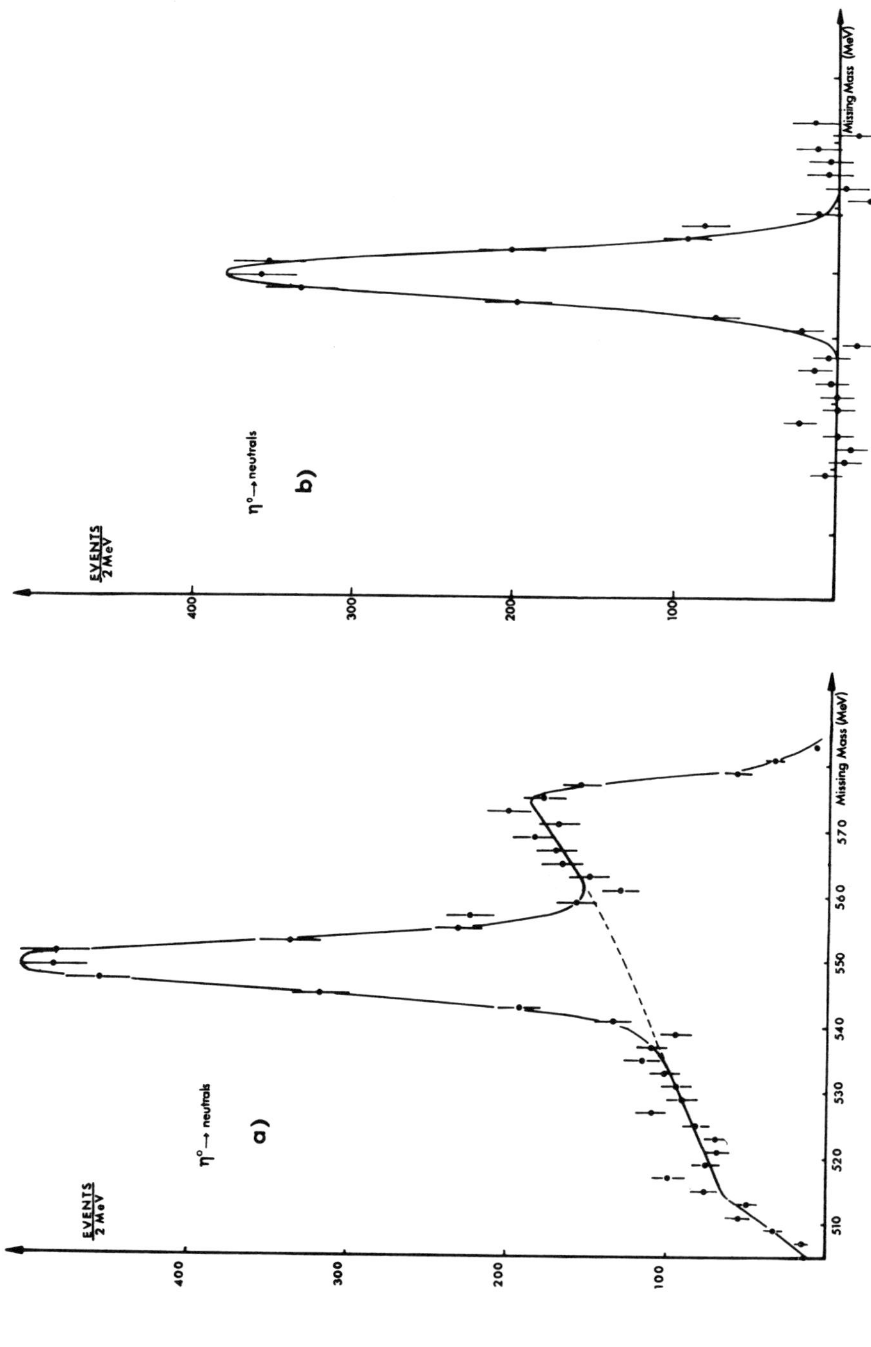

Fig. 25. – *a*) The neutron missing-mass distribution, showing the η peak obtained using the region *a* of Fig. 24. *b*) The same η peak after background subtraction.

TABLE III. – *Contributions to the* FWHH *mass resolution.*

	(MeV)
Beam momentum spread ($\pm$ 0.75%)	6.0
Target dimensions ($\varnothing$ 4·10 cm^3)	4.4
Angular spread of the beam ($\pm$ 8 mrad)	1.2
Timing θ ($\pm$ 0.1 ns)	1.0

position corresponds to 0.75 MeV change in missing mass, one obtains the neutron position resolution equal to $\pm$ (2.5 $\pm$ 0.7) cm. This is in good agreement with the value expected for neutrons when the resolution for charged particles is used for the recoil protons.

8·2. *The time-of-flight resolution.* – The time-of-flight resolution of the neutron detector is measured in a way analogous as for the positional resolution,

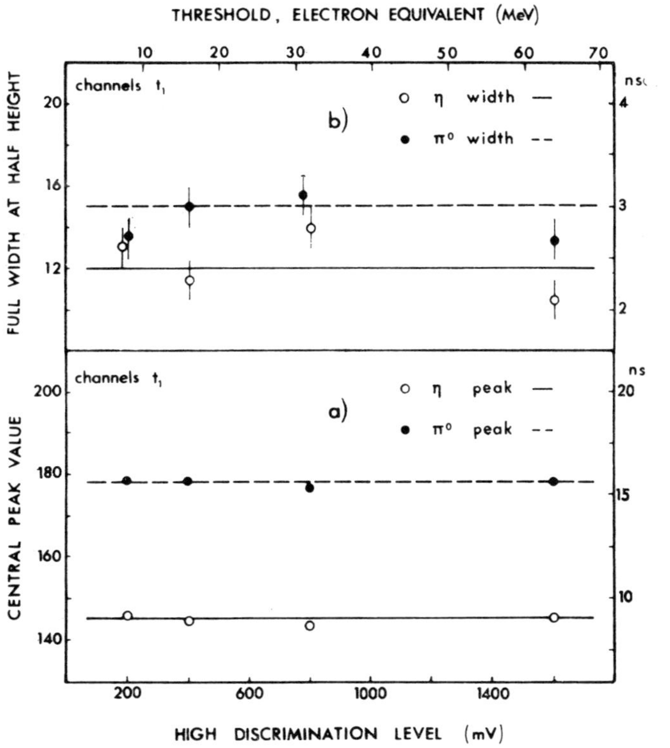

Fig. 26. – *a*) The position and *b*) the width of the η and π^0 peaks as a function of the « high » discrimination level. The abscissa scale on the top of the Figure gives the threshold in MeV electron equivalent. The ordinate on the right shows the correspondence between t_1 channels and ns.

but choosing the $(t_1\text{-}\theta_n)$ region b shown in Fig. 24. The neutron detector was placed at 25.2° on the right of the beam line, and the operating region was again defined with the help of the kinematic-selection trigger (see Subsect. **6**˙3.2). In this way, the part of the isomass curve studied is nearly perpendicular to the t_1-axis. Thus an η peak in the t_1 spectrum can be observed on a MCA; a π^0 peak can also be observed, though enlarged by a nonnegligible projection of its kinematic curve.

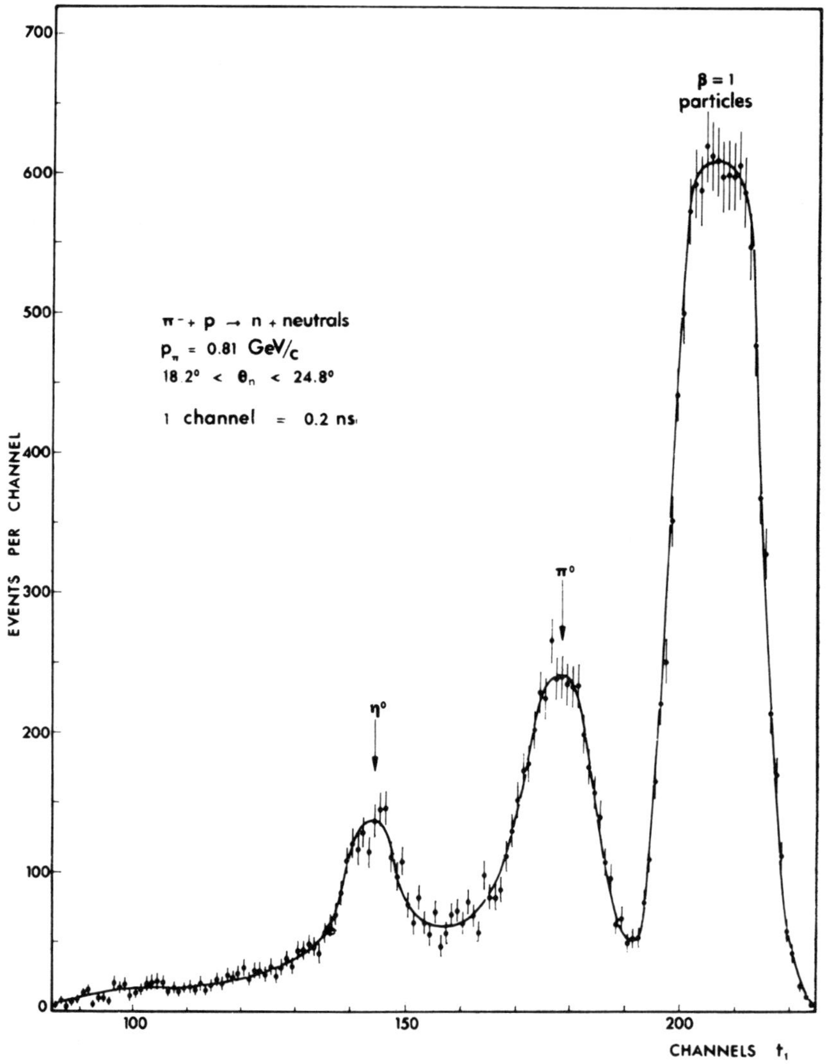

Fig. 27. – The η and π^0 peaks observed on MCA, corresponding to the region b of Fig. 24. Four spectra at different thresholds have been added.

In these conditions, the time resolution was studied at different thresholds of the « high » channel discriminators. Four measurements, with ~ 400 η events each, were done at 200, 400, 800 and 1600 mV (8, 16, 32 and 64 MeV equivalent electron energy respectively). The results are displayed in Fig. 26 *a*), where the position of the centre of the η and π^0 peaks are plotted as a function of the threshold. Figure 26 *b*) shows the FWHH of the η and π^0 peaks as a function of threshold; in both graphs one can observe that there is no important correlation between the measured quantities and the discrimination levels.

The t_1 time resolution is shown in Figure 27, where the MCA spectra of the different threshold measurements have been added up. This is justified on the basis of the results shown in Fig. 26 *a*) and 26 *b*). The observed width of the η and π^0 peaks are (2.2 ± 0.1) ns and (3.0 ± 0.1) ns FWHH respectively To these widths contribute the longitudinal dimension of the target, the projection of the kinematic curve, and the spread of the beam momentum, as

TABLE IV.

	η (ns)	π^0 (ns)
Observed widths	2.2 ± 0.1	3.1 ± 0.1
Beam momentum spread ($\pm 0.75\%$) Projection of the kinematics on t_1 axis Finite target dimensions	1.2 0.5 0.3	1.2 2.6 0.1
Measured resolution for neutrons Kinetic energy of neutrons	± 0.9 (150÷210) MeV	± 0.7 (365÷445) MeV
Time resolution for charged particles Counter thickness (18 cm) Mistiming Back-Front for nonrelativistic particles on MCA Timing t_1	0.7 1.05 0.45 0.2	0.7 0.9 0.27 0.2
Calculated resolution for neutrons	± 0.84	± 0.69

shown in Table IV. By subtraction of these contributions, the time of flight resolution is found to be ± 0.9 and ± 0.7 ns in the neutron energy ranges (150÷210) MeV and (365÷445) MeV respectively.

This value is in excellent agreement with what is expected from the calibrations with charged particles (see Table IV).

9. – The neutron detection efficiency and background.

9·1. *Previous experimental results.* – In plastic scintillator, below 10 MeV, only neutron-proton interactions produce light, so that the neutron detection efficiency can be calculated and agrees with the experimental measurements. For higher neutron energies, interactions with the carbon nuclei become significant and rescattering contributions in large detectors are not negligible.

KURZ has developed a Fortran program (11) (TOTEFF), which uses all the available cross-section data to calculate the neutron detection efficiency in the energy range 1 to 300 MeV (*). This calculation takes into account both neutron-proton and neutron-carbon reactions (see Table V), the fractional

TABLE V. – *Neutron reactions in scintillator.*

Reaction	Q value (MeV)	Threshold (MeV)	Cross-section at 20 MeV (b)	Cross-section at 100 MeV (b)	Cross-section at 200 MeV (b)
n-p	0	0	0.480	0.071	0.04
C(n, α)Be	− 5.71	0.2	0.025	0	0
C(n, n′ 3α)	− 7.26	7.9	0.285	0.106	0.10
C(n, p)B	−12.59	13.6	0.080	0.108	0.10
n-C total	—	—	1.550	0.480	0.30

TABLE VI.

Authors	Plastic trade name	Dimensions (cm)	Energy range (MeV)	Threshold proton energy (MeV)
WIEGAND *et al.* (12)	—	15×20×20	4÷ 76	4
GATTI *et al.* (13)	NE 102	10 ×17.6 diam.	21÷ 29	6
CRABB *et al.* (14)	NE 102A	28.6×30.5 diam.	20÷130	6
BRADY *et al.* (15)	NE 102	30 ×12.7 diam.	10÷ 70	3.5, 5.2
This experiment	—	100×108×36	70÷390	25, 48

(11) R. J. KURZ: UCRL-11339 (1964).

(*) TOTEFF does not take into account the inelastic pion production which takes place above 300 MeV neutron kinetic energy.

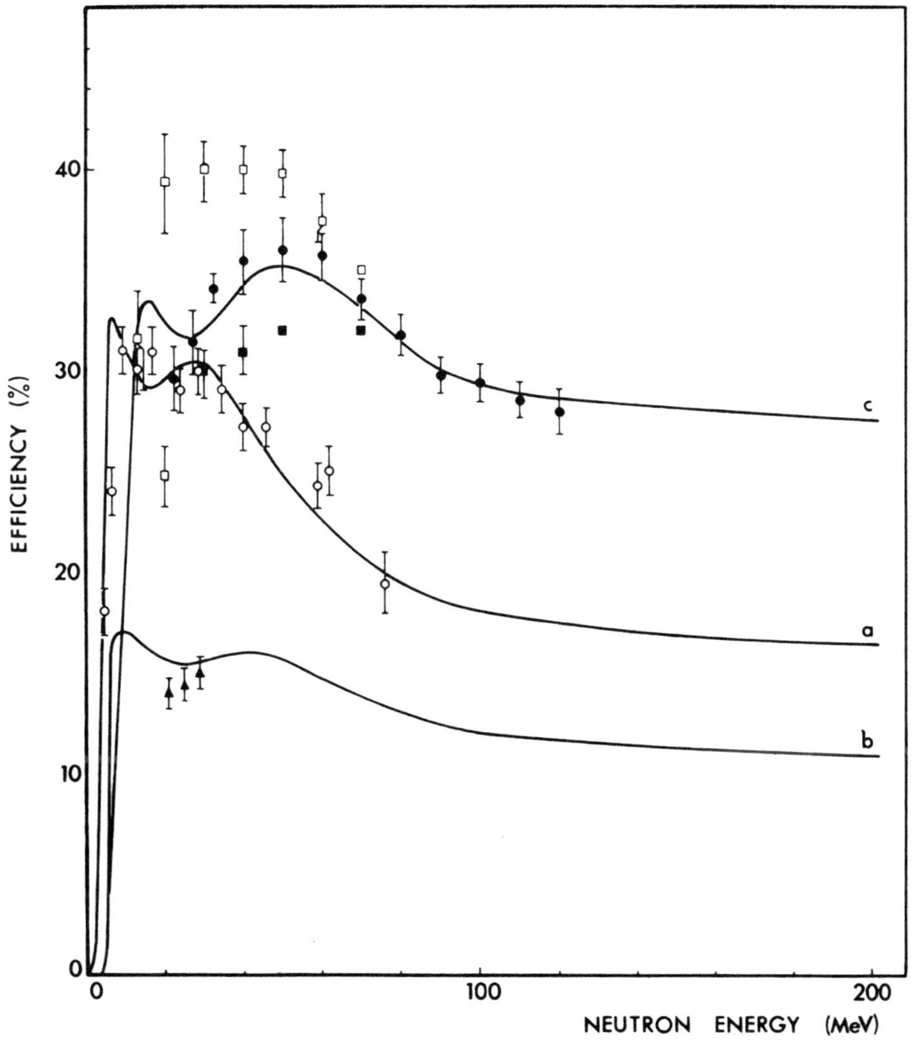

Fig. 28. – Experimental neutron-detection efficiencies for different thresholds (in MeV proton equivalent) and different geometries. Curves *a*, *b* and *c* represent the results of program TOTEFF for the measurements of WIEGAND *et al.* [12], GATTI *et al.* [13] and CRABB *et al.* [14], respectively. With the exception of the results of BRADY *et al.* [15] this Figure is taken from ref. [14]. • CRABB *et al.* (28.6 cm, 6 MeV), ○ WIEGAND *et al.* (15 cm, 4 MeV), ▲ GATTI *et al.* (10 cm, 6 MeV), □ BRADY *et al.* (30.5 cm, 3.5 MeV), ■ BRADY *et al.* (30.5 cm, 5.5 MeV).

[12] C. E. WIEGAND, T. ELIOFF, W. B. JOHNSON, L. B. AUERBACH, J. LACH and T. YPSILANTIS: *Rev. Sci. Instr.*, **33**, 526 (1962).

[13] G. GATTI, P. HILMANN, W. C. MIDDELKOOP, T. YAMAGATA and E. ZAVATTINI: *Nucl. Instr. Meth.*, **29**, 77 (1964).

[14] D. G. CRABB, J. G. MCEWEN, E. G. AULD and A. LANGSFORD: *Nucl. Instr. Meth.*, **48**, 87 (1967).

[15] F. P. BRADY, J. A. JUNGERMANN, J. C. YOUNG, J. L. ROMERO and P. J. SYMONDS: *Nucl. Instr. Meth.*, **58**, 57 (1968).

energy resolution and the mean threshold. The rescattering effects are included by integration over all neutrons which are produced in the final state of primary reactions, which have not produced sufficient light to be detected. The uncertainty in the calculated value of the efficiency is estimated to be $\pm 10\%$, which is of the same order as the uncertainty in the total nonelastic n-C cross-section.

Several experiments ([12-15]) have been performed with different geometries of neutron detectors and different threshold in the energy range 4 to 130 MeV. Table VI gives a summary of the interesting features of these experiments. Notice that CRABB *et al.* ([14]) have studied the uniformity of the efficiency over the whole surface of their detector and found that at 1 cm from the edge the detection efficiency for 120 MeV neutrons decreases only by 11% with respect to the mean efficiency.

The neutron detection efficiencies measured in these experiments ([12-15]) are shown in Fig. 28 and are compared with the calculated efficiency ([11]). On the basis of these data, one can conclude that for neutron energies above 100 MeV and when the threshold effects can be neglected, the neutron detection efficiency is nearly constant with energy and is of the order of 1% per cm. Moreover, GALAKTIONOV *et al.* ([16]) have measured the neutron efficiency in the energy range 1 to 2 GeV and their results, extrapolated to our geometry (36 cm scintillator thickness), give an efficiency of about 30%. The results indicate that the calculations of KURZ ([11]) are in good agreement with experiments ([12-14]), or not inconsistent ([15]).

9'2. *Efficiency as a function of neutron energy.* – The detection efficiency has been measured via the study of the reactions

$$\pi^- + p \to \eta + n \quad \text{at} \quad p_\pi = 0.81 \text{ GeV/c and } 1.52 \text{ GeV/c}\,,$$

$$\pi^- + p \to \omega + n \quad \text{at} \quad p_\pi = 1.52 \text{ GeV/c and } 2.12 \text{ GeV/c}\,.$$

The method consisted in the comparison of the number of η and ω observed with the rate expected from the known production angular distributions ([17,18]) and from the acceptance of the detector (*).

([16]) YU. V. GALAKTIONOV, L. G. LANDSBERG and V. A. LYUBIMOV: *Nucl. Instr. Meth.*, **39**, 351 (1966).

([17]) W. B. RICHARDS, C. B. CHIU, R. D. EANDI, A. C. HELMHOLZ, R. W. KENNEY, B. J. MOYER, J. A. POIRIER, R. J. CENCE, V. Z. PETERSON, N. K. SEHGAL and V. J. STENGER: *Phys. Rev. Lett.*, **16**, 1221 (1966).

([18]) M. A. ABOLINS, O. I. DAHL, J. S. DANBURG, D. DAVIES, P. HOCH, J. KIRZ, D. H. MILLER and R. RADER: *Proceedings of the Heidelberg International Conference on Elementary Particles* (Heidelberg, 1967), p. 509 and private communication.

(*) For details, see Table VII.

TABLE VII. – *Neutron-detector efficiency.*

Neutron detector threshold (MeV)	Missing mass	p_π $\left(\frac{\text{GeV}}{c}\right)$	Decay mode	N_π	Accepted cross-section (μb)	Correction factor	Neutron kinetic energy (MeV)	$\frac{\Delta\varphi_n}{2\pi}$	Observed number of events	ε_n (%)
16	η	0.8	total	$1.55\cdot10^9$	213±18	0.72±0.02	70÷85	0.087	1133±71	13.0±1.4
		1.5	neutral	$3.0\cdot10^8$	240±20	0.52±0.05	192÷317	0.058	300±35	33.0±5.5
	ω	1.5	total	$3.8\cdot10^8$	490±30	0.83±0.03	240÷390	0.078	1220±120	25.0±3.5
		2.1	total	$5.8\cdot10^8$	102±7	0.82±0.02	120÷153	0.057	220±34	20.0±3.5
					92±7	0.83±0.02	153÷186	0.056	221±30	21.5±3.5
					81±6	0.85±0.02	186÷219	0.055	207±31	22.0±3.5
					71±5	0.85±0.02	219÷252	0.054	207±29	25.0±4.0
37	η	0.8	neutral	$3.7\cdot10^8$	375±31	0.35±0.02	54÷78	0.088	116±14	6.5±1.1
					290±24	0.38±0.02	78÷102	0.086	97±14	6.8±1.2
					240±20	0.39±0.02	102÷126	0.092	110±15	8.3±1.3
					205±17	0.40±0.02	126÷150	0.094	118±14	10.1±1.6
					183±15	0.40±0.02	150÷174	0.101	164±16	14.5±2.2
					170±14	0.31±0.03	174÷198	0.110	149±16	16.4±2.5

The experimental set-up for this measurement is shown in Fig. 8 and described in Sect. **5**.

The events were accepted in large time-of-flight intervals. A selection of smaller intervals of neutron energy was made by computer, using the encoded data, thus obtaining detailed information on the neutron detection efficiency over a wide range of neutron energies: (70 ÷ 390) MeV. The missing-mass spectra showed a signal-to-background ratio which was typically 2.5 in the η-meson region and 1.0 in the ω-meson region.

The mass distributions were fitted by the least-square method, using a smooth polynomial function to describe the background, and a Gaussian function to describe the η or the ω peak. This procedure allowed the background to be subtracted and the number of η-mesons or ω-mesons to be evaluated.

Systematic effects, such as beam contamination, random vetoing by Q, $US\bar{R}$ scaler losses and neutron absorption in the target, in the air and in the G counters, have been taken into account. Furthermore, the event losses due to charged pions from an η or an ω decay, hitting the anticoincidence counters Q, G_R and R, have been calculated by Monte Carlo.

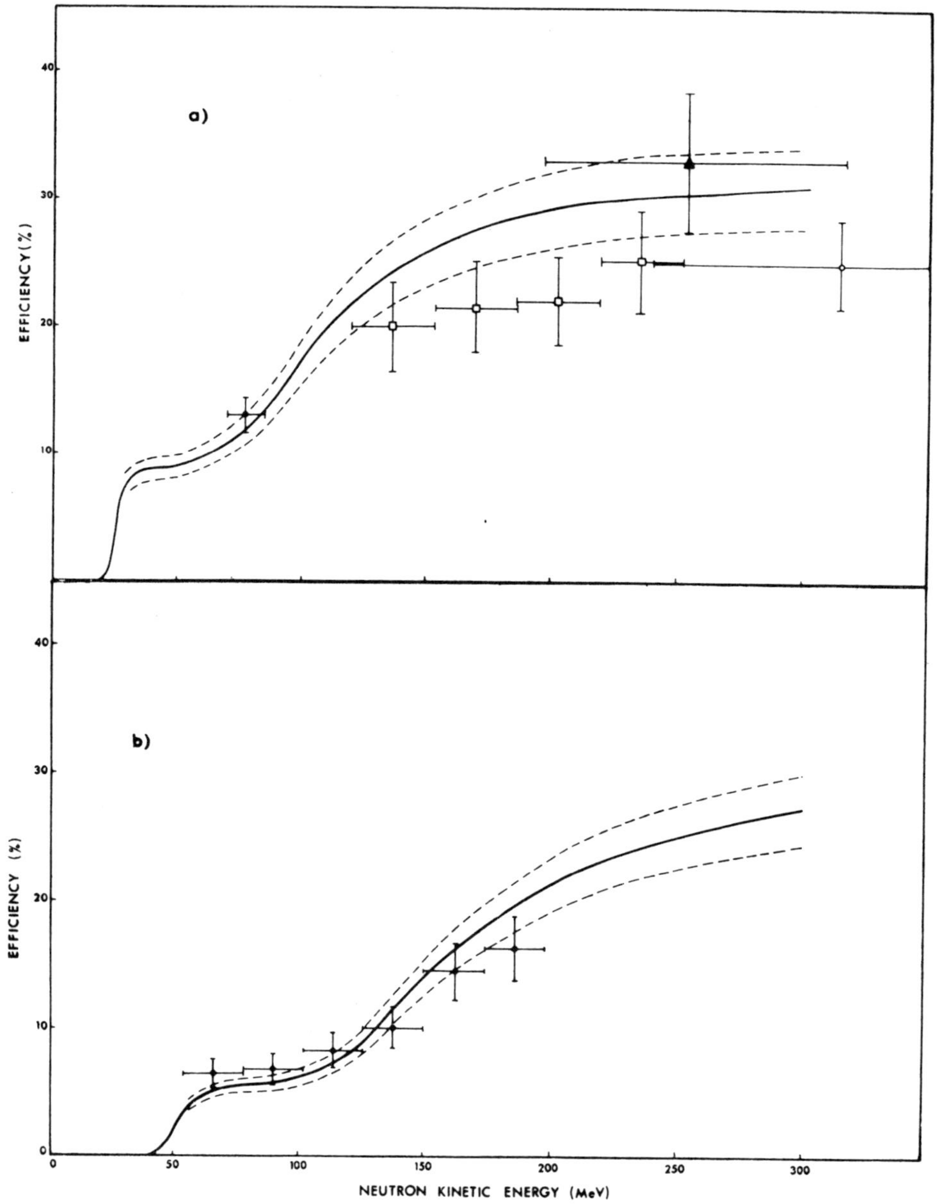

Fig. 29. – The neutron detection efficiency measured using neutrons produced in association with: *a*) η at 0.8 and 1.5 GeV/c incident pion momentum, ω at 1.5 and 2.1 GeV/c, *b*) η at 0.8 GeV/c. The measurements have been performed with a « high » discrimination level of: *a*) 16 MeV and *b*) 37 MeV equivalent electron energy. The solid curves and the dashed lines represent the results of program TOTEFF ([11]) with its error bands. • η^0 production at $p_\pi = 0.8$ GeV/c; ▲ η^0 production at $p_\pi = 1.5$ GeV/c; ○ ω^0 production at $p_\pi = 2.1$ GeV/c; □ ω^0 production at $p_\pi = 2.1$ GeV/c.

The measured neutron detection efficiency is shown in Fig. 29 as a function of neutron energy at two different thresholds. The region between the two dashed curves is the error band corresponding to the calculations done with program TOTEFF [11]. The parameters of these calculations were given by our experimental conditions, namely:

scintillator thickness = 36 cm;

scintillator density = 1.17 g/cm^3;

fractional energy resolution = $\pm 6\%$ (Subsect. **7**˙1);

a bias level of 16 MeV (Fig. 29 *a*)) and of 37 MeV (Fig. 29 *b*)) equivalent electron energy [*], the bias level being calibrated using 2.1 GeV/c pions.

The detector geometry was approximated to a cylinder (36 cm thick $\times$ $\times$100 cm diameter), with the axis along the incident-neutron direction. The average escape distance (necessary to compute rescattering effects) was calculated for uniform neutron illumination of the cross-sectional area.

Concerning the thickness used, one should remark that the experimental efficiency was obtained in two 18 cm thick counters, but that a thickness of 36 cm was assumed in the calculation; this is correct apart from cases where neutrons give charged recoil particles such that the light collection in the two 18 cm counters is not sufficient for the event to be detected, but the light emission in one 36 cm counter would be sufficient. This effect has been estimated to be of the order of 1% in the worst case and therefore can be neglected with respect to the other uncertainties.

A comparison of the measured and calculated efficiencies shows that the values obtained from η production are in agreement, whilst the values obtained from ω production are systematically low but are compatible within the large experimental errors. This behaviour may be explained by the fact that the cross-sections and angular distributions for η production [17] are much better known than those for ω production [18].

9˙3. *Efficiency and background as a function of pulse-height threshold.* – The variation of the neutron detection efficiency with the threshold of the « high » discrimination channel of the detector was studied observing η and π^0 production, in the same experimental conditions as those described in Subsect. **8**˙2. In the selected region of $(t_1\text{-}\theta_n)$-plane, the neutron energy ranged from 150 MeV to 210 MeV for the η and from 365 MeV to 445 MeV for the π^0. As we have already said, the time-of-flight spectra were taken at thresholds corresponding to 8, 16, 32 and 64 MeV equivalent electron energies.

(*) These levels correspond respectively to 25 MeV and to 48 MeV equivalent proton energy [19].

(19) H. C. EVANS and E. H. BELLAMY: *Proc. Phys. Soc.*, **74**, 483 (1959).

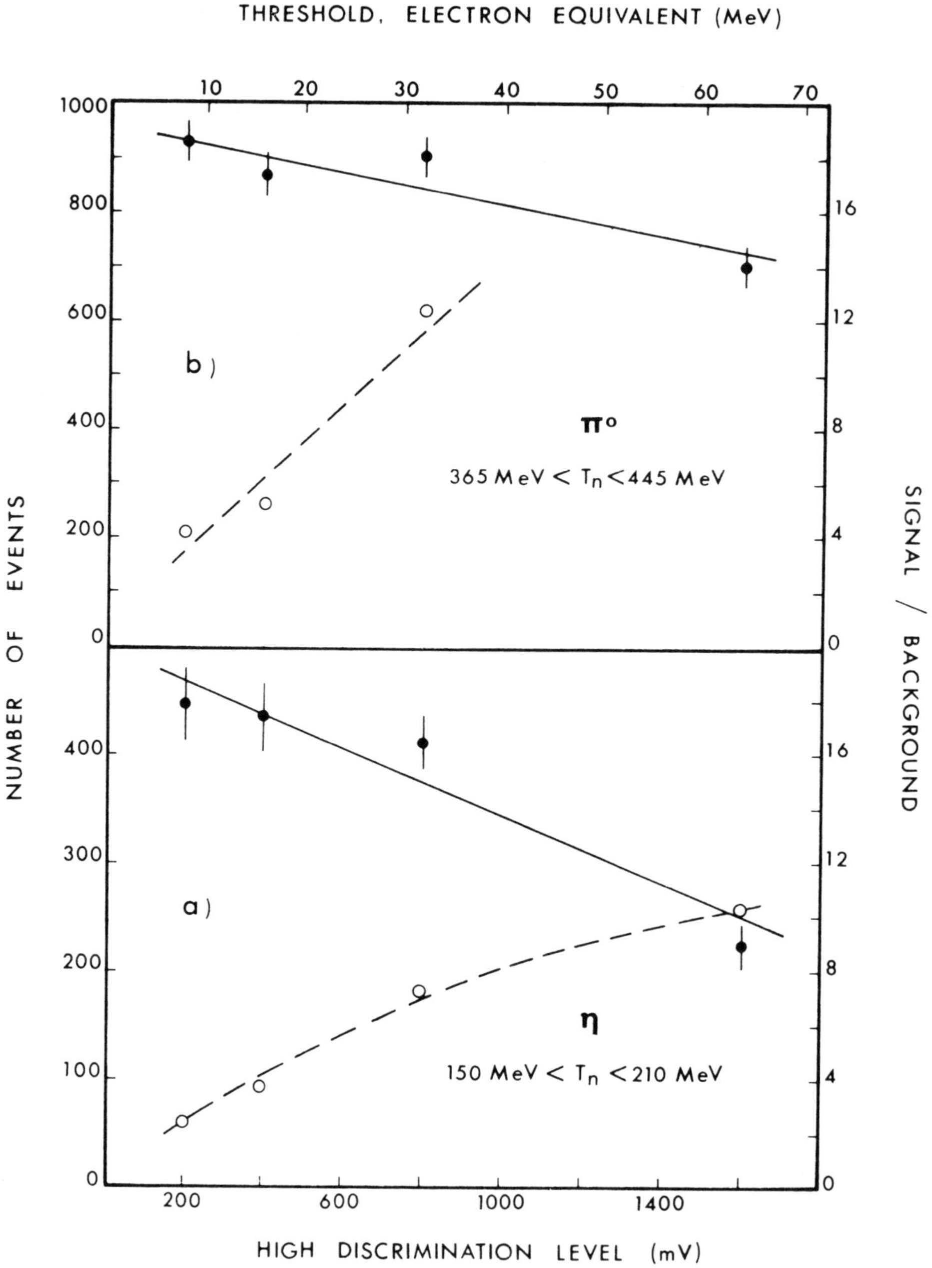

Fig. 30. – The number of observed *a*) η and *b*) π^0 as a function of the « high » discrimination level. The solid line shows the slope predicted by program TOTEFF at the corresponding neutron energies. The dashed line gives the ratio signal/background as a function of the threshold. The abscissa on the top represents the threshold in MeV electron equivalent. Notice that the π^0 signal-to-background ratio at 64 MeV threshold is compatible with zero background.

The numbers of η and π^0 produced were normalized to the same number of incident pions. The decrease of the observed event rate, when the threshold increases, is shown in Fig. 30 *a*) and *b*) for the η and π^0 data respectively. A comparison between the measured values and the predictions of program TOTEFF ([11]), represented by a continuous line, is also shown. Within the 8% statistical accuracy, the predicted and measured slopes agree. The dashed lines in Fig. 30 *a*) and Fig. 30 *b*) show the variation of the signal-to-background ratios.

These curves allow a choice of the best detection threshold, depending on whether the best absolute detection efficiency or the best signal-to-background ratio is needed for a given physical problem.

10. – Conclusions.

The neutron detector described in this paper has been extensively used to study neutral resonances produced in π^-p interactions ([1-5]).

The large volume of scintillator, in the particular geometrical arrangement chosen, allowed a mean detection efficiency of about 25% in the range $(70 \div 390)$ MeV neutron kinetic energy for a laboratory solid angle of 0.14 sr at 4 m radial distance.

With the techniques adopted, calibrations with charged particles could be easily performed in a few hours using a low beam intensity.

A very interesting feature of this instrument is the accuracy achieved in locating incident particles, which is ± 1.4 cm for charged particles, and ± 2.5 cm for neutrons. An instrument of this type can be, in some cases, competitive with spark chambers, if one considers the big advantage of a fast response.

The accuracies achieved for the time-of-flight measurement were ± 0.35 ns for charged particles and less than three times worse for neutrons.

With these accuracies on the neutron time of flight and angle, the missing mass was determined with a good resolution, which depended of course on the kinematics of the reactions studied, and was, for example, of ± 4 MeV for η, ± 10 MeV for ω, and ± 15 MeV for φ.

* * *

We wish to thank our technicians Messrs. J. Berbiers and O. Polgrossi for their skilful assistance.

The neutron counters were constructed at the Istituto di Fisica dell'Università di Bologna, and we would like to thank Messrs. W. Chiarini, G. Lelli, F. Massera and G. Pacchioni for their hard work and their friendly collaboration.

RIASSUNTO

Si descrive un rivelatore di neutroni a grande accettanza ed alta efficienza. La superficie ed il volume sensibili del rivelatore sono rispettivamente 2.16 m² e 0.78 m³. Il rivelatore consiste di ventiquattro elementi di scintillatore plastico, ciascuno avente le dimensioni $(100\times18\times18)$ cm³. Il grande volume di scintillatore, nella particolare configurazione geometrica scelta, permette di ottenere una efficienza media di rivelazione di circa il 25% nell'intervallo di energia cinetica del neutrone $(70\div390)$ MeV, coprendo un angolo solido di 0.14 sr nel laboratorio a una distanza radiale di 4 m. Con le tecniche adottate, calibrazioni con particelle cariche possono essere fatte in poche ore usando un fascio di bassa intensità. Una caratteristica interessante di questo strumento è la precisione ottenuta nella localizzazione di particelle incidenti; essa è di ±1.4 cm per particelle cariche e di ±2.5 cm per neutroni. Le precisioni ottenute nella misura del tempo di volo sono ±0.35 ns per particelle cariche e ±0.7 ns per neutroni. Con queste risoluzioni nel tempo di volo e nell'angolo del neutrone, l'incertezza sul difetto di massa risulta essere ±4 MeV per il mesone η, ±10 MeV per il mesone ω, ±15 MeV per il mesone φ.

Новый нейтронный детектор высокой эффективности с большим дианазоном захвата для исследования недостающих масс.

Резюме авторами не представлено.

D. BOLLINI, *et al.*
21 Maggio 1969
Il Nuovo Cimento
Serie X, Vol. **61** A, pag. 125-172

D. Bollini, A. Buhler-Broglin, P. Dalpiaz, T. Massam, F. Navach,
F.L. Navarria, M.A. Schneegans et A. Zichichi

UN DÉTECTEUR DE NEUTRONS POUR LA SPECTROMÉTRIE DE MASSES MANQUANTES

From
Revue de Physique Appliquée *4, 301 (1969)*

1969

REVUE DE PHYSIQUE APPLIQUÉE TOME 4, JUIN 1969, PAGE 301.

UN DÉTECTEUR DE NEUTRONS POUR LA SPECTROMÉTRIE DE MASSES MANQUANTES

D. BOLLINI, A. BUHLER-BROGLIN, P. DALPIAZ, T. MASSAM, F. NAVACH, F. L. NAVARRIA, M. A. SCHNEEGANS et A. ZICHICHI,

C.E.R.N., Genève, Suisse,

Istituto di Fisica dell'Università di Bologna, Italie,

Istituto Nazionale di Fisica Nucleare, Sezione di Bologna, Italie,

Centre de Recherches Nucléaires, Strasbourg, France.

Résumé. — Nous décrivons un détecteur de neutrons de grand volume sensible ($2 \times 0{,}39$ m³ de scintillateur plastique) capable de mesurer avec précision les coordonnées du point d'interaction du neutron détecté ainsi que son temp-de-vol. Employé comme spectromètre de masses manquantes, il permet d'observer par exemple le méson η avec une résolution de $\pm$ 4,2 MeV.

Abstract. — A large (2×0.39 m³ plastic scintillator) neutron detector capable to measure with high accuracy the coordinates of the neutron interaction point as well as its time-of-flight is described. As a missing mass spectrometer, it allows to observe for example the η meson with a mass resolution of $\pm$ 4.2 MeV.

I. Introduction. — L'étude des résonances neutres produites dans la réaction :

$$\pi^- + p \rightarrow n + B^0 \qquad (1)$$

est rendue possible par l'emploi d'un détecteur de neutrons. La connaissance du moment du pion incident, associée à la détermination du moment et de l'énergie du neutron révélé, permet l'identification de la masse manquante B^0. Un détecteur de neutrons répondant à ces exigences sera décrit.

II. Description physique du détecteur. — Le détecteur de neutrons est constitué par douze compteurs à scintillations de $100 \times 18 \times 18$ cm³. La surface de révélation est de 1,08 m². L'épaisseur utile est de 36 cm. Un grand compteur à scintillation de $145 \times 117 \times 2{,}5$ cm³, placé devant le détecteur de neutrons, est mis en anticoïncidence; il garantit que la particule détectée est neutre. Chacun des douze compteurs est vu par deux photomultiplicateurs XP 1040 placés à leurs deux extrémités. Le détecteur, suspendu à un treuil et monté sur roues, est mobile dans le sens vertical et horizontal.

III. Les fonctions du détecteur. — La réaction (1) est produite en bombardant une cible d'hydrogène liquide par des pions négatifs de grande énergie; dans l'état final, nous avons la résonance B^0 et le neutron émis à un angle azimutal θ_n avec une énergie E_n. La révélation du neutron est rendue possible par la détection des particules chargées produites dans les réactions (n, p), C(n, n′ 3α) et C(n, p)B. La détermination de l'angle azimutal θ_n est réalisée par la mesure du point d'interaction du neutron dans le détecteur. La lumière produite par l'interaction du neutron au sein du scintillateur atteint les deux photomultiplicateurs PM_1 et PM_2 situés aux deux extrémités du compteur aux temps t_1 et t_2. La mesure de la différence du temps-de-vol de la lumière $|t_1 - t_2|$ détermine le point d'interaction et par conséquent l'angle d'émission du neutron.

L'énergie du neutron, mesurée par son temps-de-vol, est connue par la détermination de la différence du temps $|t_0 - t_1|$ ou $|t_0 - t_2|$. t_0 est le temps de passage du pion dans un compteur à scintillation placé à 2 m devant la cible. $|t_0 - t_1|$ est composé du temps-de-vol du pion jusqu'à la cible, du temps-de-vol du neutron de la cible au point d'interaction et du temps-de-vol de la lumière du point d'interaction respectivement au photomultiplicateur PM_1 ou PM_2.

IV. Quelques caractéristiques électroniques. — Le signal issu de la treizième dynode des photomultiplicateurs PM_1 et PM_2 est fortement discriminé et indique l'existence d'une réaction nucléaire dans le scintillateur. L'information « temps » est portée par le front de montée du signal d'anode, faiblement discriminé et mis en forme. La présence simultanée des deux signaux d'anode et de dynode est requise pour déclencher le processus d'analyse des différences de temps $|t_1 - t_2|$, $|t_1 - t_0|$ et $|t_2 - t_0|$. Celles-ci sont déterminées par la mesure de la surface de superposition des impulsions mises en forme (t_1, t_2), (t_1, t_0) et (t_2, t_0); le résultat de la mesure, mis sous forme analogique, est transféré sur bande magnétique.

Pour améliorer la résolution en temps de vol du neutron, sans pour autant sacrifier l'efficacité de détection, l'épaisseur utile est physiquement et électroniquement divisée en deux et l'information « temps » est donnée par le compteur qui détecte en premier l'interaction du neutron. En outre, des sélections cinématiques sont possibles en choisissant électroniquement le domaine angulaire et le temps de vol les plus intéressants pour le neutron émis dans la réaction (1).

V. Performances du détecteur pour des particules chargées. — *i)* La résolution en énergie de nos compteurs est de $\pm$ 6 %; *ii)* la longueur d'atténuation du scintillateur employé ([1]) est égale à (385 $\pm$ 50) cm. L'indice de réfraction est égal à 1,56 $\pm$ 0,03, la densité est égale à 1,17 $\pm$ 0,01 g/cm^3; *iii)* la résolution spatiale est de $\pm$ 1,4 cm avec une linéarité parfaite de la réponse; *iv)* la résolution en temps est de $\pm$ 0,35 ns; *v)* la non-uniformité de réponse du compteur est inférieure à 5 %; *vi)* la stabilité de la réponse des douze compteurs sur une période d'une semaine est meilleure que 0,2 ns.

([1]) Tous les compteurs de neutrons ont été construits à l'Institut de Physique de Bologne.

VI. Performances du détecteur pour des neutrons. — L'utilisation de la réaction $\pi^- + p \rightarrow n + \eta$ donne une calibration physique du détecteur. L'observation du méson η avec une résolution en masse de ($\pm$ 4,2 $\pm$ 0,2) MeV (1/2 L.T.M.H.) confirme que la résolution spatiale du détecteur pour des neutrons est de l'ordre de ($\pm$ 2,5 $\pm$ 0,7) cm. L'observation de η au moyen du spectre de temps de vol (t_1) corrobore la résolution en temps de $\pm$ 0,35 ns. Finalement, le confrontement du nombre total de mésons η observés avec le nombre théoriquement attendu permet de calculer l'efficacité de détection du détecteur pour des neutrons. L'efficacité ainsi déterminée est en bon accord avec les prévisions des calculs de Kurz [1]; elle est quasi constante et voisine de 25 % pour des neutrons de 100 MeV à 400 MeV et pour un seuil de détection de 25 MeV pour les protons.

VII. Conclusion. — L'excellente résolution spatiale et temporelle de ce détecteur de neutrons, alliée à une grande acceptance et efficacité, a permis une utilisation fructueuse dans de nombreuses expériences [2].

BIBLIOGRAPHIE

[1] Kurz (R. J.), UCRL-11339, 1964.

[2] Buhler-Broglin (A.) *et al.*, *Nuovo Cim.*, 1967, **49 A**, 183.

Bollini (D.) *et al.*, *Nuovo Cim.*, 1968, **56 A**, 531.

Bollini (D.) *et al.*, *Nuovo Cim.*, 1968, **56 A**, 1173.

Bollini (D.) *et al.*, *Nuovo Cim.*, 1968, **57 A**, 404.

Bollini (D.) *et al.*, *Nuovo Cim.*, 1968, **58 A**, 289.

Bollini (D.) *et al.*, *Nuovo Cim.*, 1969, **60 A**, 541.

A. Zichichi

AN APPARATUS OF THE NBC TYPE AND THE PHYSICS RESULTS OBTAINED

From
Annals of Physics 66, n. 2, 405 (1971)

1971

Reprinted from ANNALS OF PHYSICS

Vol. 66, No. 2, August 1971
Printed in Belgium

An Apparatus of the NBC Type and the Physics Results Obtained

A. ZICHICHI

Istituto Nazionale di Fisica Nucleare, Bologna, Italy and
Istituto di Fisica dell'Università, Bologna, Italy

Received October 13, 1970

DEDICATION

Amos de Shalit was known primarily as a theoretical physicist, but he was much more than that; his interests knew no frontiers, and in our restricted world of physics, he was not only stimulated by the possible theoretical significance of experimental results, but he took a deep and personal interest in all facets of physical research, including the financial and technical problems of experiments. He realized that the multipurpose, electronically triggered set-up was a most promising way of reducing the cost and time needed to perform experiments in those fields where bubble chambers were inefficient, and he was enthusiastic about its possibilities.

In gratitude for his interest in our work, and on behalf of all physicists who worked on the project, I dedicate this paper to the memory of Amos de Shalit, as a tribute to his scientific and human qualities, which cannot be forgotten by those who had the privilege of knowing him.

NOMENCLATURE

BC	bubble chamber
BR	branching ratio
BW	Breit–Wigner formula
C	charge conjugation
C.M.M.	current mixing model
d	depth of the EM shower detectors
$E_{c.m.}$	total energy in the center of mass system
EM	electromagnetic
FWHM	full width at half maximum
g	coupling constant
GWSR	generalized Weinberg spectral function sum rule
I	isotopic spin
J	spin
K_n	neutron kinetic energy
m, M	mass of a particle (used also as invariant mass of a system of particles)
M^0	neutral nonstrange meson state

M.M.M.	mass mixing model
MPNBC	multipurpose non bubble chamber
N	number of events
$\mathcal{N}$	nucleon
NBC	nonbubble chamber
p	linear momentum
P	parity
PCAC	partial conservation of the axial vector current
PS	protonsynchrotron
$P(\chi^2)$	probability of a χ^2 distribution
r.l.	radiation length
R, R'	branching ratio
t	four momentum transfer to the nucleon
T	time reversal
t_n	neutron time-of-flight
t_1, t_2	neutron time-of-flight plus the transit of the light in the scintillator as seen from side 1 and 2
Y	hypercharge
β	relative velocity, v/c
Δp	error on the linear momentum
$\Delta\theta$	angular range
$\Delta\sigma$	integrated differential cross section over a fraction $\Delta\Omega$ of the solid angle
Γ	width of a resonance or decay rate (used also as experimental resolution)
θ	original Sakurai mixing angle
θ_D	angular setting of the EM shower detectors around an horizontal axis
θ_G	generalized mixing angle
θ_n	neutron emission angle in the laboratory system
$\theta_n{}^*$	neutron emission angle in the center of mass system
θ_N	baryonic mixing angle
θ_0	central angle of the neutron detectors
θ_W	angular setting of the EM shower detectors around a vertical axis
θ_Y	hypercharge mixing angle
σ	total cross section
ω_1	$SU(3)$ singlet
ω_8	eight component of an $SU(3)$ octet

Summary

A nonbubble chamber (NBC) setup, especially designed to perform a series of experiments, is described. The setup consists of a neutron missing-mass spectrometer coupled with a system of large electromagnetic shower detectors. In spite of its large dimensions (the sensitive surface and volume are 2.16×10^4 cm^2 and 7.8×10^5 cm^3, respectively), the neutron spectrometer has a time resolution of ± 0.70 nsec FWHM, with ~25 % detection efficiency in the range (70–390 MeV) neutron kinetic energy. The time equalization between the various components of the neutron spectrometer has been established to within ± 0.1 nsec. At present this is the most powerful and the most accurate high-energy neutron detector. The electromagnetic shower detector is based on the principle of simultaneous measurements of the spatial development of the electromagnetic cascade and of its energy release. This is obtained with nine elements of lead foil — spark chamber — plastic scintillator, all sandwiched together. The sensitive surface and volume of the electromagnetic shower detector are 1.45×10^4 cm^2 and 7.2×10^5 cm^3, respectively.

The setup has been used to perform a series of 14 experimental investigations. Ten of these have already produced conclusive results which are either published or submitted for publication:

1. Evidence against the existence of the S^0 meson;
2. an accurate measurement of the branching ratio $\eta \to$ neutrals/$\eta \to$ total;
3. an accurate determination of the branching ratio $\omega \to$ neutrals/$\omega \to$ total;
4. study of ϕ production in $\pi^- + p$ interactions at 2.1 GeV/c;
5. evidence for the new decay mode $\phi \to e^+e^-$ and measurement of the cross section $\sigma[\pi^- + p \to (\phi \to e^+e^-) + n]$;
6. evidence for the new decay mode $\omega \to e^+e^-$ and measurement of the cross section $\sigma[\pi^- + p \to (\omega \to e^+e^-) + n]$;
7. points (5) and (6) above allowed the first direct determination of the $(\omega - \phi)$ mixing angle;
8. evidence for the new decay mode $X^0 \to \gamma\gamma$;
9. cross section measurements for X^0 production and value of the branching ratio $X^0 \to$ neutrals/$X^0 \to$ total;
10. evidence for a structure in the neutral A_2 mass region.

    The data of the following three experiments are being analysed:
11. measurement of the branching ratio $X^0 \to \gamma\gamma/X^0 \to$ total;
12. study of the $\rho - \omega$ interference in the reaction $\pi^- + p \to \pi^+ + \pi^- + n$ at 1.67GeV/c incident pion momentum;
13. study of the $(\pi^0\pi^0)$ system in the invariant mass region from 400–950 MeV.
14. Finally, the last experiment is still under way:
    the apparatus is at present running for the detection of the decay mode $\phi \to \eta + \gamma$ and for the measurement of the branching ratio $\phi \to \eta + \gamma/\phi \to$ total.

The setup described and the experimental results obtained show that it is possible, in the field of NBC physics, to reduce the cost and the time needed for an experiment provided that multipurpose NBC setups are used.

1. Introduction

From the point of view of the technique used, experimental physics in high-energy laboratories has developed along two main directions: bubble chamber (BC) and nonbubble-chamber (NBC) devices.

A typical difference between BC and NBC setups has been the fact that BC setups have been used for many experiments while NBC setups have usually remained in the limited domain of apparatus able to do only one experiment. This is, of course, a serious handicap for NBC setups because, as in any large scale enterprise, experimental high-energy subnuclear physics has been faced with the problem of organizing facilities in order to maximize the number of experiments performable with a single apparatus, in order to reduce the cost of each investigation.

The most popular instrument for such a scientific policy has, of course, been the bubble chamber, whose pictures, taken under different conditions of incoming beam particles and momenta, represent an ideal tool for physicists of various laboratories, even very far away from the place where the instrument is actually working, to perform their investigations. On the other hand, a BC setup carries all negative features of a large scale project.

On the contrary, if a multipurpose NBC (MPNBC) setup is designed to perform a series of experiments, it will have the power of a large scale project, while possessing a high degree of flexibility: its basic properties should, in fact, be such that important improvements can be incorporated without destroying the basic elements of the original setup. Moreover, the great advantage of a nonbubble chamber experiment (i.e., the small dimensions of the research group involved, with the consequent high impact between physicists and the experimental setup) should of course remain in an MPNBC apparatus.

To sum up, an MPNBC setup should have the following properties:

(i) to be complementary to a BC device for the physics program which can be performed with it;

(ii) to be able to be constantly improved upon without drastic alterations to its vital components;

(iii) to require the services of a small group of physicists ($\lesssim 10$) for actual operation of data taking.

The purpose of this paper is to give a review of a setup especially designed to fulfill the above-mentioned requirements, and to report on the experimental results obtained so far. These results represent, in fact, the best proof that this MPNBC setup has the desired properties.

The setup consists of a neutron spectrometer coupled with an electromagnetic (EM) shower detector. Both the neutron spectrometer and the EM shower detector have large dimensions and high resolution either in position determination (neutron spectrometer) or in particle identification (EM shower detector).

The instrument, in its original components, was designed, studied and tested by the physicists who used it for the actual experiments. In particular, the 24 large plastic scintillator counters ($18 \times 18 \times 100\ \text{cm}^3$) which constitute the neutron detector were constructed at the Institute of Physics of the Bologna University. The construction of these scintillators in the Institute has allowed a very close look at their properties. For example, special melting procedures made it possible to achieve an optimum optical uniformity and transparency of the scintillator. The detailed study of various elements allowed the choice of the best geometrical configuration to achieve an optimum uniformity in the efficiency of response over the whole volume of the counter.

2. The MPNBC Setup

2.1. *General Description*

A schematic diagram of the MPNBC setup is shown in Fig. 1. It consists of the following:

(i) A system of "beam-defining counters" $\check{C}US\bar{R}$. $\check{C}$ is a gas Čerenkov counter to select different particles present in the primary beam; U is an important plastic scintillator counter used in the timing of the neutron; S is a very thin plastic scintillator, fast coincidence counter (0.05 cm thick, in order to reduce the interactions outside the H_2 target as much as possible); $\bar{R}$ is an anticoincidence counter to remove beam halo.

(ii) Two EM shower detectors called TOP and BOTTOM. In front of them there are the coincidence counters M, and the thin-plate spark chambers K which, for the sake of clarity, have all been omitted in Fig. 1 (but are shown in Fig. 7). When looking for γ-rays, a lead converter (0.5 cm thick) is placed before the coincidence counters, which are used to detect the conversion products of the γ-rays.

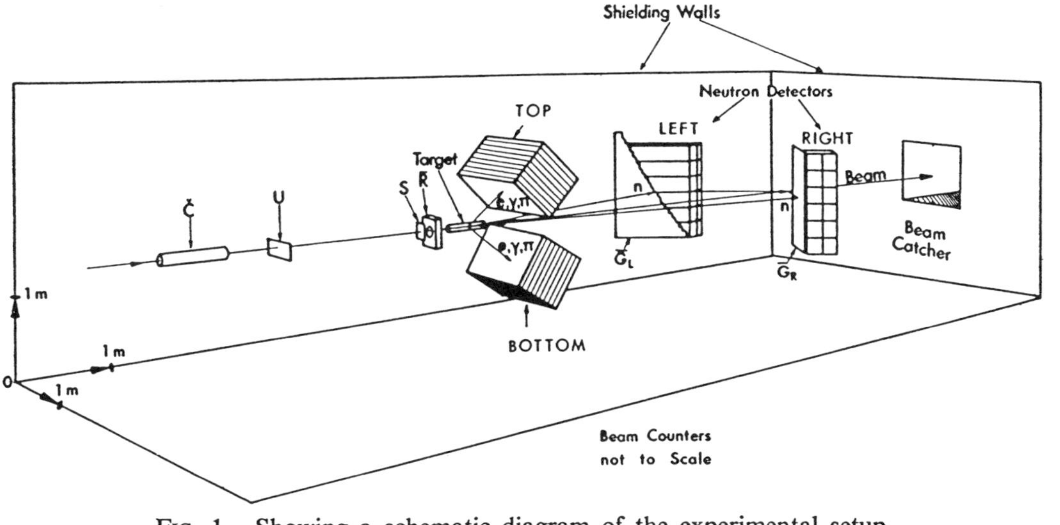

FIG. 1. Showing a schematic diagram of the experimental setup.

(iii) Two neutron detectors, called LEFT and RIGHT with anticoincidence counters G_L and G_R in front of them to reject charged particles impinging in the neutron counters. These two identical neutron detectors have a sensitive surface and volume equal to 2.16 m^2 and 0.78 m^3, respectively.

(iv) Depending on the experiment, two types of target are used. In the case of a pure missing-mass experiment a 10 cm long, 4 cm diameter H_2 target is used to achieve a good mass resolution. When the EM shower detectors are used as well, it is possible to reconstruct the production vertex to ± 1.5 cm and therefore the target is 40 cm long and 5 cm in diameter. A veto counter, not shown in Fig. 1, is placed behind the target in order to anticoincide noninteracting primary beam particles.

Both the neutron and the EM shower detectors are mounted in such a way that they can easily be placed in any position along the beam. This is an essential feature for calibration purposes and also to perform different experiments.

2.2. *The Neutron Detectors*

A neutron detector (Fig. 2) is made of 12 elements of plastic scintillator, each having dimensions $(100 \times 18 \times 18)$ cm^3. Each element is viewed by two *XP*-1040 photomultipliers placed on its two small faces. The principle of neutron detection

FIG. 2. Showing a photograph of the experimental setup. Notice the possibility of an easy displacement of both the neutron detectors (on the rails) and the EM shower detectors (mounted on a turntable). It is thus possible to choose the better configurations for each process to be investigated and to calibrate the detectors in the direct beam. This is the first example of what is now "standard" in experimental subnuclear physics: very large acceptance detectors.

is based on the scintillation light emitted by the charged products of the interaction of a neutron with the plastic scintillator nuclei. An interesting feature of this instrument is the accuracy achieved in locating incident particles; this accuracy is ±1.4 cm for charged particles, and ±2.5 cm for neutrons. The accuracies achieved for the time-of-flight measurement are ±0.35 nsec for charged particles and ±0.7 nsec for neutrons. This time and positional resolution for neutrons is strongly related to the possibility of the time equalization between all the 24 counters which could be achieved to ±0.1 nsec. An example of this time-

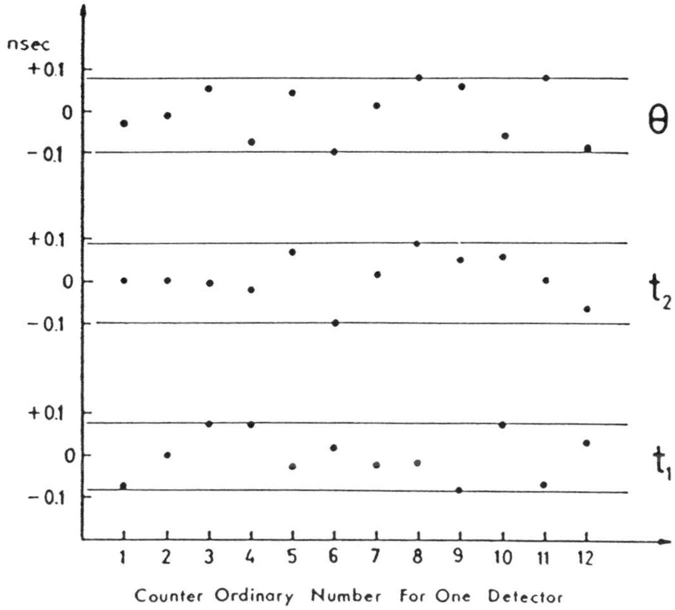

FIG. 3. Relative timing of the 12 elements of the neutron detector RIGHT. The abscissa indicates the identification number of an element, the ordinate indicates the relative timing.

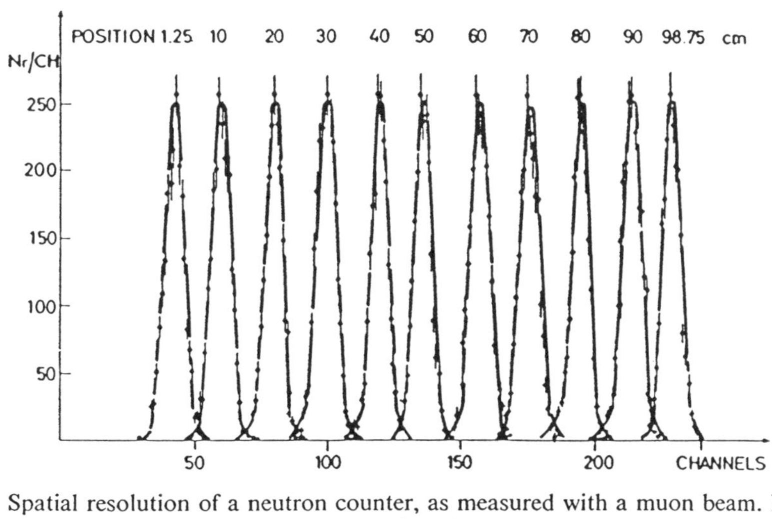

FIG. 4. Spatial resolution of a neutron counter, as measured with a muon beam. Each peak in this spectrum corresponds to a given position of the beam-defining telescope along the neutron counter.

equalization is shown in Fig. 3, where t_1 is the time difference between the (U) signal and the signal from any photomultiplier at one side of the neutron counter, t_2 corresponds to signals from the other side of the neutron counter, and θ is the difference between the two, obtained electronically.[1]

Typical data on position resolution and linearity of the neutron counters are shown in Figs. 4 and 5, respectively. In Fig. 4 the curves are labeled with the distance from the edge of the counter. The spatial resolution is shown to be all along the counter (including the extreme edges) ± 1.4 cm for charged particles. In Fig. 5 the ordinate is the distance from one edge of the counter, and the abscissa

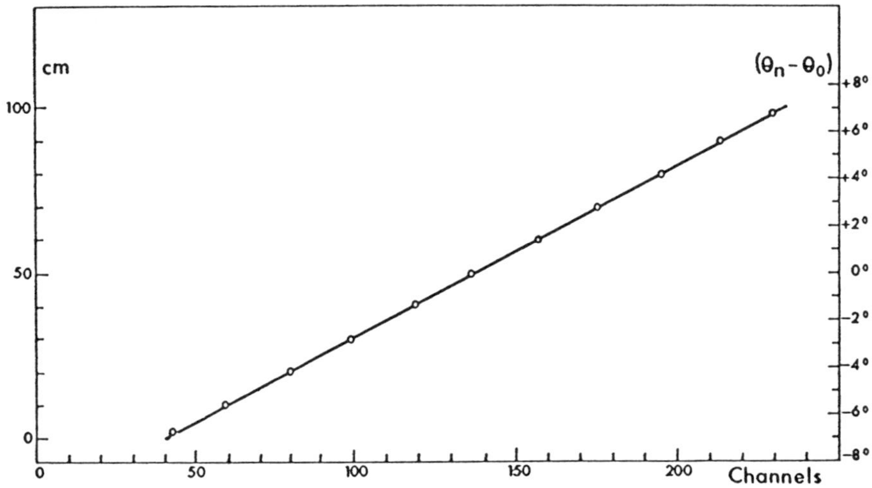

FIG. 5. Position calibration of the neutron counters, showing the position of the muon beam as a function of the channel numbers. The ordinate is also labeled with the neutron scattering angle, corresponding to 4 m distance for the counters.

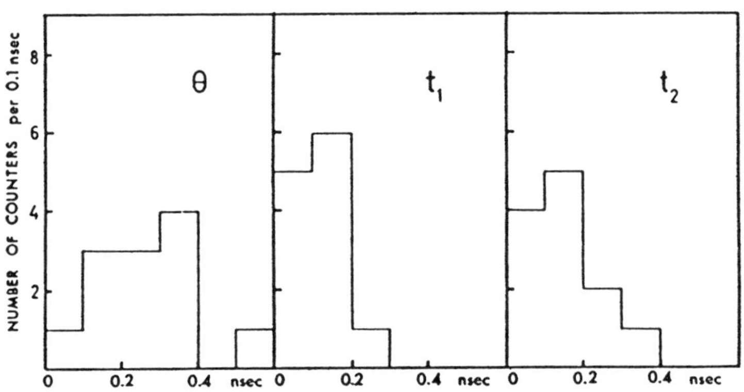

FIG. 6. Neutron counter stability over a week: The time variation for θ, t_1, and t_2 signals is plotted for each counter.

[1] For more details on this instrument we refer the reader to Bollini *et al.* [1].

is the channel number in which the peak corresponding to a certain position [as showsn in Fig. 4) falls. The counter is seen to be linear.

Notice that there is a total of 24 elements. For all of them the above calibrations were repeated periodically in order to check the correct performance of the apparatus. For example, the neutron detector stability over a week is shown in Fig. 6, where the time variation for θ, t_1, and t_2 signals is plotted for each element of the neutron detector. The time stability of the neutron detector is remarkably good.

The neutron detectors are mounted on rails (Fig 2) so that they can be positioned at any angle depending on the reaction studied. According to the requirement of good mass resolution or of large accepted solid angle the neutron detectors can be set, for example, at 6 m or 4 m distance from the target center, other positioning obviously being possible as well.

2.3. *The* EM *Shower Detectors*

These detectors are used for electron and γ detection. One detector (Fig. 7) consists of nine elements, each element being made of a piece of lead followed by a two-gap spark chamber and a plastic scintillation counter. The first layer of lead is 1.72 radiation lengths thick; the other layers are each 0.86 radiation length thick. The over-all detector thickness is half a meter. Before the first lead layer there are two thin-plate spark chambers K (a total of 12 gaps) to allow precise kinematical

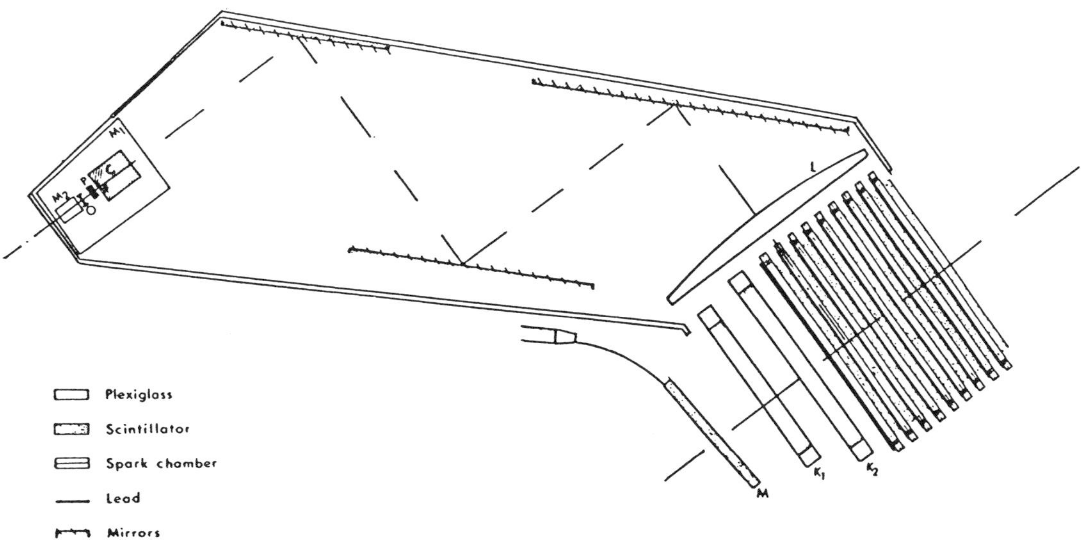

FIG. 7. Side view of the EM shower detectors. M is a scintillation counter, K_1 and K_2 are thin-plate spark chambers. Each EM shower detector consists of nine layers of lead, spark chamber and scintillator sandwiched together. When the detector is used for γ-ray detection a 0.89 r.l. lead foil for γ-conversion (not shown in the Fig.) is placed in front of the M counters.

reconstruction of the events. Thus, a long H_2 target can be used without losing accuracy in the missing-mass measurement by the neutrons. The two detectors (see Fig. 2) may be rotated independently about a horizontal and a vertical axis through the center of the H_2 target.

Figure 8 shows a calibration of one of the two EM shower detectors. The response of the telescope is plotted as a function of the energy of the electron beam. We see

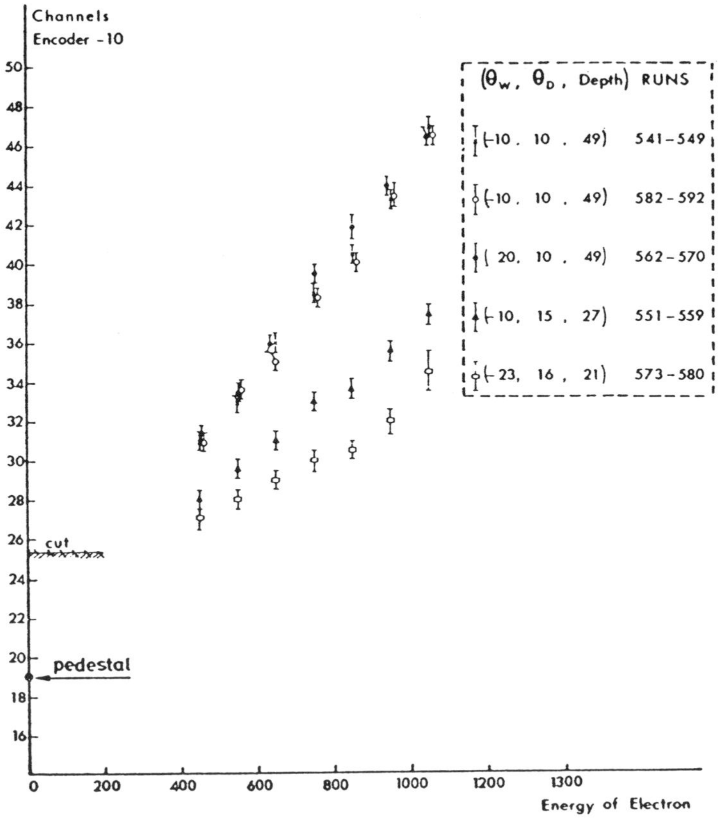

FIG. 8. Calibration of one of the EM shower detectors. Total scintillation counter pulse-height vs electron energy for various settings.

that the instrument is linear. The three sets of points in the upper curve correspond to measurements made at different times (given by the run number), and to three different positions of the beam in the detector. Near the extremes of the detector, the pulse-height decreases and the calibrations are parametrized according to the maximum depth of the detector available to the shower. The two lower curves are two of these edge-effect calibrations.

Figure 9a shows a complete efficiency calibration at fixed energy. The purpose of this figure is to show that the EM shower detectors TOP and BOTTOM had very similar characteristics. The open circles refer to the bottom detector and the full circles to the top detector. The electron energy for these two sets of points is 1050 MeV. Figure 9b shows a family of curves corresponding to 170 calibration points taken at energies from 1.05 GeV down to 0.45 GeV.

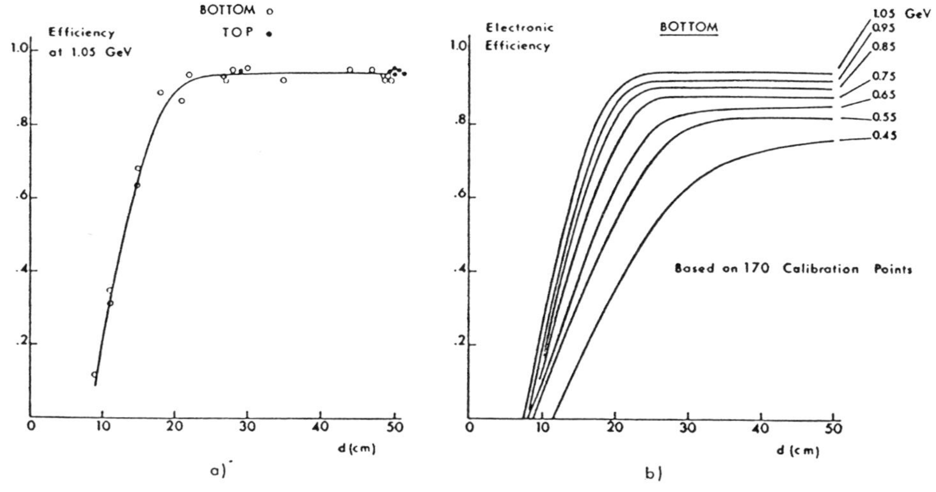

FIG. 9. Efficiency calibration of the EM shower detector as a function of the depth available to a shower: (a) Response of TOP and BOTTOM counters at 1.05 GeV. (b) Response of BOTTOM counter at different energies. The curves correspond to 170 calibration points.

The calibrations have been repeated with a γ-ray beam and the result is that the detector response is essentially the same. As mentioned above, for γ-ray detection a 0.5 cm Pb foil is placed in front of the M counters.

2.3.1. *π-e Discrimination*

When looking for rare events such as the (e^+e^-) decays of vector mesons (see Section 3.5) produced in strong interactions, a high rejection power against hadronic background is needed.

Typical figures for π and e identification in one of the two EM shower detectors are shown in Table 1. From as low as 400 MeV up to 1100 MeV, one can reject pions with a power of $\sim 3 \times 10^{-4}$. For each particle and each momentum there are three numbers: the electronic efficiency; the picture analysis efficiency; and their product, the overall efficiency.

The picture analysis is very important because it allows the elimination of charge exchange of pions, which, being accepted by the trigger, is the greatest source of trouble when one wants to distinguish a pion from an electron. From 400 to 1100 MeV, the power of the telescope against pions is practically the same, and the efficiency for electron detection is very good—between 70 % and 80 %.

To summarize, each of the two telescopes has $\sim 3 \times 10^{-4}$ rejection against pions, giving a product $\sim 10^{-7}$, which is the rejection factor for charged $\pi\pi$ pairs

TABLE I

Efficiency in the electromagnetic shower detector BOTTOM

Momentum in MeV/c	Particle	Electronic efficiency	Picture analysis efficiency	Overall efficiency
400	e	$(77.5 \pm 2.2)\,\%$	$(89.0 \pm 2.2)\,\%$	$(69.0 \pm 2.6)\,\%$
	π	$(6.3 \pm 0.2)\,\%$	$(0.43 \pm 0.2)\,\%$	$(2.7 \pm 1.6)\,10^{-4}$
1100	e	$(94.0 \pm 1.5)\,\%$	$(88.0 \pm 2.0)\,\%$	$(83.0 \pm 2.3)\,\%$
	π	$(17.6 \pm 0.6)\,\%$	$(0.16 \pm 0.16)\,\%$	$(2.8 \pm 2.8)\,10^{-4}$

and any other sort of charged multipion events. It is this rejection power that allows the study of rare events such as (e^+e^-) decays of strongly interacting particles.[2]

2.4. *Reactions Investigated*

The setup is designed to study reactions of the general type

$$a + b \to \text{neutron} + \text{missing mass} \begin{cases} \nearrow \text{charged particles} \begin{cases} \nearrow e^{\pm} \\ \searrow \pi^{\pm} \end{cases} \\ \searrow \text{neutral particles} \to \gamma\text{'s} \end{cases}$$

where a and b can be any sort of incident and target particles.

Actually, all experiments performed so far have used as initial state π^- and protons. In the final state the neutron has always been measured, while the electromagnetic shower detectors have been used only for those investigations where it was necessary to identify the nature of the decay products of the "missing mass," i.e., π's, e's or γ's. The basic reaction investigated is therefore

$$\pi^- + p \to n + M^0, \tag{1}$$

where M^0 indicates any neutral meson state. For each neutron produced in reaction (1), the neutron detector measures its time-of-flight t_n and its emission angle θ_n, thus allowing a determination of the M^0 mass. The mass resolution depends on the kinematical region accepted in the plane (t_n, θ_n) (see Fig. 10) and

[2] Full details of the EM shower detectors can be found elsewhere [2].

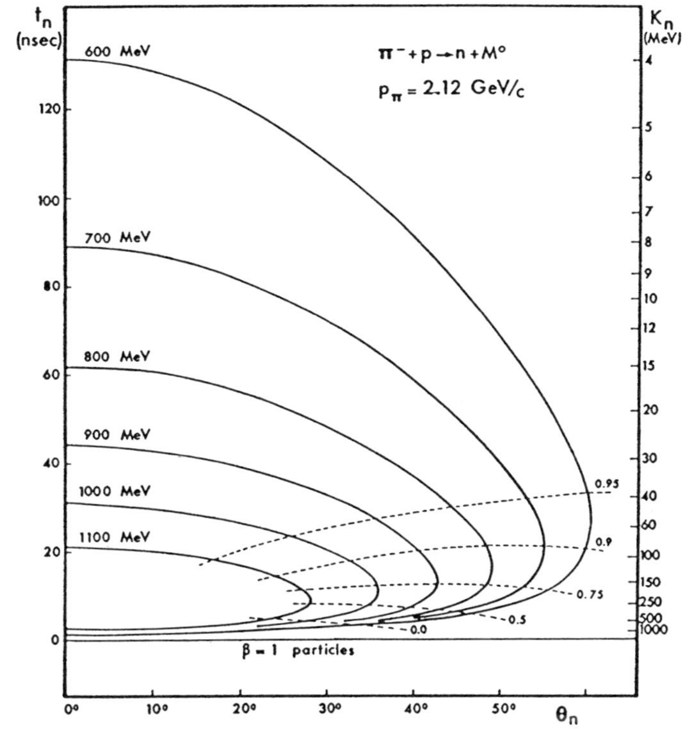

FIG. 10. The neutron time of flight t_n over a 4 m path is plotted vs the neutron emission angle θ_n in the laboratory system. The ordinate on the right refers to the neutron kinetic energy. The kinematic curves are labeled by the corresponding neutron missing masses. The dashed lines indicate constant values of $\cos\theta_n^*$ ($\theta_n^* \equiv$ neutron emission angle in the c.m. system).

on the distance of the neutron detector from the target. For example, with the neutron detectors set at 6 m, the mass resolution is ± 3.5 MeV in the η mass region at 0.8 GeV/c incident π^- momentum, ± 6.5 MeV in the X^0 mass region at 1.6 GeV/c, and ± 7.5 MeV in the A_2 mass region at 3.16 GeV/c.

3. Review of the Physics Results Obtained

3.1. *Evidence Against the Existence of the S^0 Meson* [3]

This is the first experiment which was performed. The interest in the existence of a neutral meson in the ρ-mass region was and is still of the greatest importance in order to solve the old problem of forward–backward asymmetries [4, 5] in the charged and neutral ρ-meson decays.

This asymmetry would, in fact, destroy the interpretation of the charged and neutral ρ-meson in terms of a unique particle if it were not for the obvious possi-

The reproduction stops at page 417 of the original review paper and does not include the physics results since this volume refers to the invention and implementation of new instruments only, not to the physics results obtained.

Section 3

THE LAA PROJECT AND ITS ACHIEVEMENTS

The purpose of the LAA project was to prove, on the basis of prototypes, the feasibility of essential components for an ideal detector to operate in future multi-TeV hadron colliders, at the highest energy and luminosity. A brief summary and the reproduction of a review paper on this new venture in subnuclear physics technology are reported in this section, together with a list of all the published results.

Horst Wenninger
CERN, Geneva, Switzerland

THE LAA PROJECT: A BRIEF SUMMARY

THE LAA PROJECT: A BRIEF SUMMARY

Horst Wenninger
CERN, Geneva, Switzerland

The LAA Project was implemented at CERN in 1986 (ref. CERN Council Resolution, December 1986) and Prof. A. Zichichi presented its research and development programme in an open presentation at CERN in June 1987 and subsequently to the CERN Research Board.

The LAA Project consisted of eleven sub-projects, covering all aspects of a detector lay-out, i.e.:

- high precision tracking
- calorimetry
- large area muon detection devices
- leading particle detection
- data acquisition and analysis (computing)

The LAA Project was open to all interested physicists and engineers. The goal was to prove, on the basis of prototypes, the feasibility of experimental detector set-ups for future multi-TeV hadron colliders. The machine parameters – equivalent to a 10% model of Eloisatron – were defined as:

- beam energy: > 10 TeV (LHC beam energy is 7 TeV)
- luminosity: 10**32 to 10** 33 cm-2 s-1 (LHC luminosity 10**34)
- bunch crossing time: 25 ns (LHC 25 ns)

Special attention was therefore paid to (i) radiation hardness, (ii) rate capability, (iii) hermeticity and (iv) momentum resolution of proposed detector assemblies and R&D proposals.

The LAA Project was a great success involving 250 participants and 50 institutions. 35 professional staff were recruited at CERN for the LAA Project and many of them are now involved in LHC experiments. The main achievements of the LAA Project are described by the Project Leader, Prof. A. Zichichi, in a CERN report (CERN/LAA/91-1), first published in March 1991 and reported in several open Meetings and Conferences. The reproduction of one of these reports follows.

A. Zichichi

THE MAIN ACHIEVEMENTS OF THE LAA PROJECT

From
Physics Up to 200 TeV, 327 (1991)

PLENUM PRESS
NEW YORK AND LONDON
1991

THE MAIN ACHIEVEMENTS OF THE LAA PROJECT

A. Zichichi

CERN, Geneva

Switzerland

I. INTRODUCTION

I.1. General philosophy of the LAA Project

The LAA Project [1, 2, 3, 4, 5, 6] is an intensive programme to develop new High Energy Physics experimental techniques. The ultimate goal is to prove, on the basis of prototypes, the feasibility of essential components for a detector to operate in a future multi-TeV Hadron Collider.

The LAA Project is open to all physicists and engineers who are interested in participating.

This project represents a unique opportunity for Europe to have a leading role in the advanced technology for High Energy Physics.

I.2. The basic data

Let me remind briefly the origin of the LAA Project and its basic features. As already explained in Ref. 5 and 6, at present no one knows how to build an (e^+e^-) collider even at 1 TeV, while the conceptual design of a collider for (pp) at 100 TeV already exists (the ELN Project [7]). On the contrary, no one knows how to build detectors for a 100 TeV (pp) collider, while a detector for a 1 TeV (e^+e^-) machine is within the present-day technological possibilities. This is illustrated in Fig. I.2.1.

Physics Up to 200 TeV
Edited by A. Zichichi, Plenum Press, New York, 1991

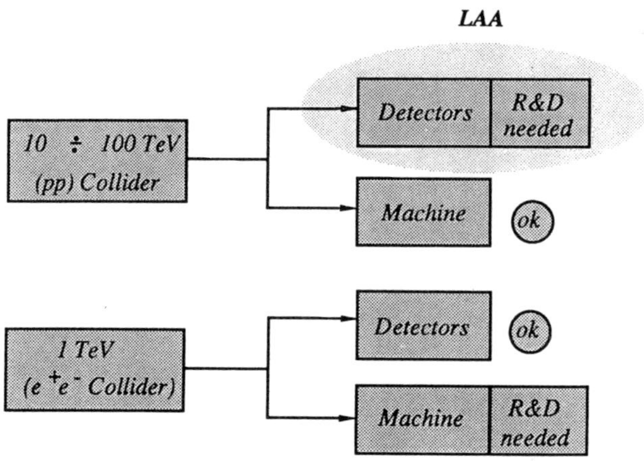

Fig. 1.2.1. The present status of Colliders and Detectors.

One of the main problems for detectors comes from the requirement that the average number of events per bunch crossing, $< n >$, must be one if the missing energy is to be used as a signature for rare events. Another limiting factor towards very high luminosities is the radiation dose the detectors can withstand.

The following machine parameters:

$$\Delta t_b \sim 100 \; ns, \text{ and}$$

$$L_{pp} \sim 10^{32} \; cm^{-2}s^{-1},$$

are well within reach from a technological point of view; on the other hand, the total (pp) cross-section is expected to be:

$$\sigma_{pp} \cong 100 \; mb = 10^{-25} cm^2.$$

These three values together produce $< n > \sim 1$. But what is wanted is $< n >= 1$ at higher luminosities. And therefore the radiation levels, the minimum bias event rate and the Δ_{tb} reach prohibitive figures:

$$\begin{array}{rcccc} L = 10^{33} & \rightarrow & 10^{34} & \rightarrow & 10^{35} cm^{-2}s^{-1}, \\ \downarrow & & \downarrow & & \downarrow \\ \Delta t_b = 10 & \rightarrow & 1 & \rightarrow & 0.1 \; ns. \end{array}$$

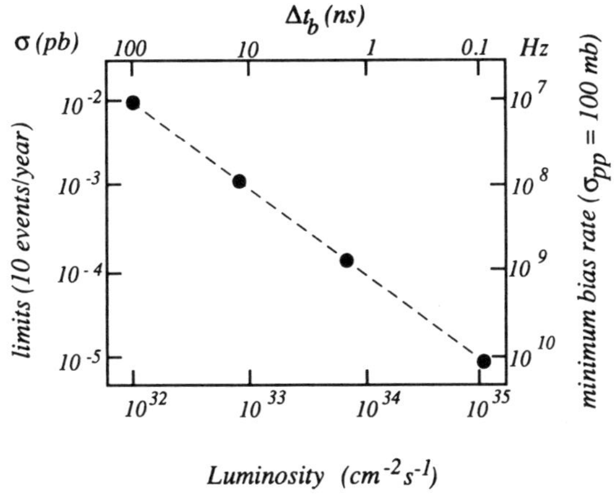

Fig. 1.2.2. Discovery limit for rare events, total rate and bunch spacing (for $<n>=1$) as a function of the Luminosity.

In conclusion, the main items where R&D is needed in order to cope with the challenging experimental environment of the next generation of (pp) colliders are:

1. RADIATION HARDNESS,
2. HERMETICITY,
3. RATE CAPABILITY,
4. TRACK & SPACE RESOLUTION,
5. ENERGY RESOLUTION,
6. MOMENTUM RESOLUTION,
7. TIME RESOLUTION,
8. PARTICLE IDENTIFICATION,

and these are the basic parts of the LAA Project.

It is very important to notice that, up to now, the first three items were not of such a great relevance.

The potentials for new discoveries in a multi-TeV (pp) collider are summarized in Fig. I.2.2. The "magic" limit of $10^{-40}cm^2$ in the cross-section limit can be reached only if the luminosity is pushed to high levels: $L = 10^{33} - 10^{34}cm^{-2}s^{-1}$ or even higher.

Figure I.2.3 shows the correlation between Luminosity, bunch spacing, and radiation dose under the $<n>=1$ condition.

Moreover, (pp) interactions at 100 TeV are expected to produce hundreds of particles per event. All these particles must be observed, measured (in terms of energy and/or momentum), and, possibly, identified with the best achievable precision.

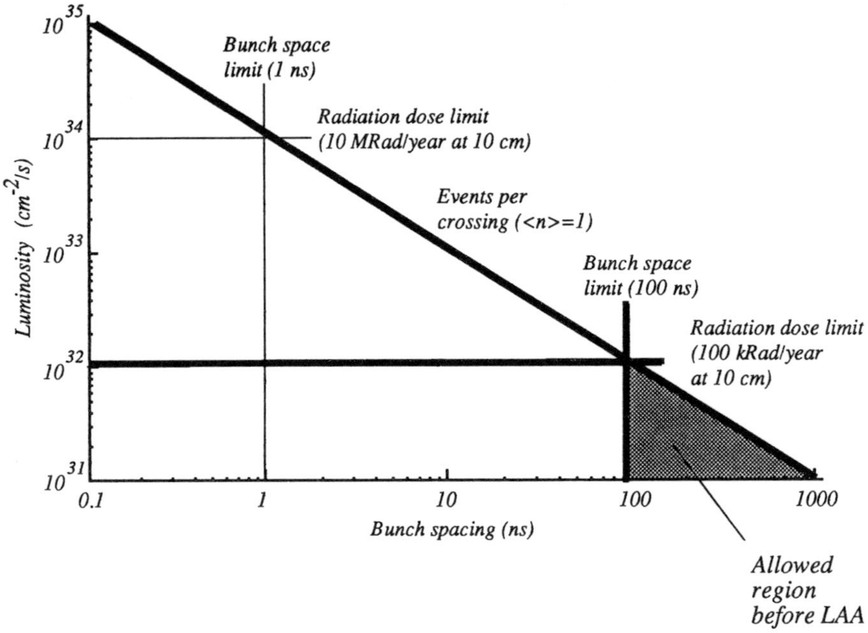

Fig. 1.2.3. Limits on Luminosity as function of bunch spacing. Also shown is the radiation dose level at different Luminosities. The grey region was allowed before LAA.

I.3. The Physics at the new generation of Hadron Colliders

What Physics are we facing with the new generation of Hadron Colliders? There are four basic theoretical problems to be solved in the multi-TeV ($10 \div 200$ TeV) energy domain:

i) The **family** problem.
ii) The **hierarchy** problem.
iii) The **proliferation** problem.
iv) The **compositeness** problem.

The questions which arise when translating these theoretical problems into physically observable phenomena are the following:

i) Do **new, heavier, quarks** and **leptons** exist?
ii) Are there **other intermediate vector bosons?**
iii) Do **Higgs bosons** exist?
iv) Do **supersymmetric partners** exist?
v) Are **quarks** and **leptons composite?**
vi) Would some **unexpected exotic process** occur?

In order to go from theory to "down-to-earth" physics, all detectors operating in the multi-TeV domain should detect and measure:

i) **electrons** and **photons**;
ii) **muons** and other **lepton-like long-lived particles**;
iii) **neutrinos** and other **non-interacting particles** (i.e. missing objects);
iv) **leading protons** (for hermeticity and new physics);
v) **hadrons** and **jets**, with and without leptons inside.

I.4. Present structure of the LAA project

The LAA project consists of eleven main parts (Fig. I.4.1):

1. HIGH PRECISION TRACKING

Three components:

a) Gaseous detectors.
b) Scintillating fibres.
c) Microstrip GaAs.

2. CALORIMETRY

Three components:

a) High precision electromagnetic.
b) Compact EM+Hadronic.
c) "Perfect" Calorimetry.

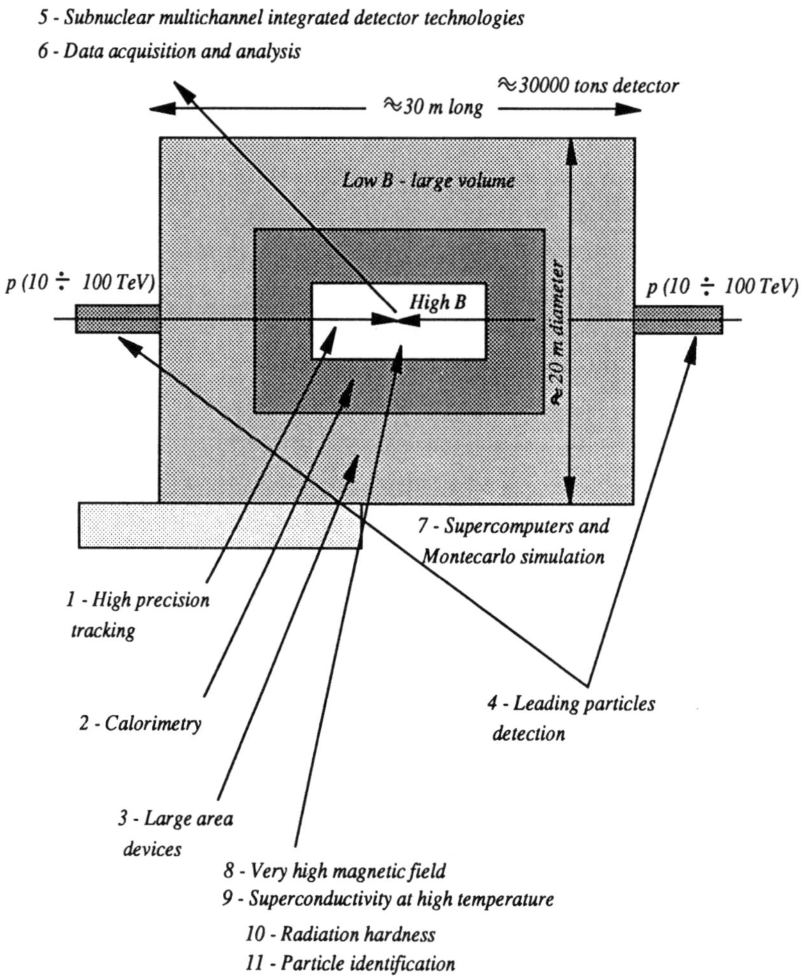

Fig. I.4.1. The eleven components of LAA.

3. LARGE AREA DEVICES

Two parts:

a) Construction .

b) Alignment.

4. LEADING PARTICLE DETECTION

5. SMIDT (Subnuclear Multichannel Integrated Detector Technologies)

Two parts:

a) Microelectronics.

b) New, Radiation-resistant Technologies.

6. DATA ACQUISITION AND ANALYSIS

Three parts:

a) Real Time Data Acquisition.

b) FASTBUS RISC computer.

c) Fine-grained Parallel Processor.

7. SUPERCOMPUTERS AND MONTECARLO SIMULATIONS

8. VERY HIGH MAGNETIC FIELDS

9. SUPERCONDUCTIVITY AT HIGH TEMPERATURE

10. RADIATION HARDNESS

11. PARTICLE IDENTIFICATION

I.5 Participants

The following Physicists, Engineers and Technicians contribute to the activity of the LAA Project:

D. Acosta[39)], J. Alberty[48)], J. Alsford[38)], C. Alvisi[2)], G. Ambrosi[37)], F. Anghinolfi[7)], F. Anselmo[7)], G. Anzivino[20)], M. Arneodo[44)], R. Arnold[12)], F. Arzarello[2)], P. Aspell[7)], R. Ayad[48)], L.E. Barberio[48)], G. Bari[2)], T. Barillari[5)], M. Basile[2,3)], R. Battiston[37)], S.P. Beaumont[21)], U. Becker[29)], L. Bellagamba[2)], M. Bénot[7)], P. Benvenuto[20)], J. Berbiers[7)], J. Berdugo[7)], F. Bergsma[7)], R. Bertin[7)], N. Bingefors[46)], D. Bisello[33)], R.K. Bock[7)], A. Bonanno[5)], D. Boscherini[2)], M. Bosteels[7)], R. Bouclier[7)], M. Bramhall[38)], G. Bruni[2)], P. Bruni[2)], S. Buontempo[7)], V. Buzuloiu[48)], L. Calôba[11)], M. Camarca[5)], M. Campbell[7)], G. Cara Romeo[2)], M. Caria[4)], R. Casaccia[20)],

H. Castro[48], S. Ceresara[17], G. Charpak[7], E. Chesi[7], M. Chiarini[48],
J. Christiansen[7], E. Christofel[12], L. Cifarelli[2,31], F. Cindolo[20],
F. Ciralli[48], E. Colavita[5], F. Coninckx[7], A. Contin[7], M. Costa[48],
F. Crea[5], I. Crotty[7], G. D'Ali[2,34], C. D'Ambrosio[7], S. D'Auria[2],
M. Dardo[45], C. Del Papa[2,3], G. Della Gatta[13], S. De Pasquale[2],
R. De Salvo[7], P. Destruel[43], J. deWitt[40], O. Di Rosa[7], D. Dorfan[40],
E. Duchovni[47], J. Dupont[7], J. Dupraz[7], J. Egger[49], T. Ekelöf[46],
C.C. Enz[42], A. Ereditato[31], J.P. Fabre[7], R. Ferrari[36], F. Fiori[2,3], P. Ford[7],
F. Frasconi[2], M. Fraternali[36], M. French[38], M. Fuchs[7], G. Fumagalli[35,36],
K. Gabathuler[49], J. Galvez[48], R. Garré[17], J. Gaudaen[7], Y. Giomataris[7],
J.P. Girod[7], P. Giusti[2], K. Goebel[7], C. Grinnel[7], H. Güsten[24],
J.L. Guyonnet[12], T. Gys[7], F.G. Hartjes[32], W. Hao[48], D. Hazifotiadu[48],
E. Heijne[7], T. Henkes[27], A.M. Henriques[7], M. Hourican[7], G. Iacobucci[2],
G. Iuvino[36], P. Jarron[7], P. Jenni[7], J.P. Jobez[9], C. Joram[6],W. Kluge[6],
W. Krisher[7], F. Krummenacher[16], A. Kuzucu[48], I. Laakso[20],J.C. Labbé[7],
G. La Commare[7], H. Larsen[7], G. Laurenti[7], T.D. Lee[10], M. Letheren[7],
H. Leutz[7], G. Levi[37], L. Levinson[47], Q. Lin[48], L. Linssen[7], B. Lisowski[2],
A. Litke[40], M. Livan[36], C. Ljuslin[7], J.G. Lynch[21], G. Maccarrone[2],
D.McIntyre[21], A. Maio[25], L. Mapelli[7], A. Marchioro[7], A. Margotti[2,3],
M. Marino[48], T. Marsico[5], T. Massam[2], T. Matsuda[48], T. Matsuura[14],
D. Mattern[7], G. Meddeler[7], K.H. Meier[7], R. Meng[14], G. Mikenberg[47],
M.R. Mondardini[23], G. Mörk[46], B. Musso[1], R. Nania[2], C. Nemoz[7],
S. Newett[38], B. Ong[39], V. O'Shea[2], N. Ozdes[48], H.P. Paar[39], F. Palmonari[3],
G. Passardi[7], F. Pastore[36], P. Pelfer[18,19], M. Pereira[25], C. Peroni[44,45],
E. Perotto[8], V. Peskov[48], D. Piedigrossi[7], R. Pilastrini[2], D. Pitzl[40],
L. Poggioli[26], G. Polesello[7], S. Qian[20], A. Racz[7], C. Raine[21], F. Rivera[48],
F. Riccardi[30,31], A. Rimoldi[35,36], T. Ruf[6], H. Sadrozinski[40], A. Sandoval[7],
G. Sannier[7], J.C. Santiard[7], G. Sartorelli[2,3], P. Sartori[33], F. Sauli[7],
C. Scheel[7,32], M. Schioppa[48], J. Schipper[7], H. Schönbacher[7], D. Scigocki[7],
M. Scioni[2,3], J. Seguinot[9], A. Seiden[40], W. Seidl[7], J.M. Seixas[48],
A. Sharma[48], P. Sharp[38], J. Schmitz[32], I.O. Skillicorn[21], A. Simon[27],
G. Simonet[7], M. Sivertz[39], K. Smith[21], P. Sonderegger[7], M.N. Souza[11],
E. Spencer[40], L. Sportelli[5], A. Staiano[44], G.C. Susinno[5,20], S. Tailhardat[7],
M. Taufer[7], N. Tartoni[18,19], M. Tavlet[7], A.E. Terraneo[10], Z.D. Thomé[11],
R. Timellini[48], J. Tischhauser[7], J. Tocqueville[9], R.M. Turnbull[21],
V. Valencic[28], B. Van Eijk[7], G. Vanstraelen[22], V. Vercesi[7], L. Votano[20],
Y. Wang[48],N. Wauquier[7], H. Wenninger[7], R. Wigmans[7], C. Williams[7],
C. Xu[48], K. You[48], T. Ypsilantis[6,8], N. Zaganidis[7] and A. Zichichi[7]

1) Ansaldo, Genoa, Italy
2) INFN, Bologna, Italy
3) University of Bologna, Italy
4) University of Cagliari, Italy
5) University of Calabria, Italy
6) University of Karlsruhe, FRG
7) CERN, Geneva, Switzerland
8) CNR, Pisa, Italy
9) Collège de France, Paris, France
10) Columbia University, New York, USA
11) COPPE, Rio de Janeiro, Brazil
12) CRN, Strasbourg, France
13) Digital Italia, Rome, Italy
14) DESY, Hamburg, FRG
15) Draper Laboratories, Cambridge, Mass., USA
16) EPFL, Lausanne, Switzerland
17) Europa Metalli - LMI, Florence, Italy
18) INFN, Firenze, Italy
19) University of Firenze, Italy
20) INFN, Laboratori Nazionali di Frascati, Italy
21) University of Glasgow, United Kingdom
22) IMEC/ASP, Leuven, Belgium
23) Cornell University, Itaca, New York, USA
24) Kern Forschungszentrum Karlsruhe, FRG
25) LIP, Lisbon, Portugal
26) LPNHE University, Paris, France
27) MAX Planck Inst., Heidelberg, FRG
28) Mead SA, Lausanne, Switzerland
29) MIT, Cambridge, Mass., USA
30) INFN, Naples, Italy
31) University of Naples, Italy
32) NIKHEF, Amsterdam, Netherlands
33) University of Padova, Italy
34) University of Palermo, Italy
35) INFN, Pavia, Italy
36) University of Pavia, Italy
37) University of Perugia, Italy
38) Rutherford Appleton Laboratory, United Kingdom

39) University of California, San Diego, USA
40) University of California, Santa Cruz, USA
41) Senter for Industriforsk., Oslo, Norway
42) Smart Silicon System, Lausanne, Switzerland
43) University of Toulouse, France
44) INFN, Torino, Italy
45) University of Torino, Italy
46) Uppsala University, Sweden
47) Weizmann Institute, Rehovoth, Israel
48) World Lab, Geneva, Switzerland
49) Paul Scherrer Institute, Zurich, Switzerland

I.6. The Main achievements of the LAA Project

The main achievements of the LAA Project, in terms of *discoveries, records, new developments* and *inventions,* are summarized in Tables I.4.I and I.4.II. They are fully described in chapter II.

Table I.6.I. Main achievements of the LAA Project in terms of: D=Discoveries, R=Records, N=New developments, I=Inventions

	D	R	N	I
1. HIGH PRECISION TRACKING				
a) Gaseous detectors		2	1	
b) Scintillating fibres	1	2	1	
c) Microstrip GaAs			1	
2. CALORIMETRY				
a) High precision EM	2		1	
b) Compact EM+Hadronic	1	2		2
c) "Perfect" Calorimetry	1	1	1	
3. LARGE AREA DEVICES				
a) Construction				2
b) Alignment		1	1	
4. LEADING PARTICLE DETECTION		1	1	
5. SMIDT				
a) Microelectronics			2	
b) New, Radiation-resistant Technologies		1		
6. DATA ACQUISITION AND ANALYSIS				
a) Real Time Data Acquisition			1	
b) FASTBUS RISC computer		1		
c) Fine-grained Parallel Processor			1	
7. SUPERCOMPUTERS AND MONTECARLO SIMULATIONS			1	
TOTAL	5	11	12	4

Table I.6.II. The main achievements of the LAA Project

1. HIGH PRECISION TRACKING	
a) Gaseous detectors	
• Records:	1. Rad-hard wire: $\gg 1$ MRad
	2. Rate resistant: $\approx 2 \times 10^6$ particles/cm^2
• New developments:	1. Mechanically reliable MWPC (MDM)
b) Scintillating fibres	
• Discoveries:	1. PMP
• Records:	1. Smallest diameter sc. fibres: 15 μm
	2. Rad-hard fibres:$\approx$ 1 MRad
• New developments:	1. 1×1 mm^2 bundles with 900 fibres each
c) Microstrip GaAs	
• New developments:	1. GaAs particle detectors
2. CALORIMETRY	
a) High precision EM	
• Discoveries:	1. EF
	2. New inorganic scintillators
• New developments:	1. SSAC
b) Compact EM+Hadronic	
• Discoveries:	1. Intrinsic resolution better in Pb than U
• Records:	1. Hadronic energy resolution: 27% $\sqrt{E}$
	2. Rad-hard fibres: >1 MRad
• Inventions:	1. Spaghetti calorimeter
	2. e/π rejection by timing
c) "Perfect" Calorimetry	
• Discoveries:	1. CsI+TMAE adsorbed layer quantum efficiency
• Records:	1. Highest efficiency photocathode (CsI+TMAE)
• New developments:	1. Systematic studies of liquid Xe

Table I.6.II. The main achievements of the LAA Project (cont.)

3. LARGE AREA DEVICES	
a) Construction	
• Inventions:	1. Blade chambers 2. Gaseous pixel chambers
b) Alignment	
• Records:	1. Best length/angles/linearity measurements
• New developments:	1. New instruments for length/angles/linearity measurement
4. LEADING PARTICLE DETECTION	
• Records:	1. Best way to cut silicon: $< 50\ \mu m$ cracks
• New Developments:	1. Leading particle detector problems solved
5. SMIDT	
a) Microelectronics	
• New developments:	1. All components of HARP done 2. Read-out electronics for pixel detector
b) New, Radiation-resistant Tech.	
• Records:	1. Rad-hard amplifier: > 1MRad
6. DATA ACQUISITION AND ANALYSIS	
a) Real Time Data Acquisition	
• New developments:	1. Test of commercial architectures with algorithms for feature extraction
b) FASTBUS RISC computer	
• Records:	1. Highest computer power in a FASTBUS board: 50 VAX-equivalent
c) Fine-grained Parallel Processor	
• New developments:	1. ASP construction and test of chips
7. SUPERCOMPUTERS AND MONTECARLO SIMULATIONS	
• New developments:	1. Full MonteCarlo Chain

II. STATUS OF THE PROJECT

I.1. High Precision Tracking

In order to measure tracks with high precision in future Hadron Colliders, three parts are needed.

i) The *closest to the vertex* → GASEOUS DETECTORS

ii) The *surrounding one* → SCINTILLATING FIBRES

iii) *Along the beam* → GALLIUM ARSENIDE (GaAs)

This is illustrated in Fig. II.1.1, where the detectors developed within are shown in a cut view perpendicular to the beam axis.

II.1.a. Gaseous detectors

The **new development** of the MultiDrift Modules (MDM), a mechanically reliable, modular MWPC, with high redundancy and precision: $\approx 60\ \mu m$, brought us to establish two **records**:

- The most radiation resistant combination of gas (DME) and wire (stainless steel): well above 1 Mrad.
- The drift chamber most resistant to rate: $\approx 2 \times 10^6$ particles/cm^2.

Before LAA, the typical measurement precision of the standard drift chambers used for the central tracking, was in the range between 100 and 200 μm [8]. The rate capability of these chambers was, at most, about 10^5 particles/cm^2 [9] and the radiation resistance was much lower than 1 MRad [8].

The present limits of the LAA MultiDrift Modules are shown in Figs. II.1.2, II.1.3, and II.1.4: precision = 60 μm, rate capability = 2×10^6 particle/cm^2, and radiation resistance well above 1 MRad, respectively.

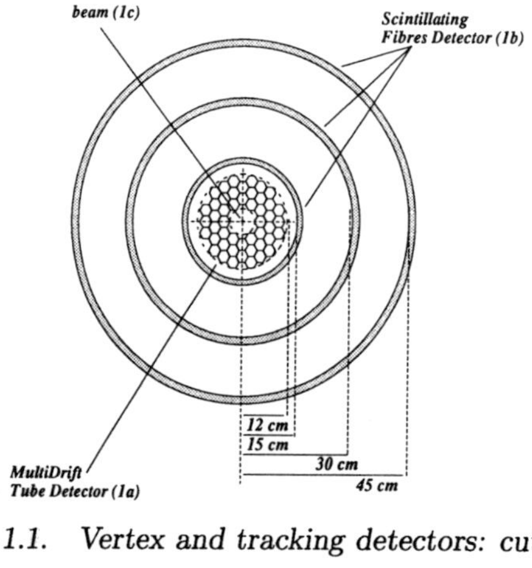

Fig. II.1.1. Vertex and tracking detectors: cut view.

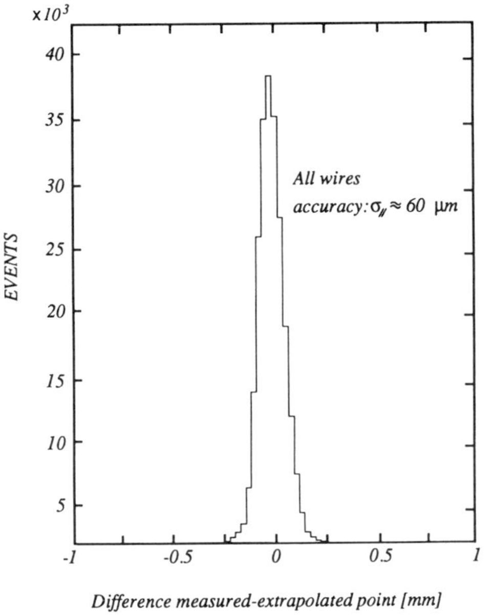

Fig. II.1.2. Position measurement perpendicular to the wires: distribution of the differences between measured and extrapolated points.

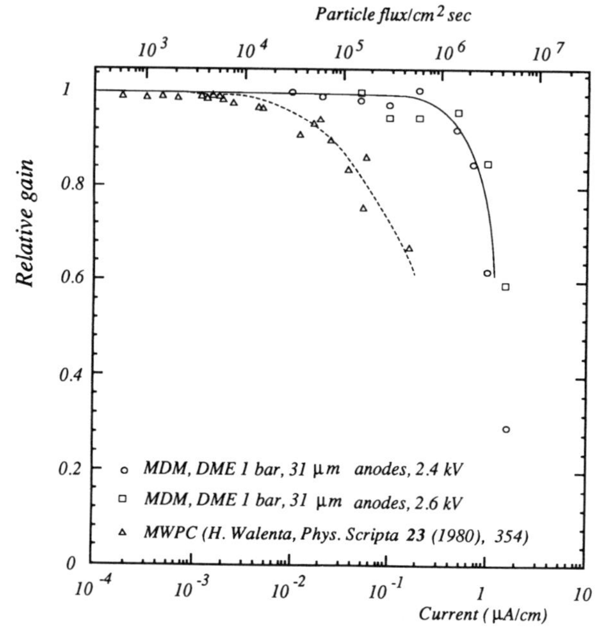

Gain drop vs wire current

Fig. II.1.3. Rate dependence of gain measured in the multidrift module with Dimethyl Ether, as a function of current and equivalent particle rate (full curve). For reference, the same measurement is shown for a conventional multiwire chamber (points on the dashed curve).

II.1.b. Scintillating fibres

Thanks to the discovery of a new scintillator with a high Stoke-shift, the PMP (see Figs. II.1.5 and II.1.6), two records were established by this component of LAA:

- The production of the smallest scintillating fibre - 15 μm diameter - with an attenuation length sufficiently large to be used in a real physics experiment. This is an improvement of nearly two orders of magnitude with respect to the situation before LAA [10].

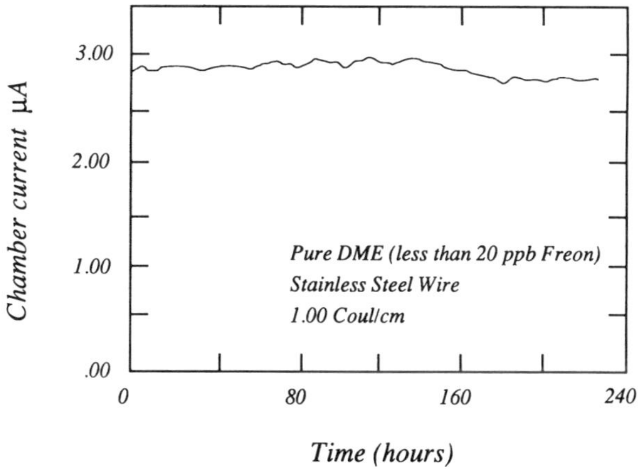

Fig. II.1.4. Radiation resistance of the modules with Dimethyl Ether using different wires and gas purities. In the best case, no degradation is observed after an irradiation corresponding to above 1 Coul/cm (= 1 MRad).

- The highest radiation resistance for small diameter fibre: about 1 MRad. Again, an improvement here by a factor 2 has been obtained by LAA [11].

Furthermore, a technological breakthrough came from the **development** of a new method for producing 1×1 mm^2 fibre bundles with about 900 fibres/bundle. These bundles are necessary, since the individual fibres cannot be handled any more at such small diameters. The bundles can be mechanically arranged to form the final tracker shells.

Figures II.1.7, 8, and 9 show the mechanical arrangement of the fibres in the fibre bundles.

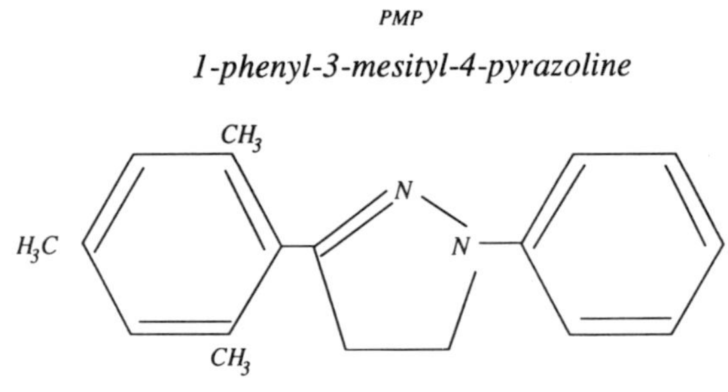

Fig. II.1.5. The PMP molecule.

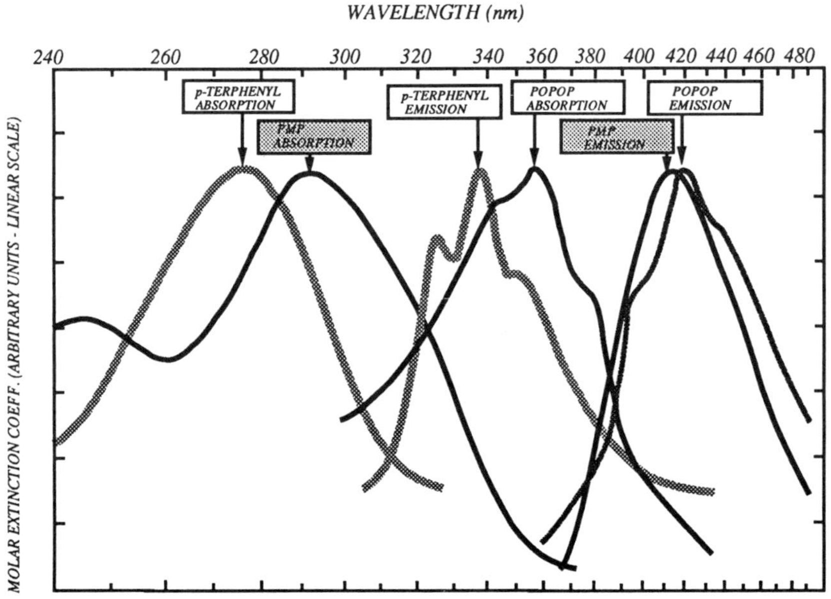

Fig. II.1.6. Absorption and emission spectra of PMP, P-TERPHENIL, and POPOP.

Fig. II.1.7. *Enlarged section of a fibre bundle containing 15 μm diameter fibres. For comparison, a female hair with 60 μm diameter is also shown.*

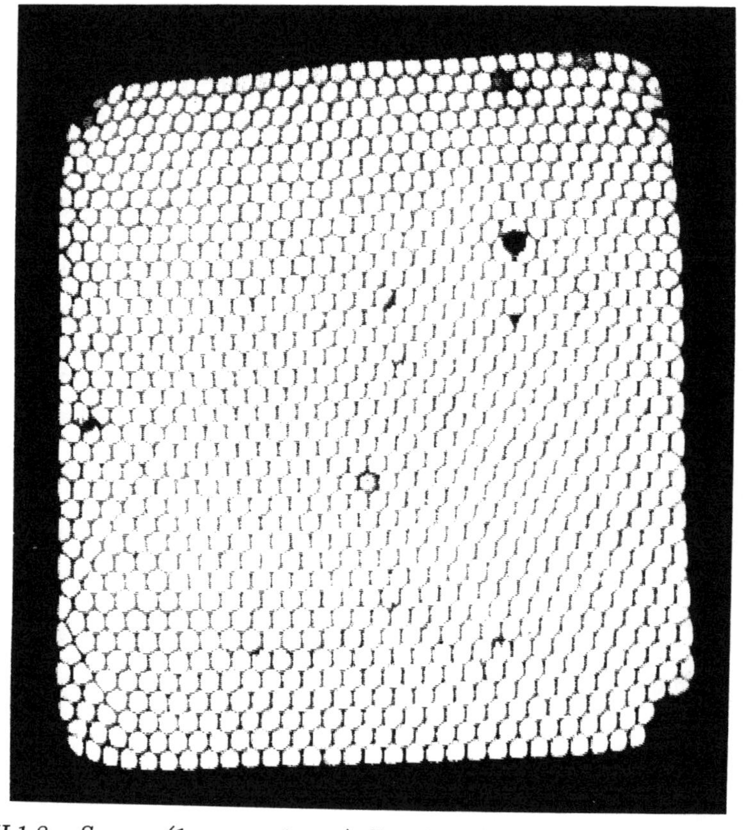

Fig. II.1.8. *Square (1 mm × 1 mm) fibre bundle containing 900 individual hexagonal fibres of 30 μm "diameter". This picture is taken through 1 m bundle length, using white light and 175 times magnification.*

Fig. II.1.9. *Magnified picture of the same fibre bundle with only one individual fibre illuminated. Fibre diameter is 30 μm. The cladding thickness separating it from the surrounding fibres is 6 μm.*

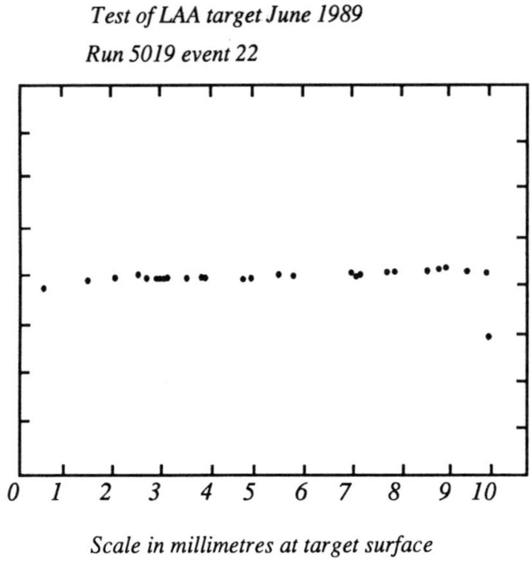

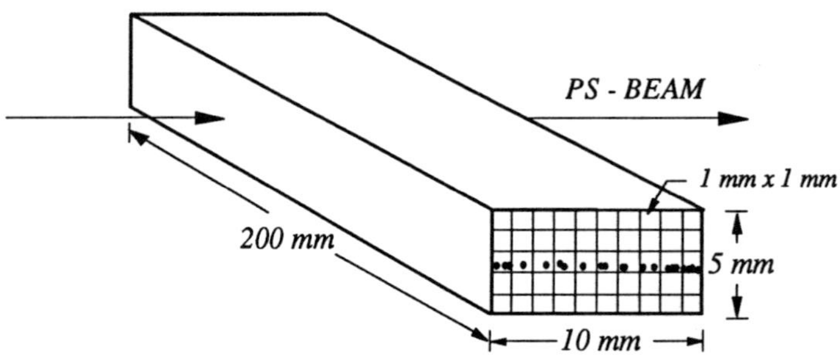

Fig.II.1.10 *Particle track seen by an assembly of multi-bundles, which represents a part of a tracker-shell. The PS-beam traverses this target like a charged particle emerging from the interaction point of a collider.*

The equivalent of one shell of the final tracker was tested in a particle beam. Figure II.1.10 shows how a single track is seen in the fibre bundle assembly.

Figure II.1.11 shows the results from 272 tracks in terms of precision and two-track resolution:

2-track resolution: 83 μm,

tracking precision: 35 μm.

Note that in classical drift chambers, the tracking precision is between 100 and 200 μm, and the two-track resolution is about 1 mm.

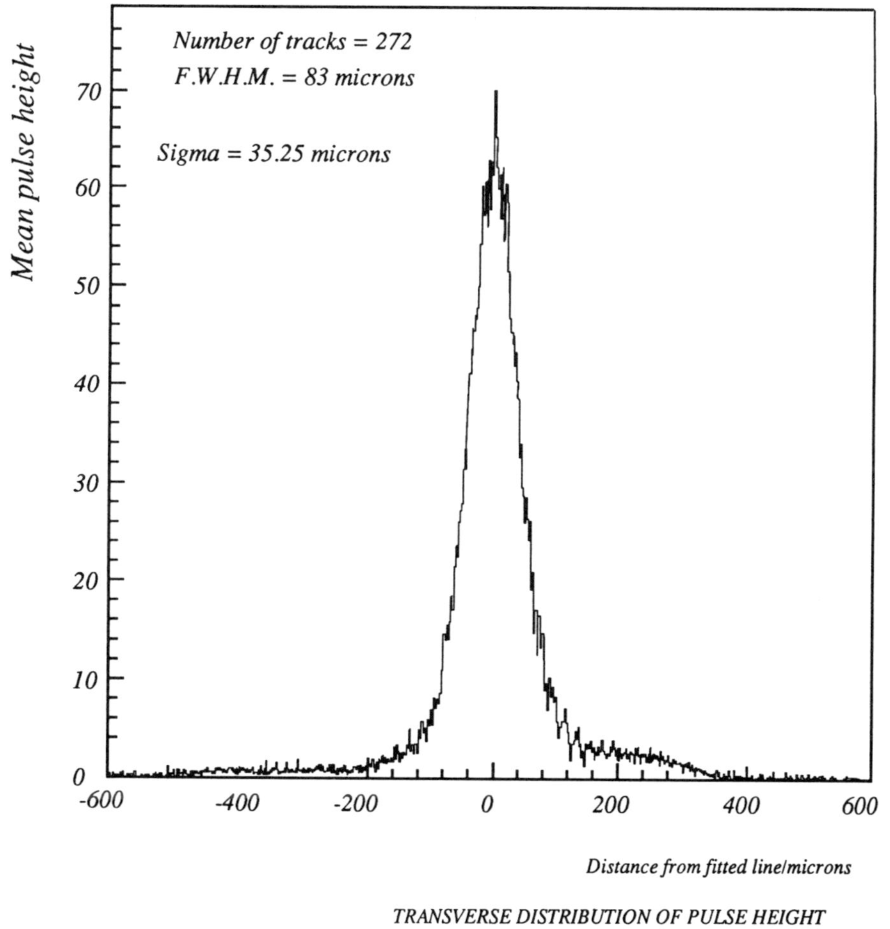

Fig. II.1.11. *Histogram of 272 particle tracks showing 83 μm two-track resolution and 35 μm spatial precision.*

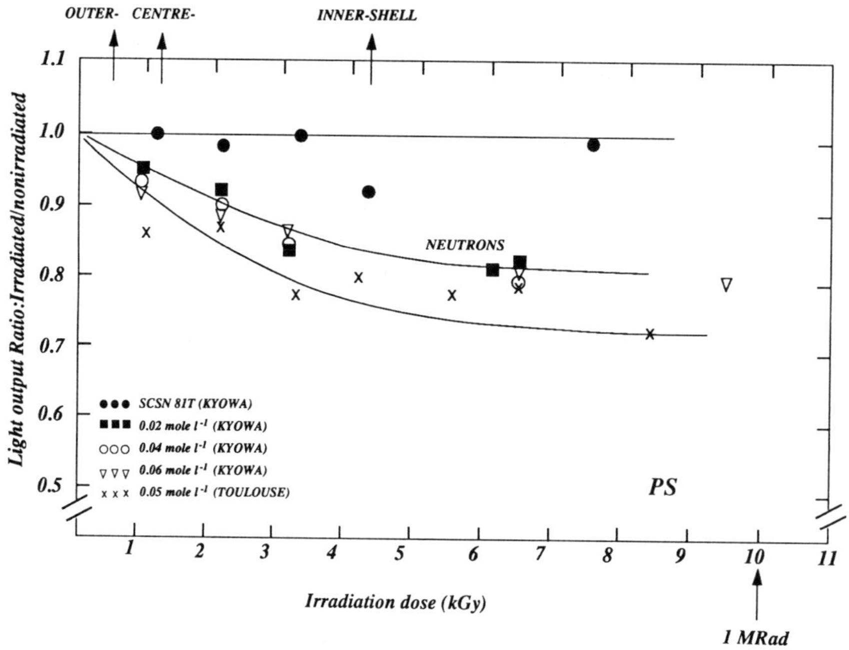

Fig. II.1.12. *Light output as a function of irradiation dose for different samples of scintillating fibres. The optimal concentration of PMP is of 0.04 model/l. The arrows show the expected level of irradiation of the three tracker shells after one year running at a luminosity of* $10^{33} cm^{-2} s^{-1}$. *The curves are free interpolation of data to guide the eye.*

The radiation resistance of the 30 μm fibres has been tested up to 1 MRad. Figure II.1.12 shows the results: at a luminosity of $10^{33} cm^{-2} s^{-1}$, and after one year running, the loss in light output due to radiation damage is of the order of 20% for the innermost tracker shell. This loss is completely acceptable for this kind of detector.

II.1.c. Microstrip GaAs

The **development** of prototype GaAs detectors by this component of LAA has shown that this material can indeed be used for building particle detectors.

The GaAs basic material for the test has been grown with the LEC (liquid-encapsulated Czochralski) method. Its relevant characteristics are shown in Table II.1.I.

Table II.1.I. Relevant properties of the GaAs as claimed by the manufacturer

DIAMETER (mm)	50.9
ORIENTATION	(100)
DOPANT	Nil
WAFER FLATNESS (μm)	< 3
WAFER THICKNESS (μm)	500 ± 25
WEIGHT (g)	5.37
RESISTIVITY ($\Omega\ cm$)	$7.5 \times 10^7 \div 1.2 \times 10^8$
MOBILITY ($cm^2\ V^{-1}s^{-1}$)	$6.8 \times 10^3 \div 6.6 \times 10^3$
CARRIER CONCENTRATION (cm^{-3})	$1.2 \times 10^7 \div 8.2 \times 10^6$
EPD (cm^{-2})	$2.1 \times 10^4 \div 3.0 \times 10^4$

We have used a 6 GeV/c hadron beam (mainly pions) to check the response of our detectors to MIPs. In Fig. II.1.13, a spectrum of energy deposited by the beam in the GaAs detector is shown. A detection efficiency curve as a function of the bias voltage is presented in Fig. II.1.14, showing that we reach $\sim$ 98% detection efficiency starting at − 40 V. It should be noticed that the spectrum presented in Fig. II.1.13 shows some edge effects of the diode. This is seen as a low number of counts that joins the pedestal to the signal spectrum. The uncertainty related to these edge effects, due to the limited acceptance of the experimental set-up, is a major contribution to the missing 2% of efficiency.

II.2. CALORIMETRY

II.2.a. High precision electro-magnetic

Through the **discoveries** of a new vapour, EF (Ethyl Ferrocene), photosensitive to the BaF_2 fast light emission (see Fig. II.2.1), and of a new family of inorganic scintillators with lower wavelength emission (ex.: $KLuF_4$, see Fig. II.2.2), the concept of SSPC (Solid State Proportional Counter) has **developed** to the SSAC (Solid State Avalanche Counter). The characteristics of these detectors are shown schematically in Figs. II.2.3 and II.2.4.

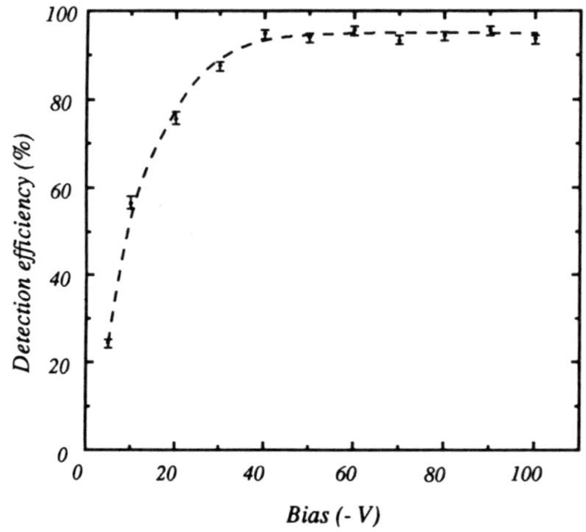

Fig. II.1.14. Efficiency of the detector as function of the bias voltage.

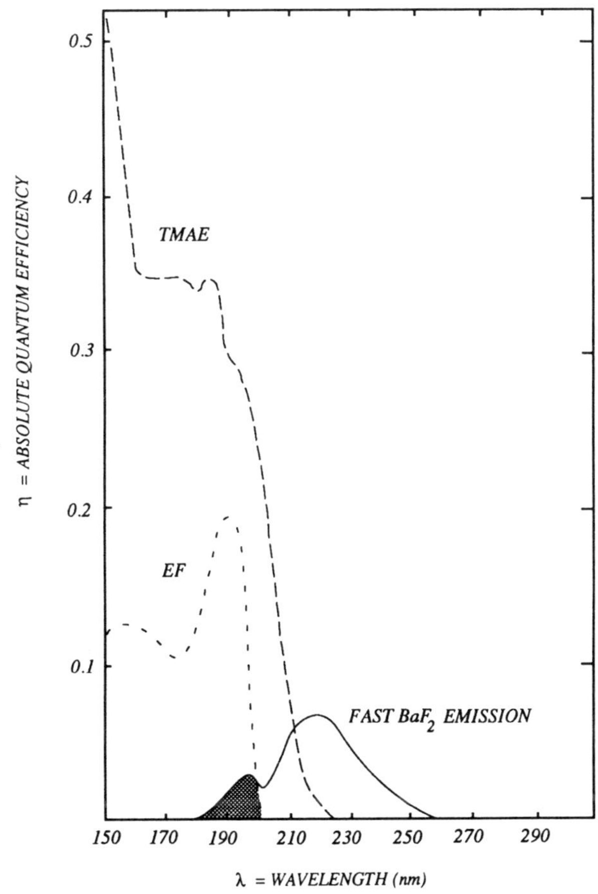

Fig. II.2.1. Ethyl-Ferrocene and TMAE quantum efficiencies. The fast (600 ps) BaF_2 emission spectrum is also shown. The dashed area of the BaF_2 emission spectrum corresponds to the overlapping component between EF and BsF_2.

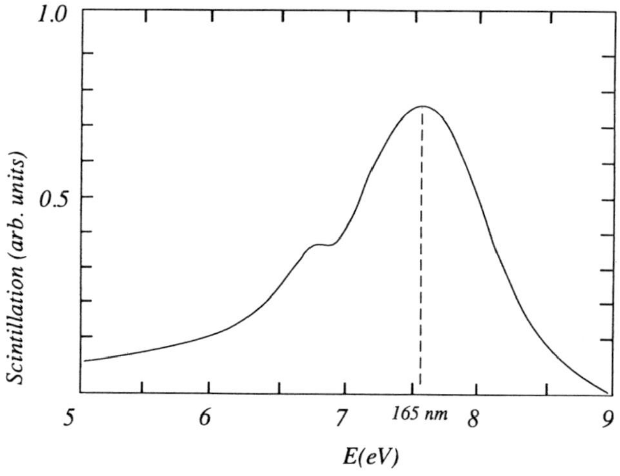

Fig. II.2.2. Luminescence emission spectra of $KLuF_4$.

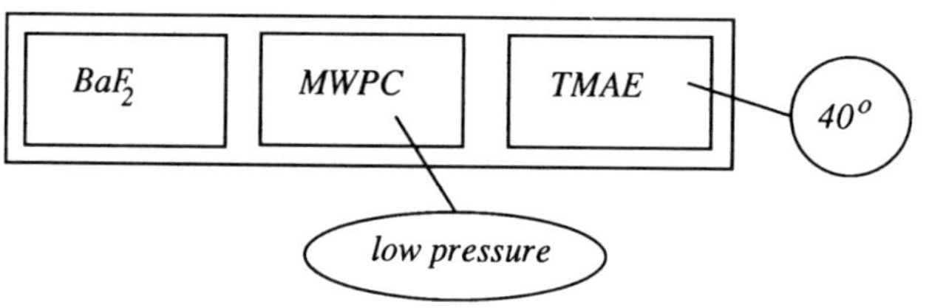

Fig. II.2.3. The essential ingredients of the "dreamt" SSPC (Solid Scintillator Proportional Counter). The aim of LAA was to get rid of low pressure and 40^0, thus achieving a counter able to work at atmospheric pressure and room temperature.

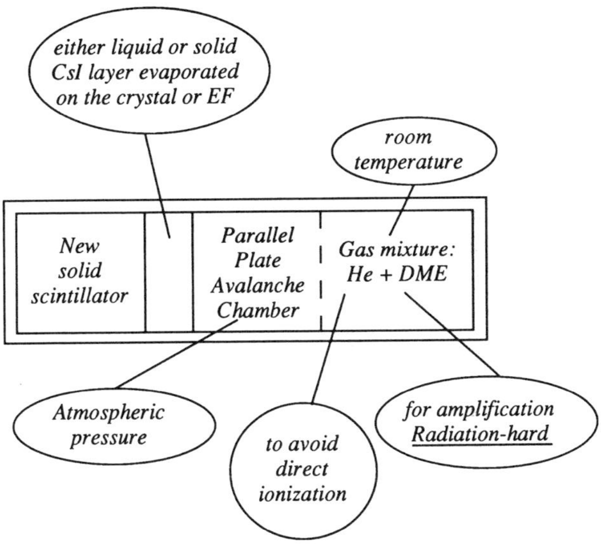

Fig. II.2.4. The Solid Scintillator Avalanche Counter.

It is interesting to know how the way was opened to the discovery of a new family of inorganic scintillators with properties similar to those of BaF_2; (they emit light in a wavelength region between 140 nm and 230 nm). This came from the complete understanding of the BaF_2 fast component light emission process (called CRL=CRoss Luminescence). The proof that the emission wavelength of the scintillators can be controlled is given by the measurement of the emission spectra of KF (potassium fluoride) when Rb^+ (Rubidium) concentration is continuously changed. Figure II.2.5 shows the CRL spectra of $K_{(1-x)}Rb_{(x)}F$ with:

$$x = 0(a), x = 10^{-3}(b), x = 10^{-2}(c), x = 10^{-1}(d), x = 1(e).$$

An example of a new scintillator is the KL_uF4 (Potassium Lutetium Fluoride), whose characteristics are better than BaF_2:

	$KLuF_4$	BaF_2
density	4.8 gr/cm^3	4.9 gr/cm^3
X_o	2 cm	2.05 cm

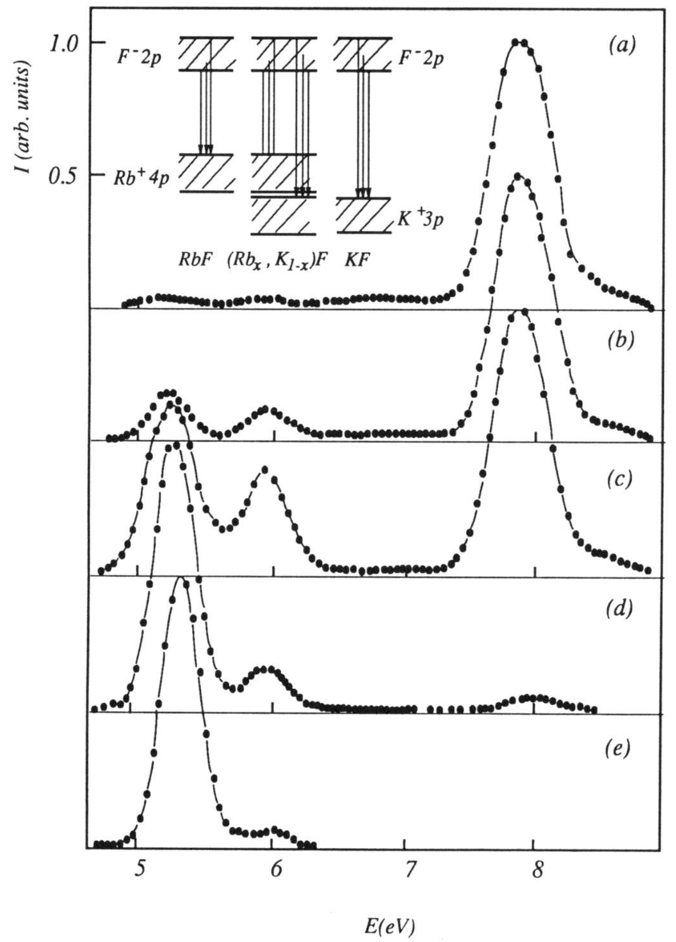

Fig. II.2.5. CRL emission spectra of $K_{1-x})RB_x)F$ solid solutions, with varying x: a) 0; B 0.001; c) 0.01; d) 0.1; e) 1.

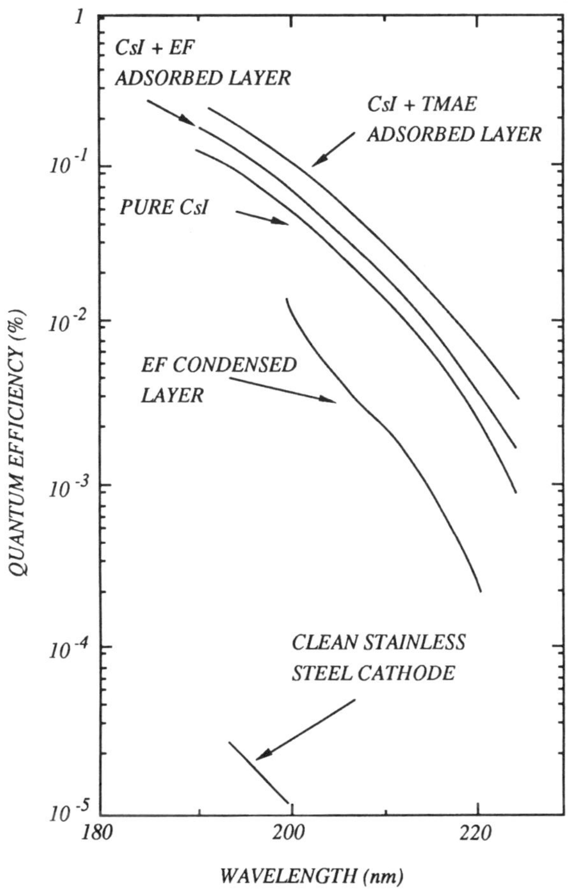

Fig. II.2.6. *Quantum efficiencies measured with a single-wire counter for CsI with TMAE adsorbed layer, CsI with EF adsorbed layer, pure CsI, EF liquid layer, and clean copper cathode.*

In addition to the points mentioned in the caption of Fig. II.2.3., the disadvantages of the "old" SSPC were:

i) the small overlap between the BaF_2 emission band and the TMAE photosensitive band;

ii) the fast ageing effects due to polymerization of TMAE in a wire chamber; and

iii) the fact that TMAE vapour is very reactive to air and corrosive for many materials. The use of liquid TMAE as a photocathode was not possible because of its low efficiency.

Another important discovery (due to V. Peskov) is the use of thin layers of photosensitive materials, either liquid (EF) or solid (CsI), deposited directly on the scintillator surface. This increases the quantum efficiency by $10^3 - 10^4$

compared to clean metallic photocathodes (Fig. II.2.6). The use of thin layers allows to eliminate the gap needed to convert the VUV light emission in the photosensitive vapour. Therefore, the newly developed Parallel Plate Avalanche Chamber (PPAC) can be used as amplification element. Conclusions: The new SSAC can operate at atmospheric pressure and room temperature.

In addition to these basic advantages the results achieved, for the new SSAC, can be summarized as follows:

- Compactness: 4 mm thick conversion gap.

  (All electrons are produced in the same superthin ($< 1\mu m$) layer of either solid (CsI) or liquid (EF), therefore the amplification signal has no jitter).
- Very fast collection time: $< 10ns$.
- Very good time resolution: $< 1ns$.
- Negligible sensitivity to direct ionization.
- Negligible sensitivity to $(\overline{B})$.
- High gain: single photoelectron detection.
- High rate capability: no space charge effect.
- No ageing due to polymerization.

The efficiency of the SSAC is equal to traditionally "dreamt" SSPC (with MWPC and TMAE) and comparable to that of a traditional PhotoMultiplier (10%). This opens up a new way towards large surface detectors.

II.2.b. Compact EM+Hadronic

A direct comparison of the results from two compensating hadron calorimeters of the lead/scintillator and uranium/scintillator sandwich type, lead to the **discovery** that the intrinsic resolution of a lead hadronic calorimeter is better than the intrinsic resolution of an uranium one.

Moreover, a deep theoretical understanding of the hadronic shower development in a variety of materials (lead, uranium, iron, silicium, etc.), has indicated that a lead/scintillator calorimeter can be compensating (i.e., it gives equal response to the electromagnetic and to the hadronic part of the shower), provided a ratio of 1:4 is kept between the two materials. This is illustrated in Fig. II.2.7, where the expected resolution of a lead/scintillator calorimeter is plotted as a function of the thickness of the lead (or of the scintillator). The experimental data obtained in the tests mentioned above are also shown in the same figure.

Both this discovery and the theoretical understanding of the shower development, have lead to the **invention** of a new kind of calorimeter: the spaghetti (lead/scintillating fibres) calorimeter.

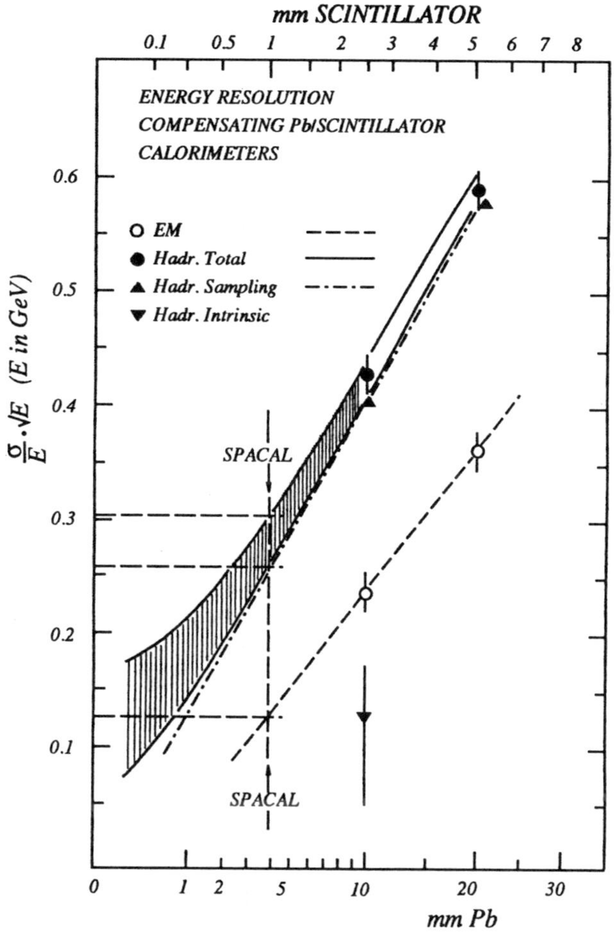

Fig. II.2.7. Energy resolution compensating lead/scintillator calorimeters.

A number of different problems have been solved during the first three years of the LAA project. Among others:

- Embedding of a very large number of fibres in lead. Figure II.2.8 shows the high uniformity of the hole dimensions and spacing obtained in prototype modules (hexagonal shape, 48.7 cm^2 cross-section, 8.6 cm apex-to-apex).
- Lateral and longitudinal uniformity of the calorimeter modules. Many details in the design of the modules have been optimized in order to obtain a "perfect" transversal (of particular importance to obtain the best energy resolution for electrons see Fig. II.2.9) and longitudinal (more important for energy resolution of hadrons see Fig. II.2.10) uniformity of light collection.
- Fibres read-out.

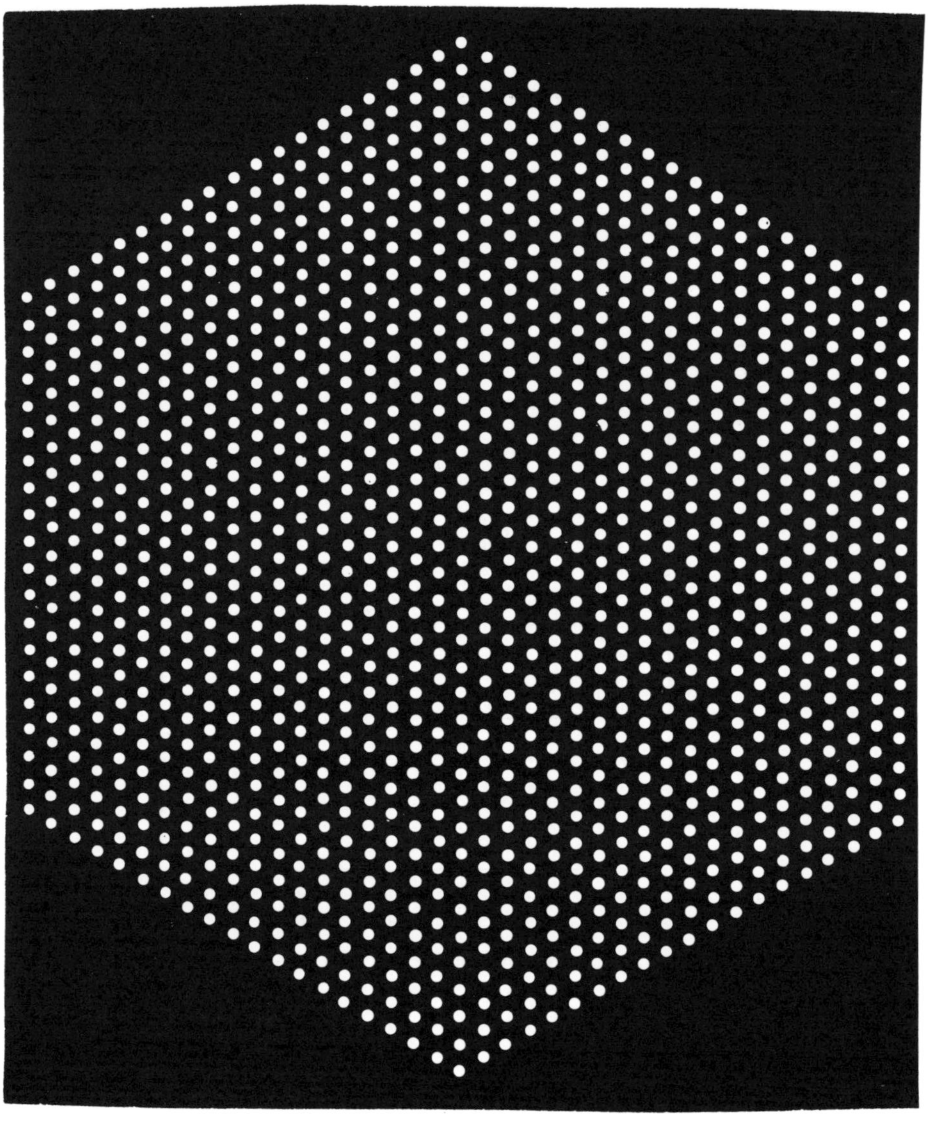

Fig. II.2.8. Front view of a spaghetti calorimeter module (48.7 cm^2 cross-section). The fibres are inserted in the holes, running parallel to the direction of the incident particles.

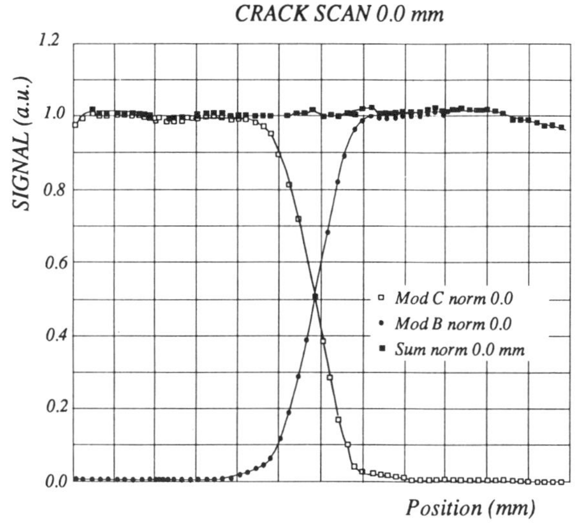

Fig. II.2.9. The calorimeter response to a collimated β-source, which is moved in small steps over the surface of two adjacent calorimeter modules. Shown are the signals in these two modules, and their sum. Results for second-generation prototype modules.

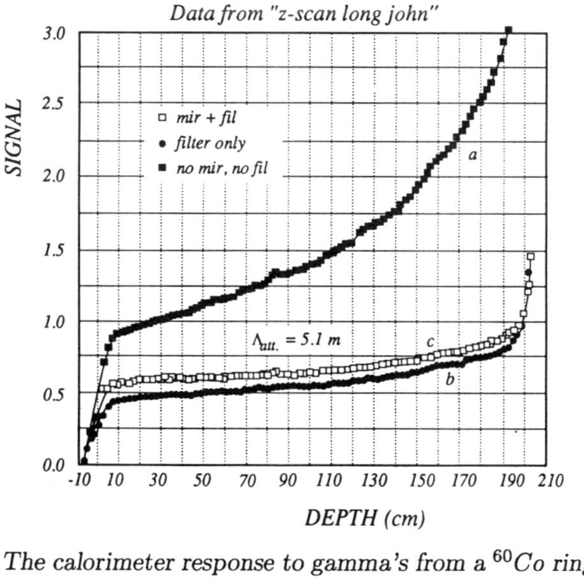

Fig. II.2.10. The calorimeter response to gamma's from a ^{60}Co ring source surrounding the module, as a function of the distance to the PM's. The measurements were done without a mirror/filter (curve a), with a yellow filter (b), and with a filter and a mirror (c).

The main motivations for using lead in place of uranium are listed in the following.

Compensating Pb/scintillator calorimeters:

- Are as compact as compensating U/scintillator, because the effective interaction length is the same for both calorimeters:

$$\sigma_{eff}^{int} \approx 20\ cm.$$

- Give better energy resolution, because the intrinsic resolution is better:

$$\left\{\frac{\sigma_{int}}{\sqrt{E}}\right\}_{Pb} = 11\% \ , \quad \left\{\frac{\sigma_{int}}{\sqrt{E}}\right\}_{U} = 19\%$$

- Give faster and better localized signals, because the neutrons from Pb have a much longer mean free path, and thus they do not contribute to the slow component (s) signal at large distances from the shower core.
- Cause less radiation damage, because the number of neutrons produced in the shower is three times smaller in Pb than in Uranium.

As to calibration, there is nothing that one can do with the U-activity that one cannot do better with removable sources inserted in a fibre slot. In particular, when such sources are used, one does not have to worry about:

- ^{235}U content fluctuations,
- plate thickness fluctuations,
- cladding thickness fluctuations ($\beta's$),
- build-up of long-lived activity in bulk material.

Moreover, lead is:

- Easy to machine.
- Non-radioactive.
- Cheap.
- Unlimitedly available.

A prototype consisting of 20 modules of the Spaghetti Calorimeter (see Fig. II.2.11), each 2 m long and with an hexagonal cross-section of 48 cm^2, was tested in high-energy particle beams from the SPS.

This prototype was used to develop the second, very important, **invention** of this component of LAA: a new method for distinguishing electrons from pions, thanks to the extremely fast (some ns) response of the lead and of the scintillating fibres. The fibres used in our calorimeter have a signal decay time of 1.5 ns.

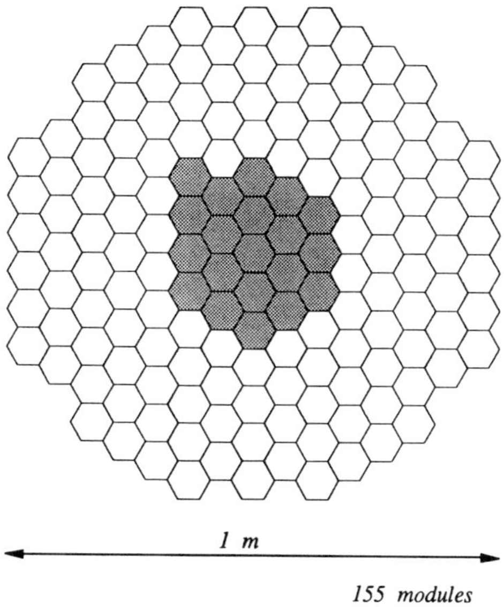

Fig. II.2.11. *Front view of the full 155-modules spaghetti calorimeter prototype. The results discussed in the text have been obtained with a smaller prototype, consisting of the grayed modules.*

This important feature paid off in an extremely interesting way. We found that the calorimeter response is so fast, that electrons and pions can be efficiently distinguished through subtle differences in the time structure of the signals, which occur on a 10 ns time scale. This is illustrated in Figures II.2.12 and II.2.13. Figure II.2.12 shows e^- and π^- signals measured with a 2 m long module. In Fig. II.2.13, the distribution of the FWFM (20%) of several hundred electron and pion signals is given. It shows a clear separation between the two types of particles, and it demonstrates that an e/π separation at the 10^{-3} level can be achieved with this completely new technique which, unlike the classical methods employing shower profiles, does not require any calorimeter segmentation whatsoever.

Another **record** of the LAA Project has also been obtained by this prototype calorimeter: the best hadron energy resolution obtained so far, i.e. $27\%/\sqrt{E}$ (test runs with a prototype for the ZEUS calorimeter gave $\sim 34\%/\sqrt{E}$ [12]). Figure II.2.14 shows the result, which, for the time being, is obtained as an extrapolation at zero shower leakage. The 20 tons full prototype, containing 155 modules, will directly measure this quantity.

A vast programme of radiation damage studies of scintillating plastic fibres has also been carried out, leading to another **record**: the development of polystyrene fibres doped with 3-hydoxyflavone (3-HF), which show only a small deterioration of the attenuation length (after recovery) up to radiation doses of 10 MRad (100 kGy), as shown in Fig. II.2.15.

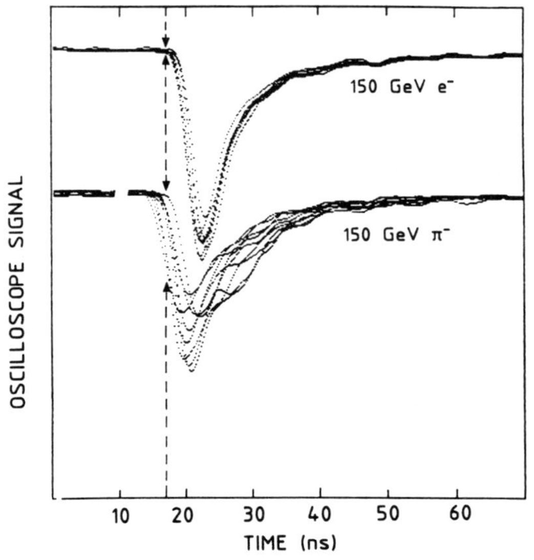

Fig. II.2.12. *Calorimeter signals from 10 different 150 GeV electron showers and from 10 different 150 GeV pion showers.*

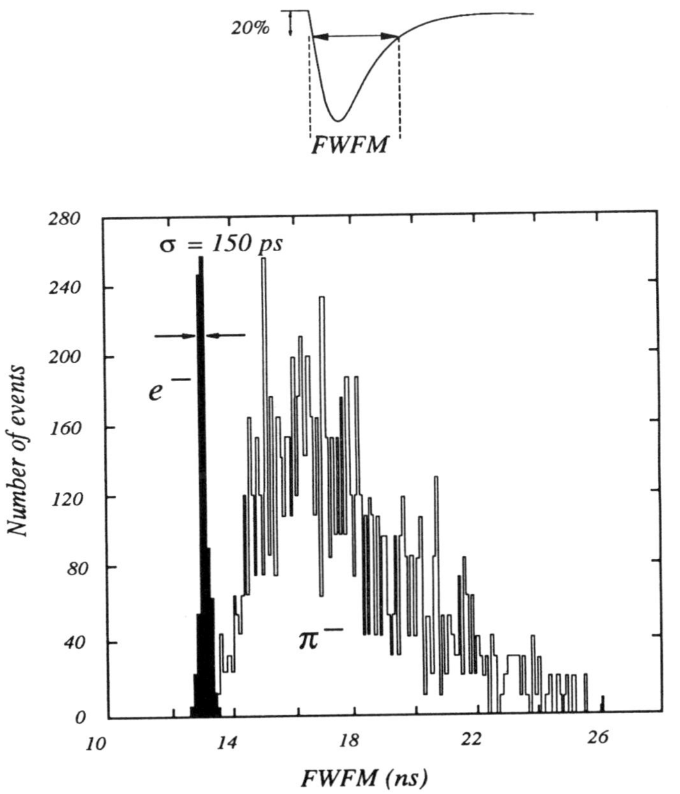

Fig. II.2.13. Distribution of the full width of the signals for 80 GeV/c pions and electrons, at 20% of the signal maximum (FWFM).

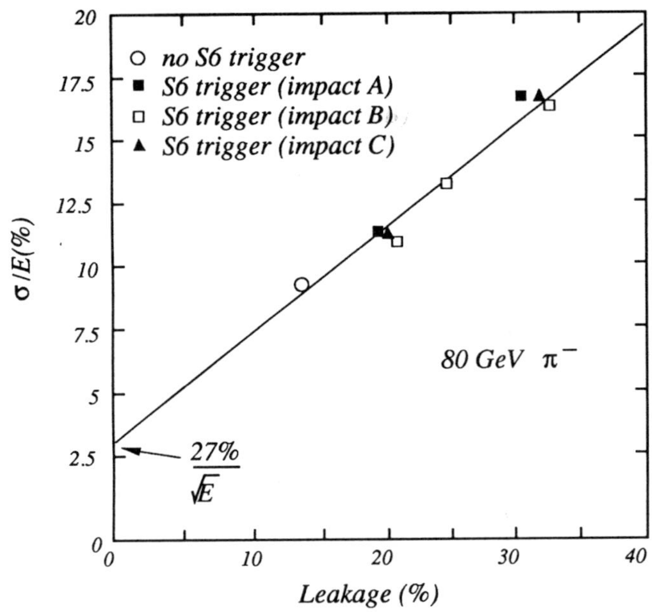

Fig. II.2.14. Energy resolution versus leakage for 80 GeV/c pions. The extrapolation at zero-leakage gives an energy resolution of $27\%\sqrt{E}$.

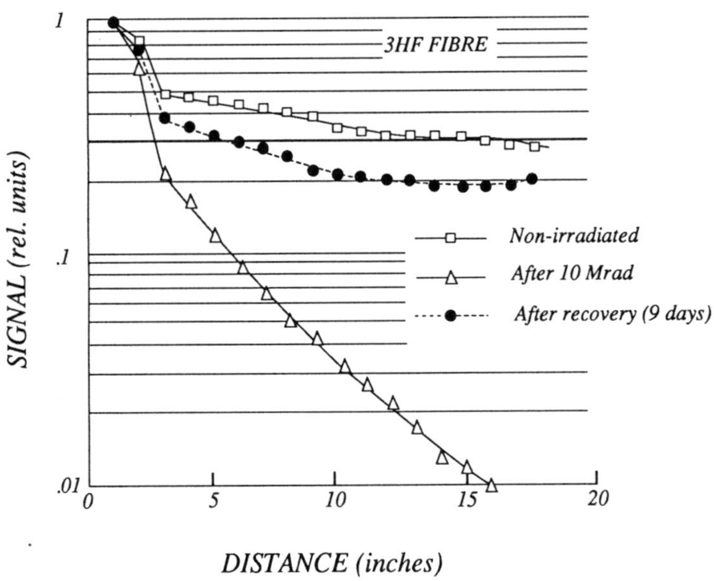

Fig. II.2.15. Attenuation curves measured on polystyrene fibres doped with 3-HF, before and after irradiation with 10 MRad.

II.2.c. "Perfect" Calorimetry

The most important **discovery** of this component of LAA is the increase in quantum efficiency of a CsI photocathode if a monomulecular adsorbed layer of TMAE is deposited on it. The experiment has been done, first by depositing a CsI reflective photocathode on a metal substrate, then by flowing (for about 1 hour) a low concentration of TMAE gas over the cathode surface. In this way a monomolecular film of TMAE is formed. This "sensitized" cathode has a higher quantum efficiency than the cathode with CsI alone. At the Xenon scintillation wavelength, this cathode has a **record** high quantum efficiency: $\approx 46\%$, as shown in Fig. II.2.16.

On the way towards the construction of the "perfect" electromagnetic calorimeter, the knowledge of the properties of liquid Xenon has been further **developed**, in terms of:

- scintillation speed;
- transmission of light;
- photon production in the scintillation process.

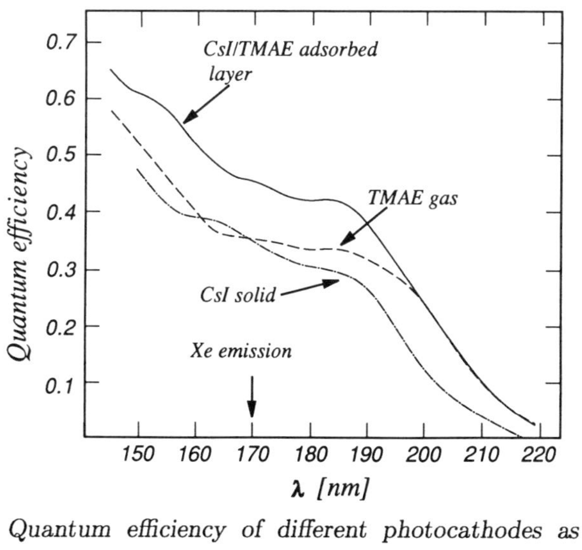

Fig. II.2.16. Quantum efficiency of different photocathodes as function of the wavelength. The liquid Xenon scintillation wavelength is $\lambda \approx 170$ nm.

A pulsed 100KV electron accelerator has been built and tested as a source for excitation of liquid xenon. A light beam from a H_2 flash lamp , with a FWHM pulse width of 15 ns, is focused onto a CsI vacuum reflective photocathode to produce the electrons which are accelerated. This device can deliver as little as 10MeV (400 electrons each with 25 KeV) or as much as 1 TeV (10^7 electrons each with 100 KeV) of excitation energy (per pulse) into the liquid Xenon, permitting the investigation of the scintillation and ionization processes over a wide range of energies.

Recently, a cleaning and condensation system was successfully inaugurated. Xenon was condensed and its scintillation lifetime in the gaseous and liquid state was measured as shown in Fig. II.2.17. The liquid lifetime is less than the observed width of 25 ns because of the lamp pulse of 15 ns. The quadrature difference gives a value of 20 ns, in agreement with other measurements of the long life triplet state of the Xe_2 molecule. The faster singlet state (3 ns) cannot be discerned with this long lamp pulse. The lifetime of the gaseous Xenon scintillation was measured to have a 60 ns component due to recombination.In conclusion, the scintillation lifetime of liquid Xenon matches very well the very short bunch spacing of multi-TeV hadron colliders.

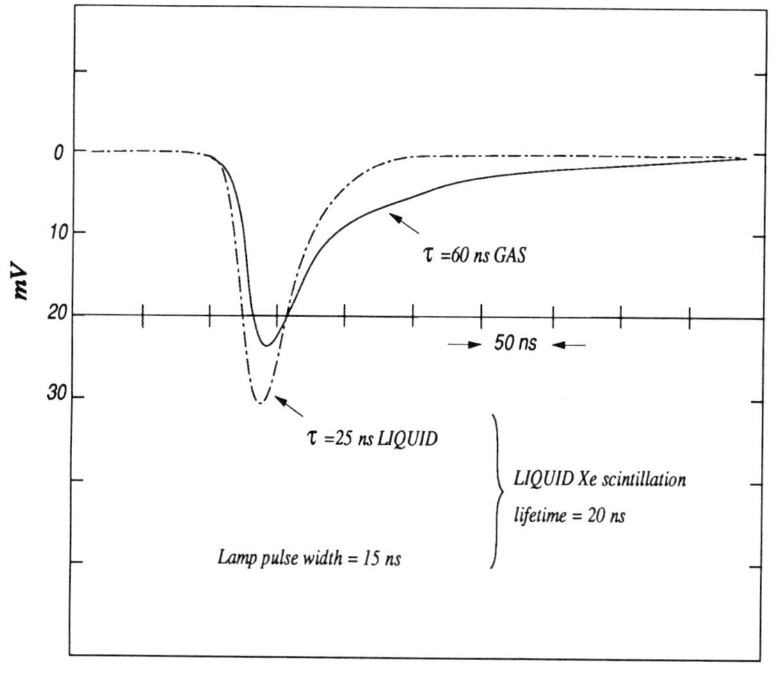

Fig. II.2.17. Liquid and gaseous Xenon scintillation light pulse duration. The scintillation was stimulated by electrons produced by a lamp with an intrinsic width of 15 ns.

The scintillation yield and transmission properties of liquid Xenon were measured by the apparatus sketched in Fig. II.2.18.

The results are shown in Fig. II.2.19. A large scintillation yield of 3×10^4 photons/MeV was measured. The quoted yield is an average with the largest pulses being two times larger (i.e. 6×10^4/MeV). The reason for this fluctuation is due to bubbling and circulation. The energy per photon is, therefore, between 33 and 16 eV, hence of the same magnitude as the ionization energy (i.e. W=15 eV). This very high photon yield promises a very good energy resolution of the final calorimeter. The transmission of a column of liquid Xenon (to its scintillation light at l=170 nm) is also shown in Fig. II.2.19. The inferred absorption length in circulating mode (i.e. during filling) is low (2.7 cm), because of bubbling. On the contrary, the photon absorption length is much greater than 20 cm in the non-circulating mode (i.e. when the liquid Xenon is left evaporating). The conclusion is that there will be no self-absorption of light in the liquid Xenon calorimeter, if this will be operated in a non-circulating mode.

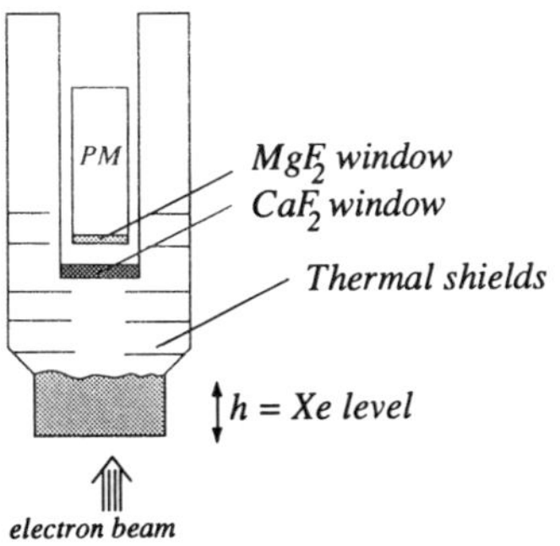

Fig. II.2.18. Principle design of the apparatus used to measure the light level and self-absorption of the liquid Xenon.

II.3. LARGE AREA DEVICES

II.3.a. Construction

The goal of this component of LAA is the development of the detectors which will cover the large area (several thousand square metres) associated with the detection of muons. The main requirements are:

- rate capability in excess of 1 Khz/cm;
- fast response time (of the order of 10 ns);
- easy construction;
- mechanical stability;
- low cost;
- easily adaptable to the geometry of the magnetic fields.

The need for a detector with unusual geometry is illustrated in Fig. II.3.1, where the principle of muon detection in the forward and backward direction is illustrated.

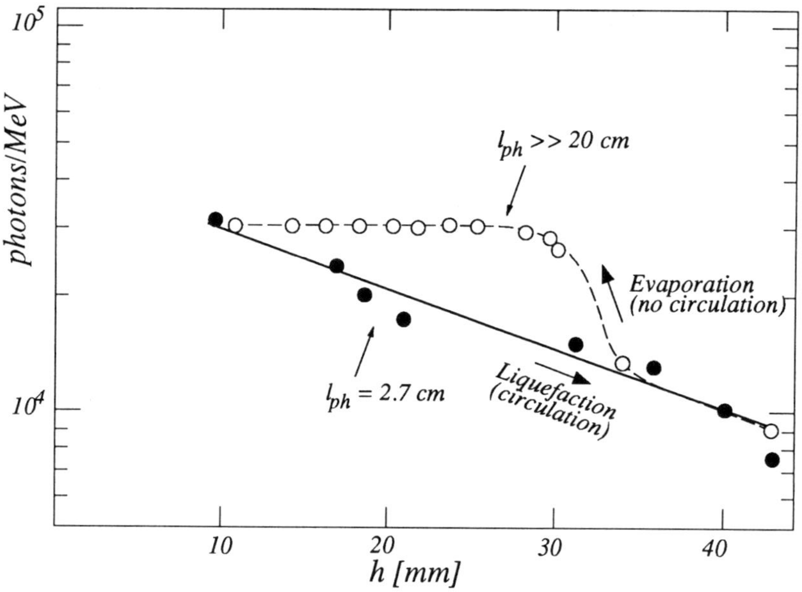

Fig. II.2.19. Absorption curves for liquid Xenon in circulating and non-circulating modes.

In these regions, a toroidal field is the only practical solution for measuring momenta. In these regions muon spectrometers require toroidal magnetic fields. In a toroidal field, a polar coordinate reading is desirable to improve trigger capability and momentum measurements. Therefore, the problem reduces to the following question: how can we build either a detector with circular cells, or a detector which can follow whatever geometry is needed? The answer to this question comes from the invention of two new types of detector:

i) A mechanically stable detector using blades instead of wires, the "blade chamber".

ii) A new wireless chamber with pixel read-out, the "gaseous pixel chambers".

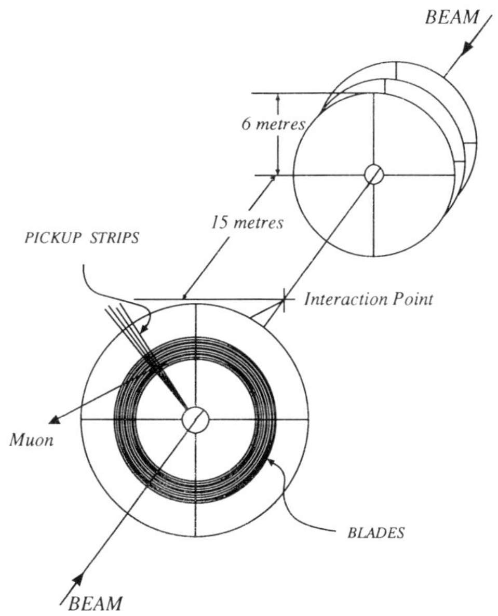

Fig. II.3.1. Geometry of the blade chambers.

Blade Chamber

The performances of a prototype "blade chamber", whose principle design is shown in Fig. II.3.2 were measured in a particle beam. The position resolution is $\sigma_\perp = 250\mu m$ transverse to the blades (Fig. II.3.3) and $\sigma_\parallel = 380\mu m$ along the blades (Fig. II.3.4). Moreover, by reading out the charge collected in the cell walls, the left-right ambiguity can be solved. This avoids the need for a double layer of chambers, as it is usually done with cell-shaped detectors. The rate limit of this new detector is 1 KHz/cm^2, matching the expected rates for muon detectors in future hadron colliders.

This new detector represents a factor 3 improvement in the precision, and a factor 10 improvement in rate capability, with respect to the classical Limited Streamer Tubes, used so far as muon detectors in big collider experiments (e.g., UA1 and ZEUS).

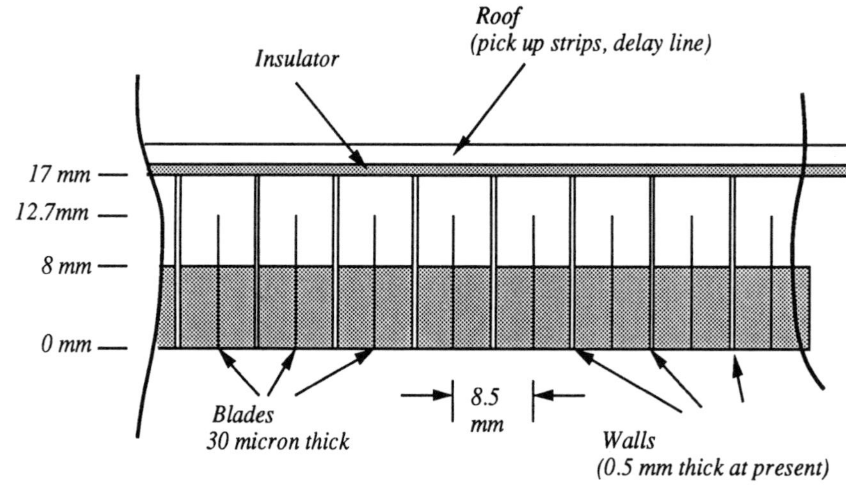

Fig. II.3.2. Cross-section of the blade chamber with insulating roof.

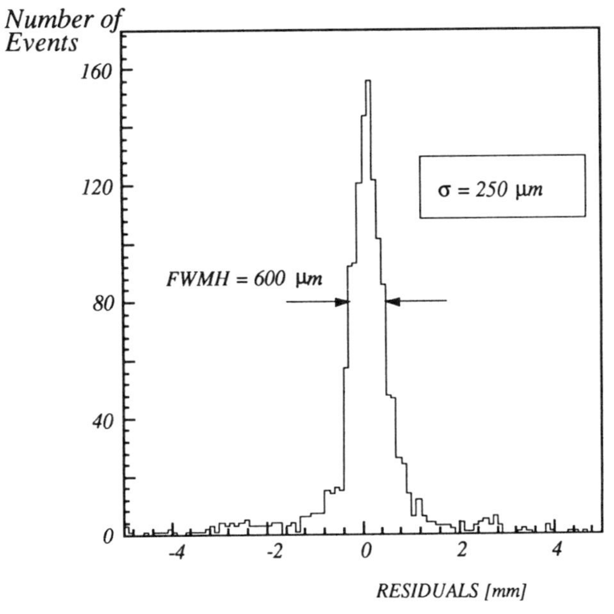

Fig. II.3.3. Spatial accuracy perpendicular to the blade from drift time measurement.

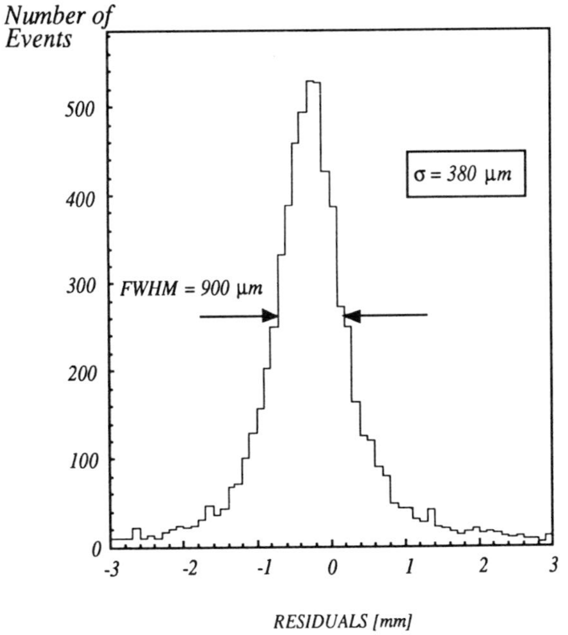

Fig. II.3.4. Position resolution along the blade using 11 mm wide pick-up strips.

Gaseous Pixel Chamber

The principle design of the gaseous pixel chamber is shown in Fig. II.3.5. It consists of a gas-filled volume delimited by a sensitive surface with pixel size of 1 cm, shown in Fig. II.3.6, and by a field-shaping roof, shown in Fig. II.3.7. The chamber works in the limited streamer mode. This device has been tested in a 6 GeV/c pion beam. Its efficiency has been measured to be 98% at an operating voltage of about 7 KV (see Fig. II.3.8). The chamber can withstand a rate of 5000 KHz/cm^2 without losses in efficiency. In conclusion, we have built a device with a very flexible geometry which can be used for triggering purposes in the whole detector.

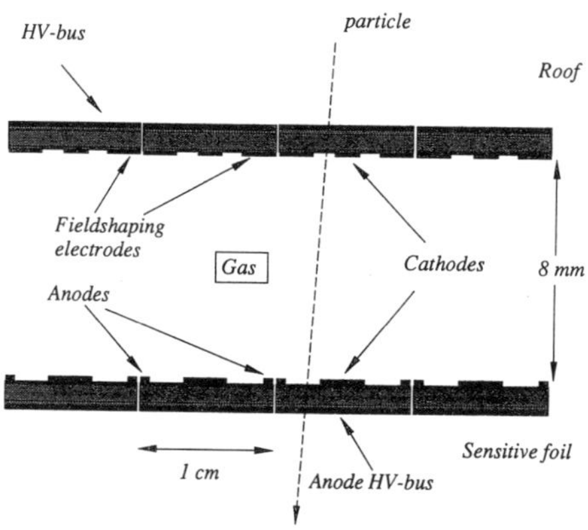

Fig. II.3.5. Principle design of the "gaseous pixel chamber".

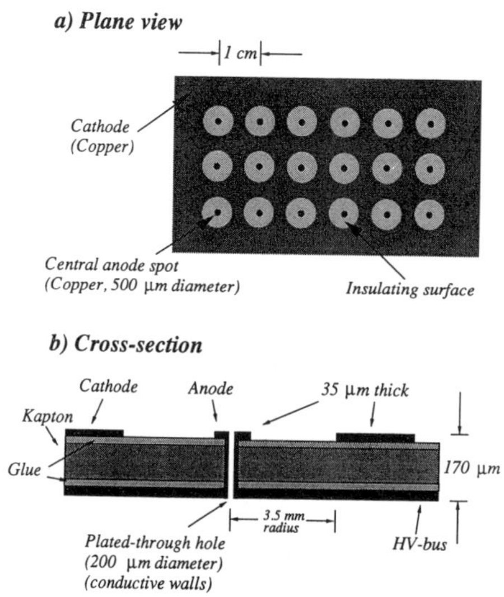

Fig. II.3.6. The "gaseous pixel chamber" sensitive surface.

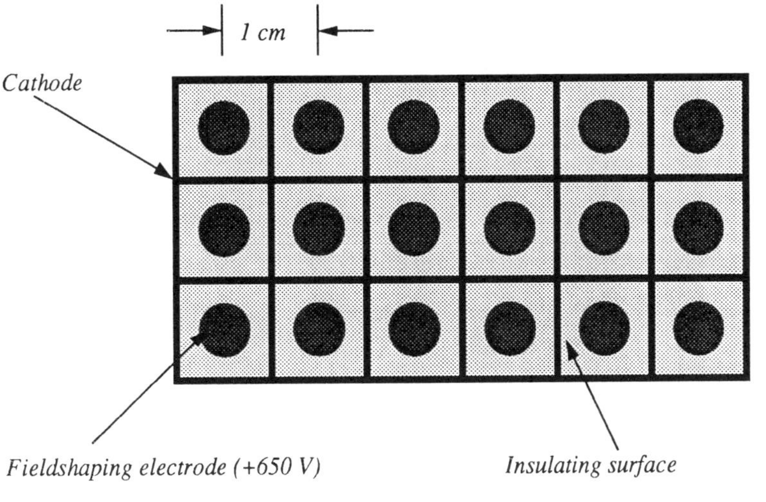

Fig. II.3.7. The "gaseous pixel chamber" roof.

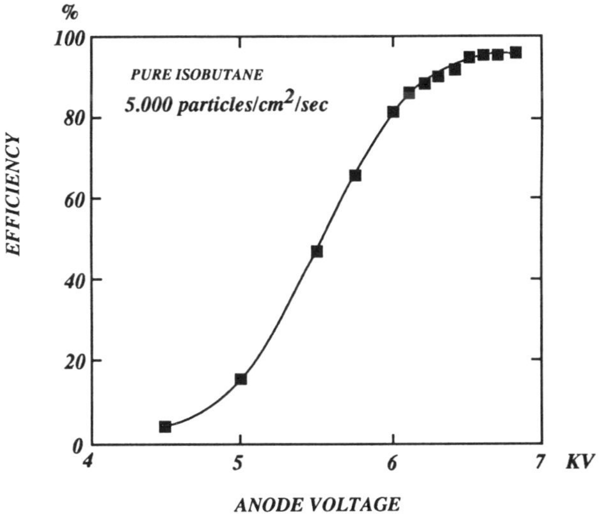

Fig. II.3.8. Efficiency versus high voltage measured in 6 GeV/c pion test beam.

II.3.b. Alignment

As already mentioned in the previous section, the muon detectors of future multi-TeV colliders will have a surface of several thousand square metres. A precision positioning of these detectors is absolutely necessary, because no detector can be better than its alignment.

Therefore, new instruments for length, angle and linearity measurements have been developed. The LAA Project has now the best set of instruments for measuring:

- lengths: 3.6 μm in 2 m;
- angles: 1 μradiant;
- linearity: 4 μm in 5.4 m.

These values, which represent a new **record** in the field of High Energy Physics, are an intermediate step towards of an overall alignment precision of 20 μm over 10 m, which is the final goal of this component of LAA. Before LAA, the most complete alignment system achieved a precision of 30 μm over 3 m [13]. In Table II.3.I, the present achievements are compared with the design goals of this component of LAA.

Table II.3.I. Design goals and achievements in aligment precision

	DESIGN GOAL	ACHIEVED
LENGTH	5 μm in 5 m	3.6 μm in 2.0 m
ANGLES	1-5 μRad	1 μRad
LINEARITY	4 - 10 μm in 10 m	4 μm in 5.4 m (needs a more powerful LASER source to increase the length)
PLANARITY	10 μm in 5 × 5 m^2	Still to be done (needs the instruments developed for linearity)

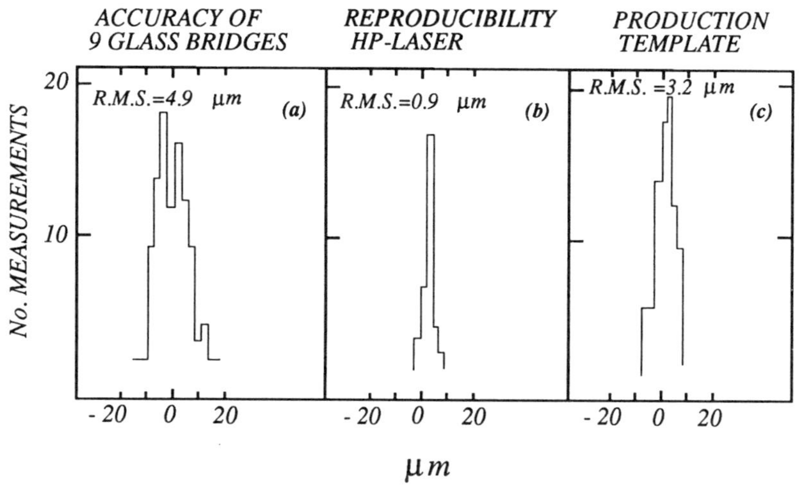

Fig. II.3.9. Error on a length standard test sample.

As an example of the measurements which are possible with the new instruments, Fig. II.3.9 shows the precision achieved with a length standard. The intrinsic error, not due to the measuring device (LASER interferometer), or to the production template, is of 3.6 μm over a length of 2 m.

II.4. LEADING PARTICLE DETECTION

The very difficult problem of designing a Leading Particle Spectrometer for a hadron collider was never tackled in the past. Therefore, all the work done by this component of the LAA Project constitutes a **new development.**

After three years of work, all instruments have been designed and built, and the various problems connected with the harsh environment of the very forward region (small space between the machine elements, radiofrequency, high background), have been solved.

In particular, one of the crucial problems was the cutting of the microstrip silicon detectors to a predefined (arbitrary) shape matching as much as possible the beam profile. This must be done because the acceptance for leading protons increases steeply for decreasing distances from the beam axis. Therefore, the detectors must be positioned as close to the beams as possible.

After several trials with different instruments, we succeeded in cutting the detectors to an elliptical shape, keeping the cracks on the surface to a **record** low level of 50 μm, i.e. 10 times better than industrial standards (see Fig. II.4.1).

Fig. II.4.1. Magnified view of the detector edge after the cutting with a diamond saw.

Some of the detectors have been received from the manufacturer (see Fig. II.4.2). Their characteristics are the following:

- Sensitive material: silicon;
- Active area: $80.96 \times 36.80\ mm^2$;
- Number of strips: 1024;
- Strip width: 25 μm;
- Strip pitch: 81.3 μm.

Fig. II.4.2. *One of the silicon microstrip detectors to be used for leading particle detection.*

Fig. II.4.3. The prototype pot, with the mechanics for positioning and for the fast extraction of the detector.

A collaboration has been established, which has allowed the LAA Leading Particle Detection group to make some research and development based on real life at the HERA (ep) collider. Part of this collaboration consisted in designing on the one hand the inner radiofrequency shielding and on the other hand the details of the interface between pot mechanics and the machine. Figure II.4.3 shows the complete prototype pot, with the mechanics for the fast extraction and very precise positioning of the detector.

Moreover, a direct measurement of the background conditions near the beams was done at the CERN SPS. The results are presented in Fig. II.4.4, where the extrapolations to a multi-TeV hadron collider are also shown. Notice that the background caused by beam tests is three orders of magnitude bigger than the stable beam conditions. Therefore, a continuous monitor of the beam conditions is absolutely necessary to preserve the leading particle detector from disastrous radiation damage.

In particular, the following circuits have been designed, produced and tested:

- *Front-end CMOS amplifiers.*

  Three different circuits have been successfully tested:
  - *A fast (100 ns) preamplifier in conventional CMOS technology.*
  - *A preamplifier with 1% tuneable gain.*
  - *A faster (15 ns) preamplifier "bipoltest" using lateral bipolar effect in SACMOS process.*

  The output signals from the above amplifiers are shown in Figs. II.5.2 (a), (b), and (c), respectively.

- *Discriminator;*
- *Analog pipeline;*
- *12-bits A/D converter.*

A partial architecture for HARP using the functional building blocks has also been **developed.** This is a 4-channel AMPLEX upgrade in SACMOS process. It has not only preamplifier building blocks, but also track/hold and multiplexing circuits, so that it can be used in conjunction with a realistic detector module.

Another new development is the read-out electronics for a silicon micro-pattern detector (size: 200 × 200 μm^2), working at 10 MHz. A layout of this circuit is shown in Fig. II.5.3.

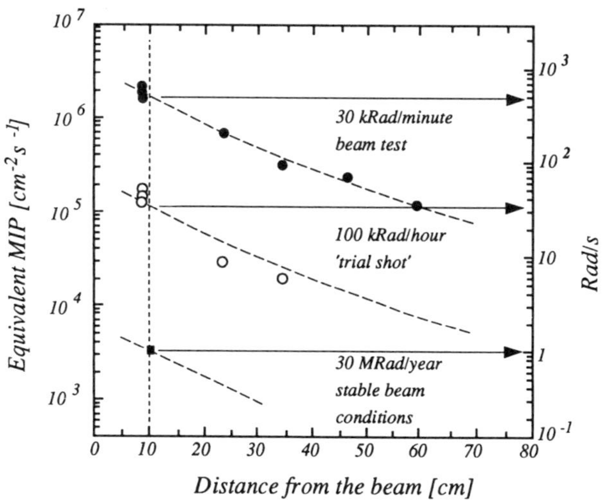

Fig. II.4.4. Left scale - results of the test at the CERN SPS ($L = 10^{29} cm^{-2} s^{-1}$) : dose rate as a function of the distance from the beam. Right scale - extrapolation to a Super Collider ($L = 10^{33} cm^{-2} s^{-1}$). (1 Rad corresponds to 3.6×10^7 $MIPs/cm^2$).

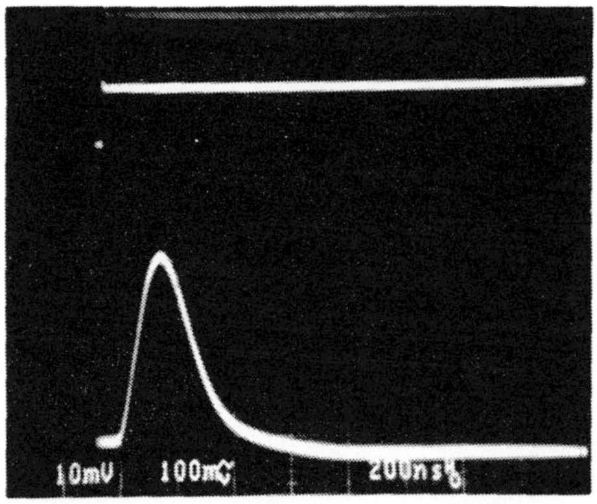

a

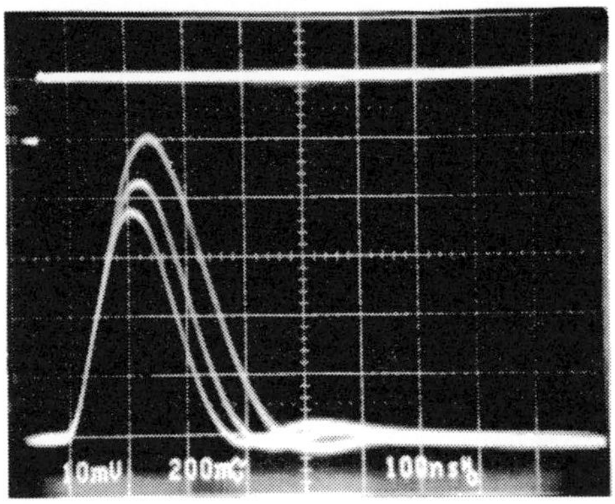

b

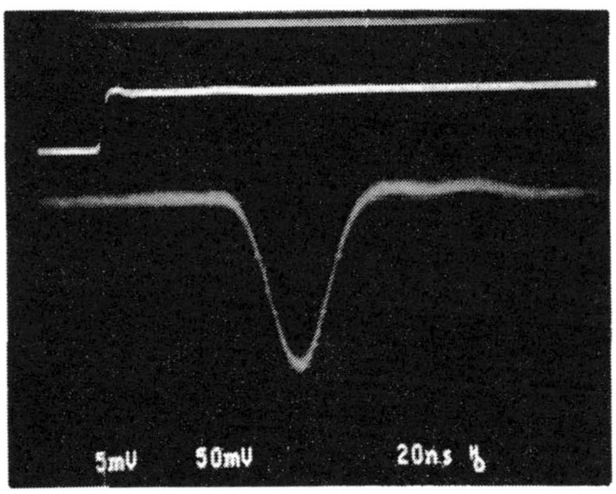

c

Fig. II.5.2. (a) *Output signal of the fast preamplifier (100 ns peaking time);*
(b) *The effect of the pilot current on the gain (top) is obtained for $I_{pilot} = 0.2\ \mu A$, the lowest (bottom) for $I_{pilot} = 0.6\ \mu A$);*
(c) *Output signal of the experimental lateral bipolar amplifier in CMOS (15 ns peaking time).*

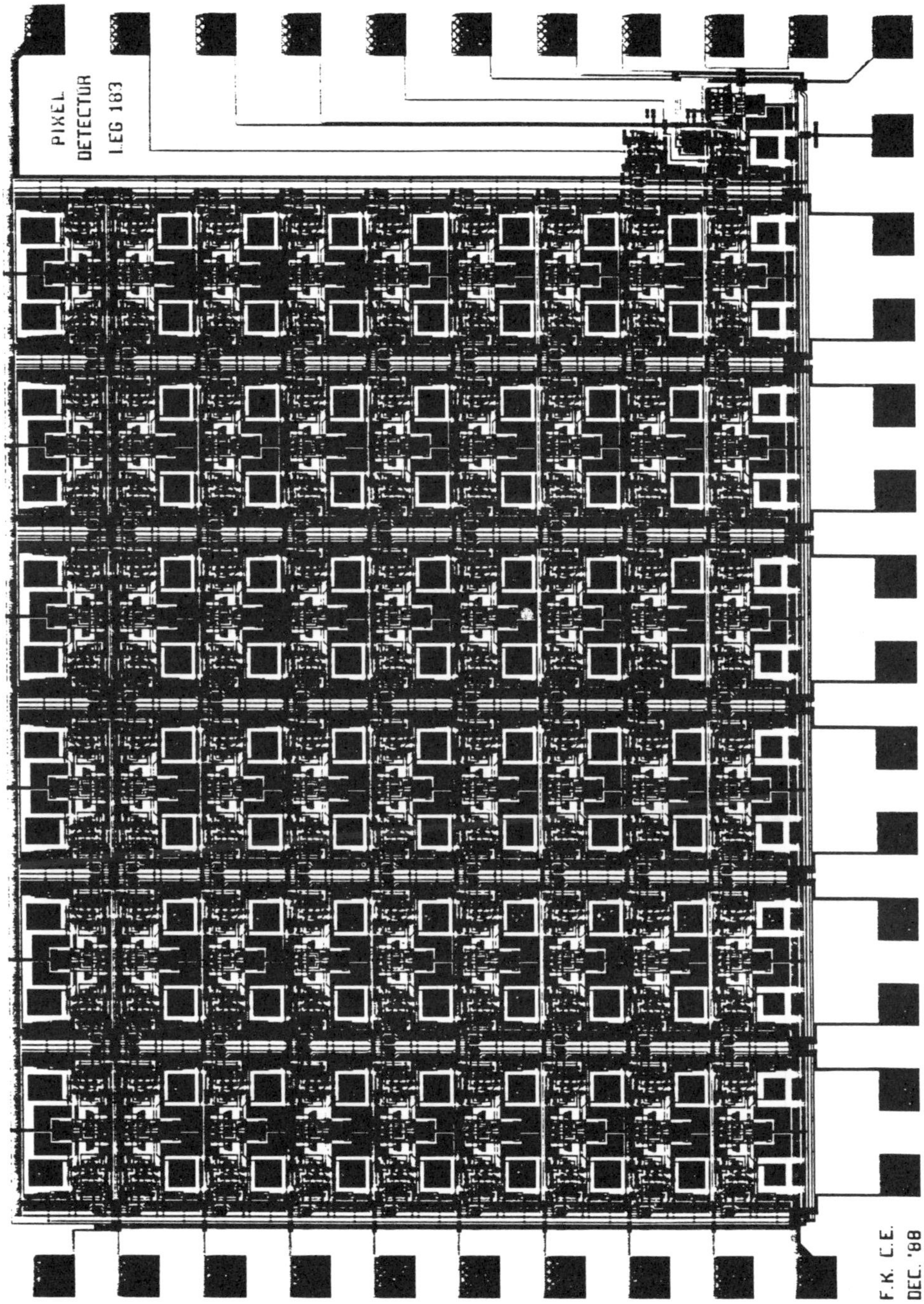

Fig. II.5.3. The read-out electronics for the pixel detector. The size of each amplifier, matching the pixel size, is $200 \times 200\ \mu m^2$

II.5.b. Radiation-hard technologies

The radiation-hard technology chosen by this component of the LAA Project is Silicon-on-Sapphire (SOS). The SOS technology is part of a larger category called Silicon-on-Insulator (SOI), meaning that the substrate onto which the CMOS integrated circuitry-structures are built is not made of bulk silicon, as in ordinary CMOS, but of sapphire. The intrinsic advantage of this arrangement is that, since sapphire is an insulator and not a semiconductor like silicon, there are less parasitic electrical couplings through the substrate plate. In particular, there can be no thyristor-type short-circuits so-called latch-ups appearing in the substrate. Such short-circuits may be provoked in bulk CMOS circuits by very strongly ionizing nuclear collisions in the substrate, caused, for example, by incident heavy ions or very high energy protons.

A first test circuit consisting of a cascode amplifier stage with bias points stabilized by current mirrors, and which is foreseen as a front-end amplifier on a microstrip detector readout chip, has been designed and fabricated.

The fabricated circuit has been irradiated with a ^{60}Co source. Fig. II.5.4 shows the measured noise, normalized to 1 before irradiation, as function of radiation dose up to 400 krad. No significant variation can be seen up to 400 krad. Further tests have confirmed that, even if the chip is irradiated up to 1.4 Mrad, the noise level remains acceptable, provided that some days of recovery are allowed at a temperature of 120^0C. This constitutes a new **record** for circuits to be used in High Energy Physics.

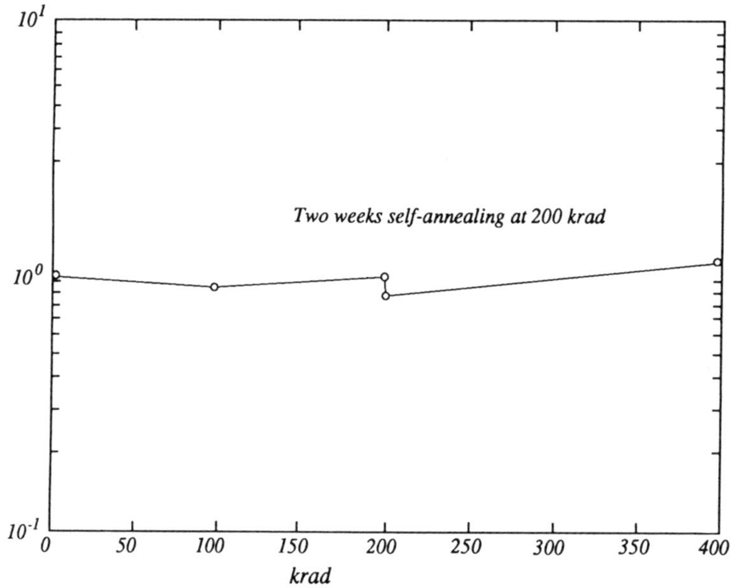

Fig. II.5.4. Measured noise as a function of radiation dose. The noise has been normalized to 1 at 0 krad.

II.6. DATA ACQUISITION AND ANALYSIS

In a future multi-TeV collider, the amount of data which has to be managed by the data acquisition system is enormous. For a luminosity ranging between 10^{33} and $10^{34} cm^{-2}s^{-1}$, the collision rate is expected to be between 10^8 and 10^9 events/s. With at least 10^6 channels per event containing significant data, coded in one byte (8 bits), we expect a total data volume of

$$10^{14} \div 10^{15} \; bytes/s.$$

Therefore, it is of paramount importance to reduce the data volume as early as possible. Ideally, this can be achieved by embedding the data reduction systems in each detector part, using simple computer architectures.

We have exposed **three selected architectures** commercially available (a Data-Cube image processing system with Sun 3 host station, a single-string 16384-processor ASP system from Brunel University, and a 48 × 48-processor GAPP system from Laben, Milano) to a set of **seven typical feature-extraction algorithms** (benchmarks): peak finder, pattern finder, calorimeter cells clustering with fixed neighbourhood, calorimeter cells clustering with variable neighbourhood, missing energy, Houghtransform for track finding, and gray level histogramming. For comparison, the same algorithms have also been run on a VAX8300. The results of the benchmarking, run either through proprietary simulators or on real hardware, are shown in Fig. II.6.1, where more or less trivial extrapolations for SIMD array sizes and technology are taken into account.

The use of commercial architectures constitutes a **new development** for the High Energy Physics community.

The conclusion is that the strong and weak points of the architectures mostly depend on the algorithms. Image processing systems are overall the easiest to cope with, and predictably the most efficient solutions: the ultimate computer performance on pixel information is obtained with one processing element per pixel.

A second path towards "intelligent detectors" consists in *embedding commercial architectures* in our environment of detectors with extremely fast response.

A second path towards intelligent detectors consists in embedding commercial architectures in our environment of detectors with extremely fast response.

The important pilot project of integrating a processor in experimental readout pipelines (typically VLSI design) has led the LAA group to defining an extension to the design of the SPARC chip. The SPARC (Scalable Processor ARChitecture, from Sun Microsystems) is an open RISC (Reduced Instruction Set Computer) processor architecture. The LAA project ultimately will intercept the SPARC at the architectural level, extending the instruction set to allow direct addressing of the Fastbus. In order to produce a usable device on a relatively short timescale,

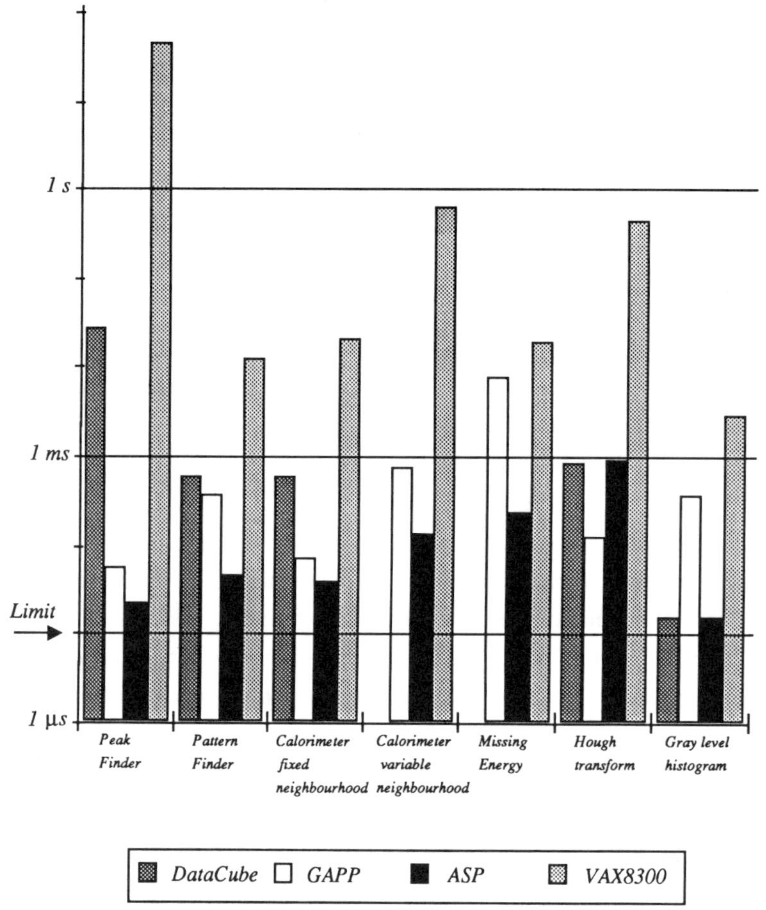

Fig. II.6.1. Performance of different computer architectures on various feature-extraction algorithms. The 10 μs limit corresponds to the maximum time allowed to the second level trigger.

timescale, an intermediate approach has now been chosen, interfacing standard commercial SPARC chips via a coprocessor to Fastbus.

This pilot project has now a finished design, for a 4-processor Fastbus board, as shown in Fig. II.6.2. The design has been fully simulated. Final negotiations with the chip manufacturer have started.

This LAA project will result in a commercial-like product allowing to put together experiment-dependent Fastbus boards with a record high computing power of 50 Vax equivalents with complete software support.

A third part of the "Data Acquisition and Analysis" component of LAA has developed an image processing workstation for physics analysis of very complex events (CCD images or data from future detectors with extremely high granularity): the TRAX-I. It consists of a VME-based workstation as a host processor, a series of hard wired image processing modules for data acquisition, background subtraction, gain equalization and image display, an Associative String Processor (ASP) with 16000 processing elements, and an array of Transputers interconnected through VME and a fast data channel.

The novelty in this architecture is the Associative String Processor (ASP) developed by R.M. Lea and colleagues at the University of Brunel. The ASP is a parallel processing computational structure comprising a string of identical Associative Processing Elements (APEs).

The TRAX-I architecture has now been completely specified, the first batch of VLSI chips for TRAX-I has been produced (64 processing elements per chip made in 2 μm CMOS, as shown in Fig. II.6.3) and tested. The LAA benchmarks described above have been implemented on the ASP simulator: at present the TRAX-I corresponds, for specific tasks, to 200 to 3000 VAX8300. And this is a **record**.

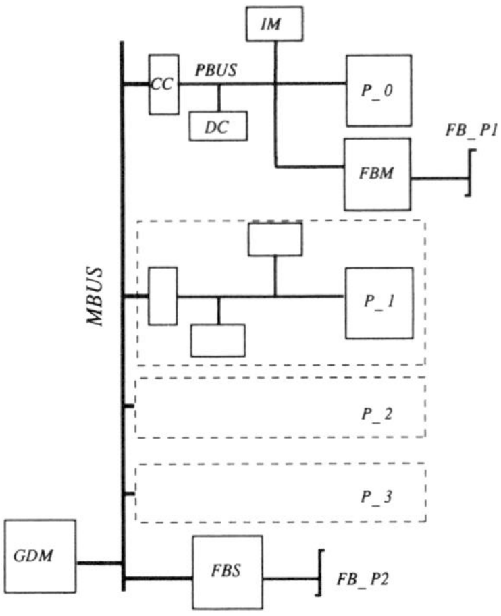

Fig. II.6.2. P_0, P_1, P_2, P_3: SPARC Processors; IM: instruction memory; FBM: Fastbus master interface; DC data cache; CC: cache controller; FBS: Fastbus slave interface, GDM: global shared memory; FB _P1,2: Fastbus backplane.

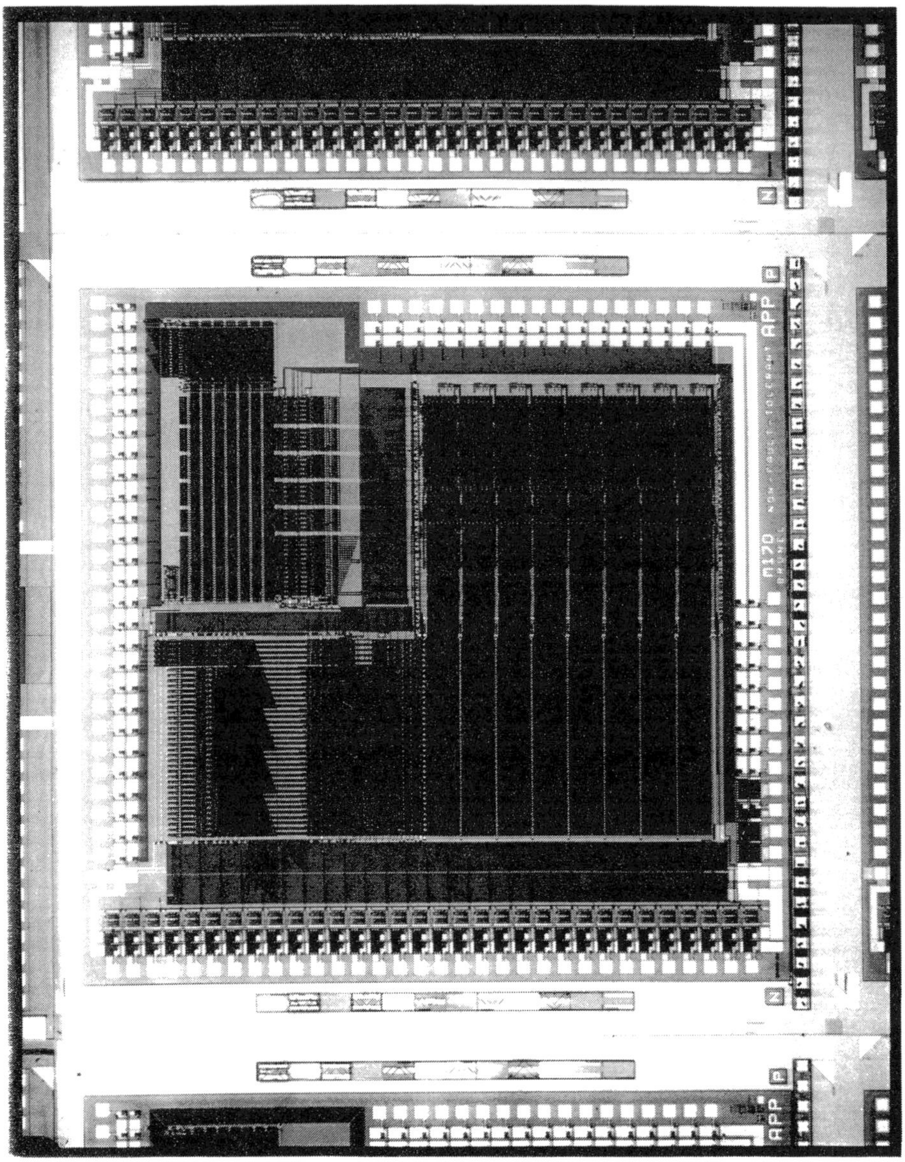

Fig. II.6.3. Picture of one VLSI chip for TRAX-I.

II.7. SUPER COMPUTERS AND MONTE CARLO SIMULATIONS

Monte Carlo Simulations for future Super Colliders are of great relevance. They are in fact needed for the prediction and/or extrapolation of the Physics scenario to very high energies, for the detector design, and for the interpretation of the experimental data.

The MonteCarlo Simulations component of LAA has proceeded along two lines:

i) *Program implementation and software tools development.*
   The software tools for utilizing a full Monte Carlo simulation chain have been developed. The hardware equipment to be used by this component of LAA has been acquired and installed and the training of a group of young physicists on various event generator programs has been completed.

ii) *Monte Carlo studies and simulations.*
   Studies and calculations on different event generators and on different effects (multihadronic production and leading effect, heavy flavours) have been carried out.

The main **development** of this component of LAA is a Full MonteCarlo Chain (FMC), illustrated in Fig. II.7.1, integrating different physics and detector simulation programs in a single, user-friendly package.

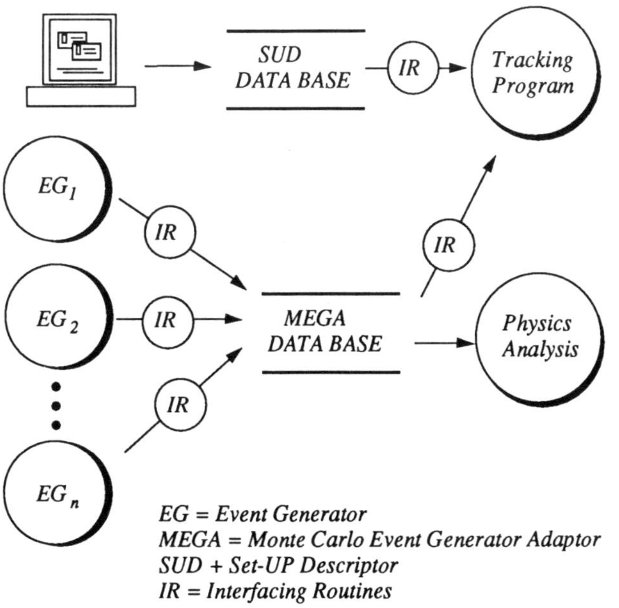

Fig. II.7.1. *Block diagram of the Full Monte Carlo Chain (FMC).*

For the first time, a modern relational data structure system, ADAMO, has been used as a framework for connecting the data from various Monte Carlo event generators existing on the market (LUND, HERWIG, ISAJET, EUROJET), to different tracking programs. Interfacing routines have been written which take into account the differences in event generator outputs and tracking program inputs. These routines, collectively named MEGA (Monte Carlo Event Generator Adaptor), use data structured into tables which are handled in a simple and flexible way. The main advantages are transparency to the user and modularity, which is suitable for the comparison of different Monte Carlo event generator outputs.

The FMC is, at present, implemented in the following computers: VAX/VMS (released), IBM/VM (released), CRAY (in preparation).

Another important software tool has been developed: the SUD (Set-Up Descriptor). SUD is a batch/interactive program to define the detector elements for the tracking program.

A new program system, which represents a natural development of the FMC, is in preparation: COSMOS (COmprehensive Super MOnteCarlo System). It will be completely modular, and it will allow the user to interactively select:

- the type of interaction;
- the centre of mass energy;
- the Feynman diagrams to be used in the interaction;
- the structure functions of the partons;
- the value of Λ_{QCD};
- the hadronization model (cluster, string, etc.);
- the level of analysis (after generation, evolution, hadronization, decay, tracking).

The logic structure of COSMOS is shown in Fig. II.7.2. The grey part has been implemented.

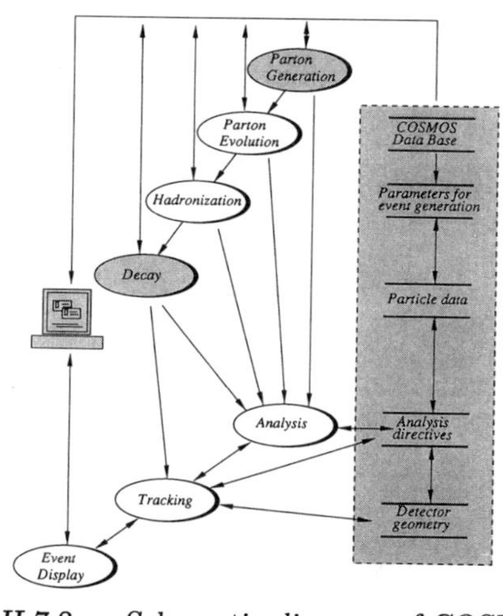

Fig. II.7.2. Schematic diagram of COSMOS.

A first detector configuration has been introduced in the FMC to simulate very high energy events. It consists in a simple tracking section (with and without a magnetic field), a Pb/scintillator calorimeter, and a muon tracker.

The complexity of the events, which are expected in a multi-TeV collider is illustrated in Figs. II.7.3 and II.7.4. These figures show different views of a $pp \to t\bar{t}X$ event at $\sqrt{s} = 16\ TeV$. The total multiplicity of the event is 483 (notice that all tracks with less than 300 MeV/c are not displayed). The top quarks (mass = 90 GeV/c^2) decays into a $\mu + \nu_\mu b$ producing a well separated muon, with a momentum of 343 GeV/c, going in the forward direction.

A theoretical study of possible effects of the baryon and lepton number violation at very high energies (order of 100 TeV), has been started by the 200 TeV Club. The picture which is under development foresees that the effects of the non-conservation of the baryon number (but conserving the difference between baryon and lepton numbers), are too small at low energies, but rapidly increase with increasing energy. These phenomena are non-perturbative results from tunneling between different vacua of the Standard Model. Figure II.7.5 illustrates the typical feature of events, very high multiplicity of particles (of the order of $1/\alpha_{\mathrm{weak}}$), expected.

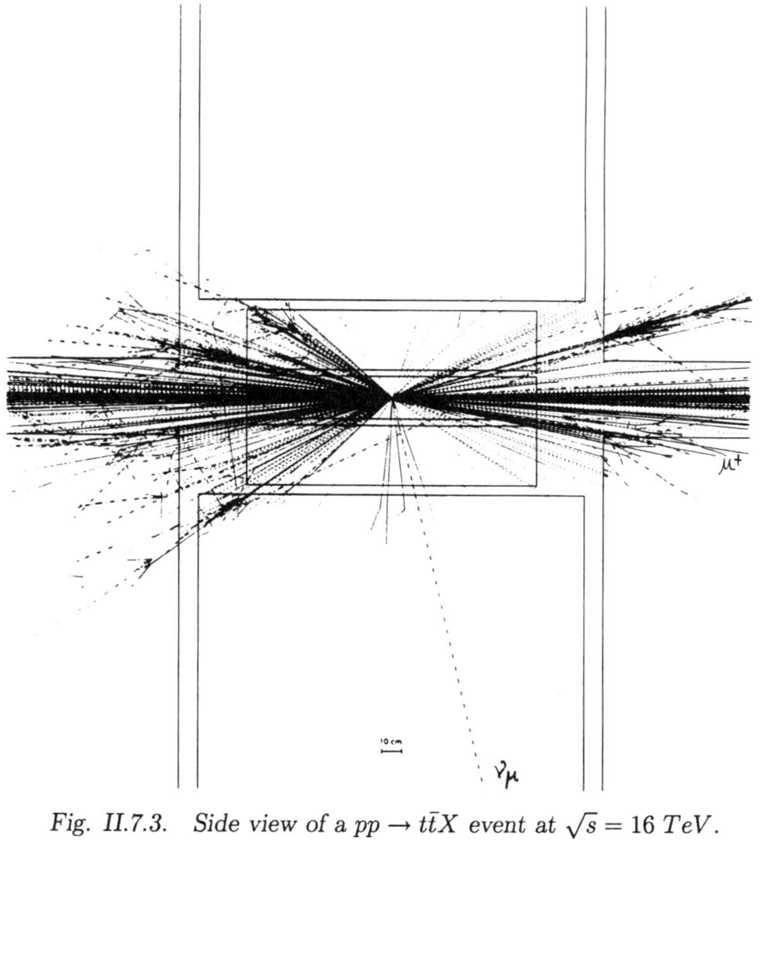

Fig. II.7.3. Side view of a $pp \to t\bar{t}X$ event at $\sqrt{s} = 16\ TeV$.

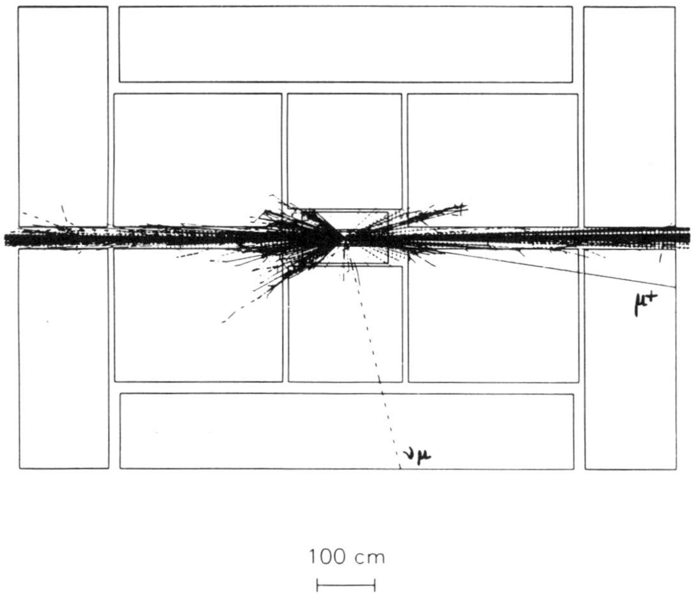

Fig. II.7.4. Same as Fig. II.7.3., but with a 4 T magnetic field in the central region.

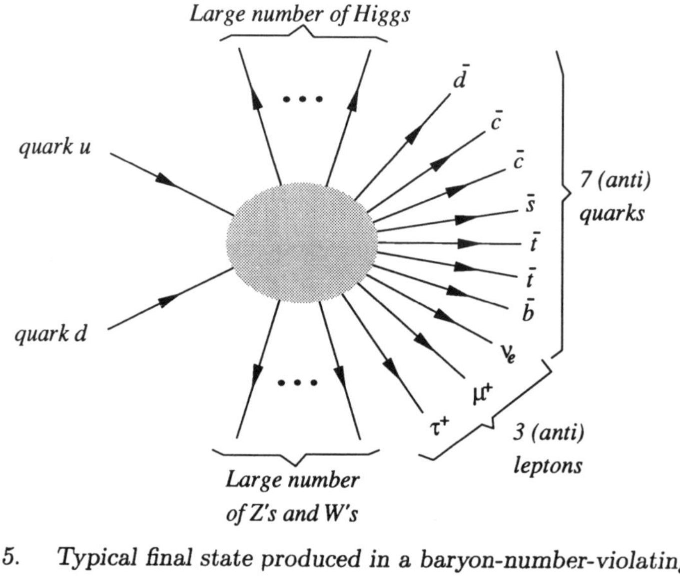

Fig. II.7.5. Typical final state produced in a baryon-number-violating event.

III. CONCLUSIONS

The results presented, i.e.: **five Discoveries, eleven Records, twelve New Developments and four Inventions,** are the best proof of the validity of the LAA Project.

References

[1] A.Zichichi, "Report on the LAA Project", Volume **1**, 15 December 1986.

[2] A.Zichichi, "Report on the LAA Project", Volume **2**, 25 June 1987.

[3] A.Zichichi, "Report on the LAA Project", Volume **3**, 19 November 1987.

[4] A.Zichichi, "Report on the LAA Project", Volume **4**, CERN-LAA/88-1, 25 July 1988.

[5] A.Zichichi, "The LAA Project", Volume **5**, CERN-LAA/88-2, 19 September 1988.

[6] A.Zichichi, "Report on the LAA Project", Volume **6**, CERN-LAA/89-1, 15 September 1989.

[7] A. Zichichi, "The ELOISATRON Project", August 1989.
A. Zichichi, "The ELOISATRON Project: Eurasiatic Long Intersecting Storage Accelerator", in *New Aspects of High-Energy Proton-Proton Collisions*, Plenum Press, 1988 (ed. A. Ali), page 1.

[8] see, for example: G.G. Hanson and M.G. Gundy, preprint SLAC-PUB-5041, July 1989, and references therein.

[9] H. Walenta, Physica Scripta, **23** (1980) 354.

[10] R.E. Ansorge et al., UA2 Collaboration, Nucl. Instr. Meth. **A265** (1988) 33.

[11] H. Schönbacker, preprint CERN-TIS/RP/201.

[12] A. Adresen et al., ZEUS Calorimeter Group, preprint DESY 89-149.

[13] L3 Collaboration, Nucl. Instr. Meth., **A277** (1989) 187.

Chairman: A. Zichichi

Secretaries: D. Brahm, S. Hsu

Discussion

– *Borden:*

Scintillation fiber tracking chambers should achieve good resolution in two dimensions (r, θ), but what about the third (z)?

– *Zichichi:*

The fiber bundles are not all parallel, they are arranged to provide high resolution in three dimensions.

– *Haas:*

Could you comment on the organizational problems involved in a collaboration of thousands of physicists? Is this environment desirable?

– *Zichichi:*

When I started to do physics, there were 10 physicists in the great Blackett group in Manchester (U.K.). Experiments were office-sized. The trend ever since has been to larger instruments and larger collaborations. This trend is inevitable if you want to do new physics. ISR made the mistake of avoiding large detectors, so this powerful collider was not fully used, and many discoveries were missed: e.g. the Upsilon.

The new particles predicted by supersymmetry are invisible, so we can only find them through missing-energy signatures. For this we require large, fast, hermetic, radiation-hard, detectors. The large dimensions in terms of people and detectors are imperative if you want to discover new phenomena in High Energy Physics. Remember, though, that there will always be the need for a single fellow with a brilliant new idea. No progress can be achieved without new ideas. The large dimensions of our enterprises should not let anybody think that new original ideas are not needed.

– *Haas:*

Are colliders the only way to find new physics?

– *Zichichi:*

Colliders were invented and implemented, first with (e^+e^-) (B. Touschek), then with (pp) (K. Johnsen). I cannot think of a better instrument. We know now that our "heavy particles" are really very light; the "truth" is at the Planck scale. Considering the consequences of our ignorance on the energy level where new physics will show up, we have to start now not only to implement LHC and SSC, but also to prepare the ground for the highest achievable energy: 200 TeV.

– *Kastening:*

The SSC has about the largest realistically imaginable ring, but is much lower than 200 TeV. How can 200 TeV be achieved? Do you anticipate stronger magnetic fields, for example, or will we use linear accelerators?

– *Zichichi:*

The Eloisatron conceptual design calls for a 300-km ring, and should reach 200 TeV, without any problem of scientific or technological nature.

– *Brahm:*

You have described a very innovative detector, probably useful for many experiments, but it is named after a fairly ordinary experiment ("Lepton Asymmetry Analyzer"). Is this a historical accident, or is this experiment particularly demanding?

– *Zichichi:*

The LAA experiment was originally planned when a 4th generation of quarks seemed likely. It was known that a very heavy quark was needed in SUSY to raise the mass of the gluino, and people did not imagine that the top was as heavy as it turns out to be. It was important, then, to determine whether a new quark was up-like (u_3) or down-like (d_4). Fortunately, scheduling problems prevented us from building the LAA then, so we continued research into detector improvements. I have since become convinced that physics must reach much higher energies, 200 TeV. The LAA project has thus become the development of a fast, hermetic, radiation-hard, detector for the next generation of colliders.

– *Ciuchini:*

We know that no decoupling theorem holds for the "low-energy" Standard Model. Do you think that high-precision measurements can replace or at least complement high-energy experiments?

– *Zichichi:*

30 years ago I worked on a high-precision experiment, the $(g-2)$ of the muon. We found that the muon is a lepton. I also did the most accurate measurement of τ_μ, i.e. of the Fermi coupling. I will never work on a high-precision experiment again! I don't think you can disentangle vacuum corrections well enough to learn much about the Standard Model from high-precision measurements. The best way to find new physics is not to worry about a high precision experiment first, but to

do a qualitatively good one in some new energy regime. Then follow it up later with high-precision measurements.

– *Kastening:*

High-precision measurements of CP violation gave ϵ'/ϵ.

– *Zichichi:*

That's true, that experiment is worth doing. But I have more interesting experiments in mind.

– *Ypsilantis:*

Would another high-precision CP-violation measurement be worthwhile, e.g. B physics?

– *Zichichi:*

It could be, but in our field we need to make choices. You must decide what priority you want to give to the many things we would like to know. My highest priority would be the search for the Higgs and the SUSY particles.

G. Anzivino, ... and A. Zichichi

THE LAA PROJECT

From
Rivista del Nuovo Cimento *13, n. 5 (1990)*

Reproduction of a few pages

EDITRICE COMPOSITORI
BOLOGNA
1990

5

La Rivista del Nuovo Cimento
della Società Italiana di Fisica
1990

A. ZICHICHI - *Project Leader*

The LAA Project

Editrice Compositori Bologna

RIVISTA DEL NUOVO CIMENTO VOL. 13, N. 5 1990

The LAA Project.

G. ANZIVINO, M. ARNEODO, F. ARZARELLO, G. BARI, M. BASILE
R. BATTISTON, U. BECKER, L. BELLAGAMBA, M. BENOT, J. BERBIERS
F. BERGSMA, R. BERTIN, R. K. BOCK, D. BOSCHERINI, R. BOUCLIER
G. BRUNI, P. BRUNI, L. CAPUTI, G. CARA ROMEO, R. CASACCIA
G. CHARPAK, M. CHIARINI, N. H. CHRIST, L. CIFARELLI, F. CINDOLO
E. COLAVITA, A. CONTIN, I. CROTTY, G. D'ALI, C. D'AMBROSIO, S. D'AURIA
M. DARDO, S. DE PASQUALE, R. DE SALVO, C. DEL PAPA, R. DOBINSON
J. DUPONT, J. DUPRAZ, T. EKELÖF, J. P. FABRE, P. FORD, F. FRASCONI
J. GAUDAEN, P. GIUSTI, K. GOEBEL, C. GRINNEL, B. GUERARD, T. GYS
E. HEIJNE, S. HELLMAN, M. HOURICAN, G. IACOBUCCI, P. JARRON, P. JENNI
L. JONES, J. KIRKBY, W. KRISHER, I. LAAKSO, J. C. LABBÉ
H. LARSEN, G. LAURENTI, T. D. LEE, H. LEUTZ, L. LONE, G. MACCARRONE
T. MASSAM, K. H. MEIER, G. MILLION, R. NANIA, V. O'SHEA, A. OLIVA
H. P. PAAR, F. PALMONARI, P. PELFER, C. PERONI, E. PEROTTO
V. PESKOV, D. PIEDIGROSSI, S. QIAN, J. C. SANTIARD
G. SARTORELLI, F. SAULI, E. SCHENVIT, J. SCHIPPER, D. SCIGOCKI
P. SHARP, H. SHÖNBACHER, G. SIMONET, P. SONDEREGGER
L. SPORTELLI, M. SUFFERT, G. C. SUSINNO, S. TAILHARDAT
A. E. TERRANEO, L. VOTANO, T. WEIDBERG, R. WIGMANS, M. WILLUTZKY
C. H. YEH, T. YPSILANTIS, A. ZICHICHI and K. ZOGRAPHOS

(ricevuto il 10 Dicembre 1989)

Preamble. – This report describes the LAA Project from its origin to its present status. The results obtained during its first year of activity are described in detail.

Let me just remind that the LAA Project [1-4] is an intensive programme to develop new experimental techniques to operate in future Hadron Colliders.

In fact, the present *status of accelerators and detectors* is as follows (see fig. 1.1).

No one knows, at present, how to design an (e^+e^-) collider in the TeV energy range, while a conceptual design of ELOISATRON [5] (the 100 TeV (pp) collider) has shown that there are no basic difficulties for the machine to be designed in all details.

The situation is *reversed for the case of detectors.*

If a multi-TeV (e^+e^-) collider were operative, we could immediately design an experimental set-up. The reason being that, due to the electroweak-photon propagator dumping, the cross-section left over is very small.

But, if a multi-TeV hadron collider were operating, no physicist in the planet would know how to perform a meaningful experiment, apart from the trivial «full screened» system. The basic motivations being the high rate of events and the strong radiation doses to which the detectors are exposed.

As discussed in subsect. **1˙2**, for the first time in the field of high-energy physics there are basic reasons to justify a jump in energy as big as possible. The

D. Acosta, ... and A. Zichichi

ADVANCES IN TECHNOLOGY FOR HIGH-ENERGY SUBNUCLEAR PHYSICS. CONTRIBUTION OF THE LAA PROJECT

From
Rivista del Nuovo Cimento 13, *n. 10-11 (1990)*

Reproduction of a few pages

EDITRICE COMPOSITORI
BOLOGNA
1990

10-11 La Rivista del Nuovo Cimento
della Società Italiana di Fisica
1990

A. ZICHICHI - *Project Leader*

Advances in Technology for High-Energy Subnuclear Physics. Contribution of the LAA Project

Editrice Compositori Bologna

Z. D. THOMÉ[11], R. TIMELLINI[48], J. TISCHHAUSER[7], J. TOCQUEVILLE[9]
V. VALENCIC[28], B. VAN EIJK[7], G. VANSTRAELEN[22], V. VERCESI[7], L. VOTANO[20]
H. WENNINGER[7], C. WERNER[48], R. WIGMANS[7], C. WILLIAMS[7], G. YPSILANTIS[6,8]
N. ZAGANIDIS[7], A. ZICHICHI[7] and K. ZOGRAFOS[7]

The following Universities, Institutions and Industries are collaborating with the various components of the LAA project:

[1] *Ansaldo, Genova, Italy*
[2] *INFN, Bologna, Italy*
[3] *University of Bologna, Italy*
[4] *University of Cagliari, Italy*
[5] *University of Calabria, Italy*
[6] *University of Karlsruhe, FRG*
[7] *CERN, Geneva, Switzerland*
[8] *CNR, Pisa, Italy*
[8] *Collège de France, Paris, France*
[10] *Columbia University, New York, N.Y., USA*
[11] *COPPE, Rio de Janeiro, Brazil*
[12] *CRN, Strasbourg, France*
[13] *Digital Italia, Roma, Italy*
[14] *DESY, Hamburg, FRG*
[15] *Draper Laboratories, Cambridge, Mass., USA*
[16] *EPFL, Lausanne, Switzerland*
[17] *Europa Metalli - LMI, Firenze, Italy,*
[18] *INFN, Firenze, Italy*
[19] *University of Firenze, Italy*
[20] *INFN, Laboratori Nazionali di Frascati (Roma), Italy*
[21] *University of Glasgow, United Kingdom*
[22] *IMEC/ASP, Leuven, Belgium*
[23] *Cornell University, Ithaca, N. Y., USA*
[24] *Kernforschungszentrum Karlsruhe, FRG*
[25] *LIP, Lisbon, Portugal*
[26] *LPNHE University, Paris, France*
[27] *Max Planck Institut, Heidelberg, FRG*
[28] *Mead SA, Lausanne, Switzerland*
[29] *MIT, Cambridge, Mass., USA*
[30] *INFN, Napoli, Italy*
[31] *University of Napoli, Italy*
[32] *NIKHEF, Amsterdam, Netherlands*
[33] *University of Padova, Italy*
[34] *University of Palermo, Italy*
[35] *INFN, Pavia, Italy*
[36] *University of Pavia, Italy*
[37] *University of Perugia, Italy*
[38] *Rutherford Appleton Laboratory, United Kingdom*
[39] *University of California, San Diego, Cal., USA*
[40] *University of California, Santa Cruz, Cal., USA*
[41] *Senter for Industriforsk., Oslo, Norway*
[42] *Smart Silicon Systems, Lausanne, Switzerland*
[43] *University of Toulouse, France*
[44] *INFN, Torino, Italy*
[45] *University of Torino, Italy*
[46] *Uppsala University, Sweden*
[47] *Weizmann Institute, Rehovoth, Israel*
[48] *World Laboratory, Geneva, Switzerland*
[49] *Paul Scherrer Institute, Zurich, Switzerland*

(ricevuto il 3 Marzo 1990)

1. – Highlights of the LAA project.

1·1. *Introduction.* – The origin of the LAA Project and its basic features have been reported [1-5].

As already explained in ref. [5], at present no one knows how to build an (e^+e^-) collider even at 1 TeV, while the conceptual design of a collider for (pp) at 100 TeV already exists (the ELN Project) [6]. On the contrary, no one knows how to build detectors for a 100 TeV (pp) collider, while a detector for a 1 TeV (e^+e^-) machine is within the present-day technological possibilities. This is illustrated in fig. 1.1.1.

The potentials for new discoveries in a multi-TeV (pp) collider are summarized in fig. 1.1.2. The «magic» limit of $10^{-40}\,cm^2$ in the cross-section can be reached only if the luminosity is pushed to high levels: $L = (10^{33} \div 10^{34})\,cm^2\,s^{-1}$ or even higher.

One of the main problems for detectors comes from the requirement that the average number of events per bunch crossing, $\langle n \rangle$, must be one if the missing energy is to be used as a signature for rare events. Another limiting factor towards very high luminosities is the radiation dose the detectors can withstand.

The following machine parameters:

$$\Delta t_b \sim 100\,\text{ns} \quad \text{and} \quad L_{pp} \sim 10^{32} \text{cm}^{-2}\,\text{s}^{-1},$$

are well within reach from a technological point of view; on the other hand, the total

LIST OF REFERENCES OF ALL LAA WORKS

LIST OF REFERENCES OF ALL LAA WORKS

(in parenthesis the number of papers)

1 — *GENERAL REPORTS (12)*

- *A. Zichichi, "The LAA Project", Report N° 1, CERN-Geneva, 15 December 1986.*
- *A. Zichichi, "The LAA Project", Report N° 2, CERN-Geneva, 25 June 1987.*
- *G. Anzivino et al., "The LAA Project", Preprint CERN-EP/87-122.*
- *G. Anzivino et al., "The LAA Project", ICFA - Instrumentation Bulletin 3, (1987).*
- *A. Zichichi, "The LAA Project", Report N° 3, CERN-Geneva, 19 November 1987.*
- *A. Zichichi, "The LAA Project", Report N° 4, Presented to the Italian Minister of Public Education Hon. Prof. Giovanni Galloni, on the occasion of his visit to CERN, July 25, 1988, CERN-LAA/88/l, 25 July 1988.*
- *A. Zichichi, "The LAA Project", Report N° 5, Presented at the International Colloquium on "Science, Culture and Peace", in honour of Victor F. Weisskopf. Geneva, 19-20 September 1988. CERN/LAA/88-2, 19 September 1988, and La Rivista del Nuovo Cimento 13, vol. 5 (1990).*
- *G. Anzivino et al., "The LAA Project", Proceedings of the XXVI Course of the "Ettore Majorana" International School of Subnuclear Physics, 1988: "The Super World III", (Plenum Press, New York-London, 1990), 329.*
- *A. Zichichi, "The LAA Project", Report N° 6, Presented at the Third LAA Open Meeting, CERN, Geneva, 27 June 1989, CERN/LAA/89-1.*
- *G. Anzivino et al., "The LAA Project: second year of activity", Proceedings of the XXVII Course of the "Ettore Majorana" International School of Subnuclear Physics, 1989: "The Challenging Questions", (Plenum Press, New York-London, 1991), 221.*
- *D. Acosta et al., "Advances in technology for High Energy Subnuclear Physics: contribution of the LAA Project", La Rivista del Nuovo Cimento 13, vol. 10-11 (1990).*
- *A. Zichichi, "The main achievements of the LAA Project", Proceedings of the XXVIII course of the "Ettore Majorana" International School of Subnuclear Physics, Erice, Italy, 1990: "Physics up to 200 TeV" (Plenum Press, New York-London, 1991) 327.*

2 — *HIGH PRECISION TRACKING AND PARTICLE DETECTION (60)*

- *R. Bouclier et al., "Recent developments of the multidrift tube", Nucl. Instr. Meth. A283 (1989) 509.*
- *M. Jibaly et al., "The aging of wire chambers filled with dimethly ether: wire and construction materials and freon impurities", Nucl. Instr. Meth. A283 (1989) 692.*
- *P. Fonte et al., "Feedback and breakdown in parallel-plate chambers", Nucl. Instr. Meth. A305 (1991) 91.*
- *P. Fonte et al., "VUV emission and breakdown in parallel-plate chambers", Nucl. Instr. Meth. A310 (1991) 140.*
- *G. Charpak et al., "Preliminary results of the study of gaseous detectors with solid photocathodes sensitive in the spectral region from 105 to 300 nm", Nucl. Instr. Meth. A323 (1992) 445.*
- *C. D'Ambrosio et al., "A new plastic scintillator with large Stokes shift", Nucl. Instr. Meth. A276 (1989) 69.*
- *H. Leutz, "Scintillating fibres for particle tracking", ICFA Instrumentation Bulletin, 6 (1989) 6.*
- *C. D'Ambrosio et al., "Supercollider SCIFI tracker", Proceedings of the Workshop on scintillating fibre development for the SSC, Fermilab, Batavia IL., USA, 14-16 November 1988.*
- *P. Destruel et al., "Plastic scintillating fibres with 50 microns core diameter: problems and solutions", Proceedings of the Workshop on scintillating fibre development for the SSC, Fermilab, Batavia IL., USA, 14-16 November 1988.*
- *J.P. Fabre et al., "Conceptual design of an optoelectronic delay line", Revue Phys. Appl., 24 (1989) 1019.*
- *H. Leutz, "Central tracking with scintillating fibres", ECFA Study Week on Instrumentation for High Luminosity Hadron Colliders, Barcelona, Spain, 14-21 September 1989, CERN 89-10, vol. 1.*
- *C. D'Ambrosio et al., "Studies on thin scintillating fibres bundle performance", ECFA Study Week on Instrumentation for High Luminosity Hadron Colliders, Barcelona, Spain, 14-21 September 1989, CERN 89-10, vol. 1.*
- *C. D'Ambrosio et al., "Measurements of reflection losses in scintillating fibres", ECFA Study Week on Instrumentation for High Luminosity Hadron Colliders, Barcelona, Spain, 14-21 September 1989, CERN 89-10, vol. 1.*
- *T. Gys et al., "Opto-electronic delay for the readout of particle tracks from scintillating fibres", ECFA Study Week on Instrumentation for High Luminosity Hadron Colliders, Barcelona, Spain, 14-21 Sept. 1989, CERN 89-10, vol. 1.*

- *C. D'Ambrosio and H. Leutz, "Vertex detection at future high luminosity hadron colliders", Perspectives for New Detectors in Future Supercolliders, eds. L. Cifarelli, R. Wigmans and T. Ypsilantis, World Scientific Publishing Co., (1991).*
- *C. D'Ambrosio et al., "Proposal for research and development on central tracking detector based on scintillating fibres", CERN/DRDC 90-29, DRDC/P4, 1990.*
- *C. D'Ambrosio et al., "Present status and future programme of scintillating fibres for central tracking", ECFA Large Hadron Collider Workshop, Aachen, Germany, 4-9 October 1990, CERN 90-10, vol. III.*
- *U. Gensch and S. Schlenstedt, "Monte Carlo simulations for central tracking with scintillating fibres", ECFA Large Hadron Collider Workshop, Aachen, Germany, 4-9 October 1990, CERN 90-10, vol. III.*
- *T. Gys et al., "Opto-electronic delay tubes", ECFA Large Hadron Collider Workshop, Aachen, Germany, 4-9 October 1990, CERN 90-10, vol. III.*
- *C. D'Ambrosio et al., "New organic scintillators with large Stokes shifts", Applied Spectroscopy, 45/3 (1991) 484.*
- *C. D'Ambrosio et al., "Scintillating fibres for central tracking of charged particles", New Technologies for Supercolliders, eds. L. Cifarelli and T. Ypsilantis, Plenum Publishing Corp. (1991).*
- *T. Gys et al., "Opto-electronic delay for the readout of particle tracks from scintillating fibres", New Technologies for Supercolliders, eds. L. Cifarelli and T. Ypsilantis, Plenum Publishing Corp. (1991).*
- *C. D'Ambrosio et al., "Reflection losses in polystrene fibres", Nucl. Instr. Meth. A306 (1991) 549.*
- *C. D'Ambrosio et al., "Organic scintillators with large Stokes shifts dissolved in polystyrene", Nucl. Instr. Meth. A307 (1991) 430.*
- *C. D'Ambrosio et al., "Scintillating fibres for central tracking of charged particles", Proceedings of the LP-HEP 91 Conference, Geneva, Switzerland, 25 July - 1 August 1991.*
- *C. D'Ambrosio et al., "RD7: central tracking with scintillating fibres, 1991 Status Report", CERN/DRDC 91-55, RD7/Status Report, 13 December 1991.*
- *C. D'Ambrosio et al, "Central tracking with scintillating fibres in a LHC detector", CERN-PPE/92-26, CERN/LAA/SF/91-25.*
- *C. D'Ambrosio et al., "Optimization of central particle tracking at future hadron colliders", Nucl. Instr. Meth. A322 (1992) 20.*
- *C. D'Ambrosio et al., "Imaging performance obtained with fused bundles of scintillating fibres", Nucl. Instr. Meth. A325 (1993) 161.*
- *T. Gys et al., "Opto-electronic delay for the readout of particle tracks from scintillating fibres", CERN/DRDC 92-42, LAA Status Report, 12 August 1992.*

- *C. D'Ambrosio et al., "Photoelectron counting with small diameter scintillating fibres", Nucl. Instr. Meth. A332 (1993) 134.*
- *C. D'Ambrosio and S. Qian, "Monte Carlo simulation and performance evaluation for the SF tracking detector", Nucl. Instr. Meth. A335 (1993) 341.*
- *H. Leutz, "Particle detection at ELOISATRON", in Supercolliders and Superdetectors, eds. W. Barletta and H. Leutz, World Scientific Co. (1993).*
- *C. D'Ambrosio, "Central tracking in high luminosity future colliders", in Supercolliders and Superdetectors, eds. W. Barletta and H. Leutz, World Scientific Publishing Co. (1993).*
- *T. Gys, "Read-out and data acquisition for tracking devices at future colliders", in Supercolliders and Superdetectors, eds. W. Barletta and H. Leutz, World Scientific Publishing Co. (1993).*
- *U. Gensch, "Tracking and pattern recognition at high luminosity colliders", in Supercolliders and Superdetectors, eds. W. Barletta and H. Leutz, World Scientific Publishing Co. (1993).*
- *D.E. Groom, "Radiation at hadron colliders: scaling to the highest energies", in Supercolliders and Superdetectors, eds. W. Barletta and H. Leutz, World Scientific Publishing Co. (1993).*
- *C. D'Ambrosio et al., "Photon counting with a hybrid photomultiplier tube (HPMT), Nucl. Instr. Meth. A338 (1994) 389.*
- *C. D'Ambrosio et al., "Further results on photoncounting with a hybrid photomultiplier tube", Nucl. Instr. Meth. A345 (1994) 279.*
- *C. D'Ambrosio et al., "RD7: Central trackng with scintillating fibres, 1993 Status Report", CERN/DRDC/93-42, RD7/Status Report, October 1993.*
- *C. D'Ambrosio et al., "New results in small diameter scintillating fibres for particle tracking", Proceedings of the Europhysics Conference on High Energy Physics, Marseilles, France, August 1993.*
- *M. Garg, "Untersuchungen an einer verzögerungsröhre zur auslese szintellierender fasern", Bergische Universität Wuppertal, WU D 93-20, April 1993.*
- *T. Gys et al., "A new position-sensitive photon detector based on an imaging silicon pixel array (ISPA-tube)", Nucl. Instr. Meth A355 (1995) 386.*
- *C. D'Ambrosio et al., "RD7: Development of small diameter scintillating fibres detectors for particle tracking", RD7 Summary Report, August 1994.*
- *C. D'Ambrosio et al., "First beam exposures of a scintillating fibre tracker readout by an ISPA-tube", Nucl. Instr. Meth A359 (1995) 618.*
- *C. D'Ambrosio et al., "Position sensitive photoncounting with an ISPA-tube", IEEE Trans. Nucl. Sci. 42 (1995) 130.*

- *C. Del Papa et al., "Results on GaAs solid-state detectors for high-energy physics", Nucl. Phys. B (Proc. suppl.), 23A (1991) 417.*
- *S.P. Beaumont et al., "Tests of GaAs detectors for minimum ionizing particles in a high radiation environment", Proceedings, eds. T. Dombeck, V. Kelly and G.P. Yost, World Scientific Publ. Co., Singapore, 1991, p.169.*
- *R. Bertin et al., "Gallium arsenide solid-state detectors for high energy physics", Proceedings of Perspectives for new detectors in future supercolliders, eds. L. Cifarelli, R. Wigmans and T. Ypsilantis, World Scientific Publ. Co., 1991, p. 37.*
- *S.P. Beaumont et al., "Gallium arsenide microstrip detectors for charged particles", Nucl. Instr. Meth. A321 (1992) 172.*
- *S.P. Beaumont et al., "GaAs solid state detectors for particle physics", Nucl. Instr. Meth. A322 (1992) 472.*
- *S.P. Beaumont et al., "Charge transport properties of undoped SI LEC GaAs solid-state detectors", Nucl. Instr. Meth. A326 (1993) 313.*
- *S.P. Beaumont et al., "Gallium arsenide detectors for minimum ionizing particles", Nucl. Phys. B (Proc. suppl.), 32 (1993) 296.*
- *C. Buttar et al., "GaAs solid state detectors for high energy physics", AIP Conf. Proc., ed. J.R. Sanford, AIP, New York, N° 272, p. 1709.*
- *F. Nava et al., "Influence of electron traps on charge-collection efficiency in GaAs radiation detectors", Nucl. Instr. Meth. A349 (1994) 156.*
- *S.P. Beaumont et al., "Gallium arsenide charge particle detectors: trapping effects", Nucl. Instr. Meth. A342 (1994) 83.*
- *L. Berluti et al., "Gallium arsenide particle detectors: a study of the active region and charge collection efficiency", Nucl. Instr. Meth. A354 (1995) 364.*
- *M. Dogru et al., "LEC SI-GaAs detectors for gamma rays", Nucl. Instr. Meth. A348 (1994) 510.*
- *S.P. Beaumont et al., "First results from GaAs double-sided detectors", Nucl. Instr. Meth. A348 (1994) 514.*
- *D. Puertolas et al, "An ISPA-camera for gamma rays", IEEE Trans. Nucl. Sci. 42 (1995) 2221.*

3 — CALORIMETRY *(79)*

- *V. Peskov et al., "Liquid and solid organic photocathodes", Nucl. Instr. Meth. A269 (1988) 149.*
- *G. Charpak et al., "Ethyl ferrocene in gas, condensed, or adsorbed phases: three types of photosensitive elements for use in gaseous detectors", Nucl. Instr. Meth. A277 (1989) 537.*

- *V. Peskov et al., "Investigation of light emission from a parallel-plate avalanche chamber filled with noble gases and with TEA, TMAE, and* H_2O *vapours at atmospheric pressure", Nucl. Instr. Meth. A277 (1988) 547.*
- *G. Charpak et al., "*BaF_2 *calorimeters with photosensitive gaseous chambers", ECFA Study Week on Instrumentation Technology for High Luminosity Hadron Colliders, Barcelona, 1989, eds. E. Fernandez and G. Jarlskog (CERN 89-10, Geneva, 1989), vol. 2, p. 588.*
- *G. Charpak et al., "New developments in calorimetry based on VUV scintillators coupled to photosensitive gaseous detectors", ECFA Study Week on Instrumentation Technology for High Luminosity Hadron Colliders, Barcelona, 1989, eds. E. Fernandez and G. Jarlskog (CERN 89-10, Geneva 1989), vol. 2, p. 593.*
- *G. Charpak et al., Proc. Int. Workshop on Liquid-State Electronics, Berlin 1988 (Hahn-Meitner-Institut, Berlin, 1988), p. 88.*
- *D. Astruc et al., Proc. Int. Workshop on Liquid-State Electronics, Berlin 1988 (Hahn-Meitner-Institut, Berlin, 1988), p. 109.*
- *G. Charpak et al., "New scintillators for photosensitive gaseous detectors", Proc. of Symposium on Particle Identification at High Luminosity Hadron Colliders, Batavia, 1989, eds. T.J. Gourlay and J.G. Morgin, Fermilab, Batavia, 1990, p. 295.*
- *V. Peskov et al., "Organometallic photocathodes for parallel-plate and wire chambers", Nucl. Instr. Meth. A283 (1989) 786.*
- *G. Charpak et al., "New developments in calorimetry based on VUV scintillators coupled to photosensitive gaseous detectors", Preprint CERN-EP/90-41, 1990.*
- *C. Charpak et al., "New read-out for* BaF_2 *calorimeters: gaseous detectors with solid photocathodes", ECFA Large Hadron Collider Workshop, Aachen, Germany, 4-9 October 1990, CERN 90-10, vol. III, p. 384.*
- *G. Charpak et al., "Liquid and solid photocathodes and their applications for fast gaseous detectors", Proc. 10th Int. Conf. on Conductivity and Breakdown in Dielectric Liquids, Grenoble, 1990, eds. P. Atten and R. Tobazeon (Lab. d'Electrostatique et Matériaux Diélectriques, CNRS and Univ. Joseph Fourier, Grenoble, 1990), p. 131.*
- *G. Charpak et al., "New photocathodes for fast gaseous detectors", Preprint CERN PPE 90-185 (1990). Presented at the Second Conference on Position-Sensitive Detectors, London, 4-7 September 1990.*
- *G. Charpak et al., "Investigation of operation of a parallel-plate avalanche chamber with a CsI photocathode under high gain conditions", Nucl. Instr. Meth. A307 (1991) 63.*
- *R. De Salvo et al., "A novel way of electron identification in calorimeters", Nucl. Instr. Meth. A279 (1989) 467.*

- *D. Acosta et al., "Results of prototype studies for a spaghetti calorimeter", Nucl. Instr. Meth. A294 (1990) 193.*
- *D. Acosta et al., "Electron-pion discrimination with a scintillating fibre calorimeter", Nucl. Instr. Meth. A302 (1991) 36.*
- *D. Acosta et al., "Localizing particles showering in a Spaghetti Calorimeter", Nucl. Instr. Meth. A302 (1991) 36.*
- *D. Acosta et al., "Effects of radiation damage on scintillating fibre calorimetry", preprint CERN PPE 91-45 (1991).*
- *D. Acosta et al., "Electron, pion and multiparticle detection with a lead/ scintillating fibre calorimeter", Nucl. Instr. Meth. A308 (1991) 481.*
- *D. Acosta et al., "On muon production and other leakage aspects of pion absorption in a spaghetti calorimeter", Nucl. Instr. Meth. A309 (1991) 143.*
- *B. Bencheikh et al., "A simple light detector gain measurement technique", Nucl. Instr. Meth. A315 (1992) 349.*
- *R. De Salvo et al., "First results on the Hybrid PhotoDiode Tube", Nucl. Instr. Meth. A315 (1992) 375.*
- *A. Contin et al., "On the calibration of the electromagnetic component of a spaghetti calorimeter at a multi-TeV hadron collider", Nucl. Instr. Meth. A315 (1992) 344.*
- *B. Bencheikh et al., "First results on the spaghetti calorimeter projective prototype", Nucl. Instr. Meth. A315 (1992) 354.*
- *R. Wigmans, "Calorimetry in the supercollider era", Proc. INFN Eloisatron Project 7th workshop: Novel features of high-energy collisions in 1-100 TeV region, Erice, 10-30 Jun. 1988 (eds. A. Ali and L. Cifarelli, Plenum Press, New York, 1989), p. 393.*
- *R. Wigmans, "The spaghetti calorimeter project at CERN", Proc. DPF summer study: High energy physics in the 1990's, Snowmass, CO, 27 Jun-Jul 1988 (ed. S. Jensen, World Scientific, Singapore, 1989), p. 798.*
- *R. De Salvo, "R&D project for calorimetry at SSC", Proc. Workshop on scintillating fiber detector development for the SSC, Batavia, IL, 14-16 Nov. 1988 (Fermilab, Batavia, 1990), p. 243.*
- *R. De Salvo, "Dream on a Supercollider Spaghetti Calorimeter", Proc. Workshop on Calorimetry for the Supercollider, March 13-17, 1989, Tuscaloosa, Alabama. Eds. R. Donaldson, M.G.D. Gilchriese, World Scientific, Singapore, 1990, p. 383.*
- *R. De Salvo, "Trigger and data acquisition prospects for the Spaghetti Calorimeter", presented at the Workshop on Perspectives for New Detectors in Future Supercolliders, Erice 17-24 October 1989. Ed. T. Ypsilantis, Plenum Press, New York, 1991, p. 85.*
- *A. Henriques et al., "Scintillating fibre calorimetry", Proc. Workshop on application of scintillating fibers in particle physics, Blossin, 3-5 Sept.*

1990. Ed. R. Nahmhauer, Inst. Hochenergiephysik, Berlin-Zeuthen, 1990, p. 25.

- *L. Poggioli et al., "Scintillating fibers", ECFA Large Hadron Collider Workshop", Aachen, Germany, 4-9 October 1990, CERN 90-10, vol. III, p. 329.*
- *C. Da Via et al., "Proposal for a new detector for small p_t physics", Proceedings of Research Directions for the Decade, ed. E.L. Berger, World Scientific, Singapore (1990) p. 526.*
- *M.R. Mondardini et al., "A new detector for small-p_t physics at LHC/SSC energies", Proceedings, Large Hadron Collider Workshop, Aachen, Oct. 1990, eds. G. Jarlskog and D. Rein. CERN 90-10, vol. 2.*
- *R. De Salvo, "Spaghetti calorimeter results and prospects", Proceedings of Research Directions for the Decade, ed. E.L. Berger, World Scientific, Singapore (1990) p. 407.*
- *R. De Salvo, "The Hybrid Photo Diode Tube", Proceedings, Research Directions for the Decade, ed. E.L. Berger, World Scientific, Singapore (1990), p. 416.*
- *D. Acosta et al., "Effects of radiation damage on scintillating fibre calorimetry", Nucl. Instr. Meth. B62 (1991) 116.*
- *R. De Salvo et al., "Hybrid Photo Diode Tube", Proceedings, New Technologies for Supercolliders, ed. T. Ypsilantis, Plenum Press, New York (1991), p. 57.*
- *C. Da Via et al., "A new detector for low P_t physics", Nucl. Instr. Meth. A323 (1992) 419.*
- *J. Orear et al., "Status of EOI-2 at the SSC; EOI2: low P_t physics at the SSC, an expression of interest", Nucl. Phys., 25B (1992) 261.*
- *D. Acosta et al., "The performance of a lead/scintillating fiber calorimeter at the LCH/SSC compatible gate widths", Nucl. Instr. Meth. A314 (1992) 431.*
- *D. Acosta et al., "Lateral shower profiles in a lead/scintillating fiber calorimeter", Nucl. Instr. Meth. A316 (1992) 184.*
- *D. Acosta et al., "Detection of muons with a lead/scintillating fiber calorimeter", Nucl. Instr. Meth. A320 (1992) 128.*
- *B. Bencheikh et al., "Results on the spaghetti calorimeter projective prototype", Nucl. Instr. Meth. A323 (1992) 398.*
- *M.R. Mondardini et al., "An imaging detector for elastic scattering at future accelerators", Proceedings, Image Processing for Future High Energy Physics Detectors, eds. V. Buzuliou and A. Zichichi, World Scientific, Singapore (1992), p. 27.*
- *V. Buzuliou et al., "Testing scintillating fibers for the spaghetti calorimeter: a feasibility study", Proceedings, Image Processing for Future High Energy Physics Detectors", eds. V. Buzuliou and A. Zichichi, World Scientific, Singapore (1992), p. 137.*

- *M.C. Lundin et al., "Images of spaghettis", Proceedings, Image Processing for Future High Energy Physics Detectors", World Scientific, Singapore (1992), p. 35.*
- *D. Acosta et al., "Effects of radiation damage on scintillating fibre calorimetry", Radiat. Phys. Chem., vol. 41, N° 1/2, p. 303 (1993).*
- *G. Anzivino et al., "Preliminary radiation damage results on a scintillating fibre and lead calorimeter", Radiat. Phys. Chem., vol. 41, N° 1/2, p. 283 (1993).*
- *H. Arnaudon et al., "The characterization and evaluation of a proximity focussed hybrid photo diode", ICFA bulletin N° 10, p. 15, December 1993.*
- *T. Aziz et al., "Lepton and photon precision physics", Nucl. Instr. Meth. A325 (1993) 23.*
- *H. Arnaudon et al., "The hybrid photodiode tube", Proceedings, EPS Conference HEP 93, Marseilles, 22-28 July 1993.*
- *C. Avila et al., "High resolution tracks in a solid SCI-FI bundle", Proceedings, Workshop on Scintillating Fiber Detectors, Notre Dame, Indiana, USA, Oct. 1993.*
- *G. Anzivino et al., "Electron-pion discrimination with projective lead/scintillating fibers calorimeters", Proceedings, IV Internat. Conf. on Calorimetry in High Energy Physics, Isola d'Elba, Italy, 19-25 Sept. 1993.*
- *G. Anzivino et al., "Results on a fully projective lead and scintillating fiber calorimeter", Proceedings of the IV Internat. Conf. on Calorimetry in High Energy Physics, Isola d'Elba, Italy, 19-25 Sept. 1993.*
- *G. Anzivino et al., "Recent developments of the HPD", Proceedings, IV Internat. Conf. on Calorimetry in High Energy Physics, Isola d'Elba, Italy, 19-25 Sept. 1993.*
- *G. Anzivino et al., "Failure modes of large surface avalanche photo diodes in high energy physics environments", Proceedings, IV Internat. Conf. on Calorimetry in High Energy Physics, Isola d'Elba, Italy, 19-25 Sept. 1993.*
- *G. Anzivino et al., "Design of a quartz fiber calorimeter for a collider experiment", Proceedings, IV Internat. Conf. on Calorimetry in High Energy Physics, Isola d'Elba, Italy, 19-25 Sept. 1993.*
- *G. Anzivino et al., "Quartz fibers for very forward calorimetry: Ultra-fast, infinitely rad-hard and shower core sensitive", Proceedings, IV Internat. Conf. on Calorimetry in High Energy Physics, Isola d'Elba, Italy, 19-25 Sept. 1993.*
- *G. Anzivino et al., "Effects of induced radioactivity in very forward calorimetry, a comparison between different techniques", Proceedings, IV Internat. Conf. on Calorimetry in High Energy Physics, Isola d'Elba, Italy, 19-25 Sept. 1993.*

- *G. Anzivino et al., "Radiation induced effects on lead and scintillating fiber calorimeters at LHC", Nucl. Instr. Meth. A346 (1994) 153.*
- *H. Arnaudon et al., "Proximity focussed hybrid photo diode characteristics evaluations", Nucl. Instr. Meth. A342 (1994) 558.*
- *A. Contin et al., "A method to identify the production vertex of intermediate-mass Higgs particles at LHC without a pointing electromagnetic calorimeter", Nucl. Instr. Meth. A337 (1994) 342.*
- *G. Anzivino et al., "Review of the hybrid photo diode tube (HPD), an advanced light detector for physics", Nucl. Instr. Meth. A365 (1995) 76.*
- *G. Anzivino et al., "Recent developments in quartz fibre calorimetry", Nucl. Instr. Meth. A357 (1995) 369.*
- *G. Anzivino et al., "Angular dependence of quartz fibre calorimeter response", Nucl. Instr. Meth. A360 (1995) 237.*
- *G. Anzivino et al., "Quartz fibre calorimetry - Monte Carlo simulations", Nucl. Instr. Meth. A357 (1995) 380.*
- *G. Anzivino et al., "Results on lead/scintillating fibers calorimetry", CERN/LAA/94-25.*
- *P. Gorodetsky et al., "Quartz fibre calorimetry", Phys. Res. A361 (1995) 161.*
- *C. Avila et al., "Elastic scattering tracking at the Tevatron", CERN/LAA/HC/94-21, presented at the 6th Pisa Meeting on Advanced Detectors, Elba, Italy, May 22-28, 1994.*
- *H. Arnaudon et al., "Caratterizzazione di un fotomoltiplicatore ibrido (HPD), submitted to "Elettroottica 94", Pavia, Italy, 1994.*
- *P. Benetti et al., "Recent developments about the hybrid photomultiplier tube", Proceedings, SPIE Conference, San Diego, July 1994.*
- *J. Seguinot and T. Ypsilantis, "Perfect Calorimetry", CERN-LAA/89-1, p. 187.*
- *T. Ypsilantis and J. Seguinot, "Liquid Xenon Calorimetry", LHC Workshop vol. 3, CERN/90-10.*
- *J. Seguinot et. al., "DRDC Proposal #P17-Liquid Xenon Calorimetry", CERN-DRDC/90-70; CERN-LAA/PC/90-17.*
- *J. Seguinot et. al., "A Totally Active Xenon (or Krypton) Electromagnetic Calorimeter for High Luminosity Hadron Colliders", Proc. "New Technologies for Supercolliders", eds. L. Cifarelli & T. Ypsilantis, Plenum Press, 1991.*
- *J. Seguinot et. al., "A Cryogenic Apparatus with On-Line Purification System for Scintillation and Ionization Studies in Noble Liquids", CERN/AT-CR/91-08.*
- *J. Seguinot et. al., "Liquid Xenon Ionization and Scintillation: Studies for a Totally Active Vector EM Calorimeter", Nucl. Instr. Meth. A323 (1992) 583.*

- *J. Seguinot et. al., "Liquid Xenon Scintillation: Photon Yield and Fano Factor Measurements", Nucl. Instr. Meth. A354 (1995) 280.*

4 — *LARGE AREA DEVICES (38)*

- *F. Bergsma et al., "Muon rates and muon interaction in matter, Perspective for new detectors in future supercolliders", Proc. 9th Workshop on Perspectives for New Detectors in Future Supercolliders, Erice, Italy, 17-24 October 1989, ed. L. Cifarelli, R. Wigmans, T. Ypsilantis, World Scientific Co., Singapore.*
- *G. Ambrosi et al., "A new gaseous detector for tracking: the blade chamber", ECFA Study Week on Instrumentation Technology for High Luminosity Hadron Colliders, Barcelona, 1989, eds. E. Fernandez and G. Jarlskog (CERN 89-10, Geneva 1989), p. 430.*
- *D. Mattern and M.C.S. Williams, "Preliminary report on a planar surface chamber working in limited streamer mode", ECFA Study Week on Instrumentation Technology for High Luminosity Hadron Colliders, Barcelona, 1989, eds. E. Fernandez and G. Jarlskog (CERN 89-10, Geneva 1989), p. 435.*
- *D. Mattern et al., "A new approach for constructing sensitive surfaces: the gaseous pixel chamber", Nucl. Instr. Meth. A300 (1991) 275.*
- *G. Ambrosi et al., "A wireless limited streamer chamber: the blade chamber", Particle World, vol. 1, no. 3 (1990) 80.*
- *G. Ambrosi et al., "A new gaseous detector for tracking: the blade chamber", Nucl. Inst. Meth. A289 (1990) 351.*
- *G. Ambrosi et al., "Detector test in a high magnetic field and muon spectrometer triggering studies on a small prototype for an LHC experiment", CERN/DRDC/90-42; DRDC/P10, 1990.*
- *F. Bergsma et al., "New results on the blade chamber", ECFA Large Hadron Collider Workshop, Aachen, Germany, 4-9 October 1990, CERN 90-10, vol. III, p. 489.*
- *F. Bergsma et al., "Conceptual design of a muon spectrometer at hadron supercolliders", ECFA Large Hadron Collider Workshop, Aachen, Germany, 4-9 October 1990, CERN 90-10, vol. III, p. 520.*
- *D. Mattern et al., "First tests of the gaseous pixel chamber fabricated on a ceramic substrate", Nucl. Inst. Meth. A310 (1991) 78.*
- *D. Antreasyan et al. "A device for particle detection at future hadron colliders: the gaseous pixel chamber", ECFA Large Hadron Collider Workshop, Aachen, Germany, 4-9 October 1990, CERN 90-10, vol. III, p. 230.*
- *D. Mattern, M.C.S. Williams, "Feasibility study of μ-tip field emission devices working as a radiation detector", CERN/LAA/HC/91-001.*

- *D. Mattern et al., "Latest results on the development of the gaseous pixel chamber", Nucl. Inst. Meth. A313 (1992) 385.*
- *M.C.S. Williams and A. Zichichi, "Operation of gaseous pixel chamber in proportional mode", Nucl. Instr. Meth. A323 (1992) 252.*
- *I. Crotty et al., "Investigation of resistive parallel plate chambers", Nucl. Instr. Meth. A329 (1993) 133.*
- *I. Crotty et al., "Results of tests of protoype resistive plate chambers", in Pavia 93, Proceedings "The resistive plate chambers in particle physics and astrophysics", p. 199.*
- *I. Crotty et al., "The non spark mode and high rate operation of resistive parallel plate chambers", Nucl. Instr. Meth. A337 (1994) 370.*
- *I. Crotty et al., "Further studies of avalanche mode operation of resistive parallel plate chambers", Nucl. Instr. Meth. A346 (1994) 107.*
- *I. Crotty et al., "Front end electronics for Resistive plate chambers", Proc. 4th Internat. Conf. on Electronics for Future Colliders, 11-12 May 1994, Chestnut Ridge, New York, USA, p. 135.*
- *I. Crotty et al., "The wide gap resistive plate chamber", Nucl. Instr. Meth. A360 (1995) 512.*
- *E. Cerron Zeballos et al., "High rate resistive plate chambers", Proc. 7th "Wire chamber" Conference, Vienna, Austria, 13-17 February 1995. Nucl. Instr. Meth. A367 (1995) 388.*
- *E. Cerron Zeballos et al., "New developments of RPC: secondary electron emission and microstrip readout", Proc. Third Internat. Workshop on Resistive Plate Chambers and Related Detectors, Pavia, Oct. 11-12, 1995. Scientifica Acta - Quaderni del Dottorato, vol. XI, anno XI, N° 1, p. 45.*
- *E. Cerron Zeballos et al., "What have we learned from a comparison between the wide gap and narrow gap resistive plate chamber", Proc. Third Internat. Workshop on Resistive Plate Chambers and Related Detectors, Pavia, Oct. 11-12, 1995. Scientifica Acta - Quaderni del Dottorato, vol. XI, anno XI, N° 1, p. 295.*
- *E. Cerron Zeballos et al., "Latest results on the performance of the wide gap RPC", Proc. Third Internat. Workshop on Resistive Plate Chambers and Related Detectors, Pavia, Oct. 11-12, 1995. Scientifica Acta - Quaderni del Dottorato, vol. XI, anno XI, N° 1, p. 317.*
- *E. Cerron Zeballos et al., "A comparison of the wide gap and narrow gap resistive plates chambers", Nucl. Instr. Meth. A373 (1996) 35.*
- *E. Cerron Zeballos et al, "New results on high rate resistive plate chambers", presented at the "Four Seas Conference", Trieste, Italy, 25 June - 1 July 1995.*
- *E. Cerron Zeballos et al., "A new type of resistive plate chamber: the multigap RPC", Nucl. Instr. Meth. A374 (1996) 132.*

- *E. Cerron Zeballos et al., "Investigation of anomalous heat production in Ni-H systems", Nuovo Cimento 109A (1996) 1645.*
- *E. Cerron Zeballos et al., "Avalanche fluctuations within the multigap resistive plate chamber", Nucl. Instr. Meth. A381 (1996) 569.*
- *E. Cerron Zeballos et al., "The multigap resistive plate chamber", Proc. 33rd INFN Eloisatron Project Workshop, Erice, Sicily, Oct. 1996, to be published by World Scientific.*
- *E. Cerron Zeballos et al., "Latest results from the multigap resistive plate chamber", presented at the Manchester Conference on position-sensitive detectors, Sept. 1996. Nucl. Instr. Meth. A392 (1997) 145.*
- *E. Cerron Zeballos et al., "Resistive plate chambers with secondary electron emitters and microstrip readout", Nucl. Instr. Meth. A392 (1997) 150.*
- *E. Cerron Zeballos et al., "Pure avalanche mode operation of a 2 mm gap resistive plate chamber", Nucl. Instr. Meth. A396 (1997) 93.*
- *E. Cerron Zeballos et al., "The micro-streamers and the micro-gap resistive plate chamber", to be published in Nucl. Instr. Meth. A.*
- *E. Cerron Zeballos et al., "The development of the multi-gap resistive plate chamber", presented at the 4th Internat. Workshop on Resistive Plate Chambers and Related Detectors, Naples, Italy, Oct. 1997.*
- *E. Cerron Zeballos et al., "Streamer free operation of a 2 mm gap resistive plate chamber with C2F5H", presented at the 4th Internat. Workshop on Resistive Plate Chambers and Related Detectors, Naples, Italy, Oct. 1997.*
- *E. Cerron Zeballos et al., "The micro-gap resistive plate chamber", presented at the 4th Internat. Workshop on Resistive Plate Chambers and Related Detectors, Naples, Italy, Oct. 1997.*
- *E. Cerron Zeballos et al., "Effect of adding SF6 to the gas mixture in a multigap resistive plate chamber", CERN/LAA-MC/98-01, submitted to the Vienna Wire Chamber Conference, Feb. 1998.*

5 — *LEADING PARTICLE DETECTION (21)*

- *H. Larsen and T. Massam, "LPS radiation tests on service electronics", ECFA Large Hadron Collider Workshop, Aachen, Germany, 4-9 October 1990, CERN 90-10, vol. III, p. 677.*
- *T. Massam, "Status of the Leading Particle Detection", Proc. 9th Workshop of the INFN Eloisatron Project, Perspectives for New Detectors in Future Supercolliders, eds. L. Cifarelli, R. Wigmans and T. Ypsilantis, World Scientific Publ. Co., (1991) p. 127.*
- *G. Anzivino et al., "The correct method to evaluate the spatial resolution in a tracking detector", CERN/LAA/LP/92-001.*

- *G. Anzivino et al., "Alignment optics for the leading proton spectrometer", CERN/LAA/LP/92-002.*
- *R. Ayad and C. Nemoz, "Background from synchrotron radiation for the ZEUS leading proton spectrometer", CERN/LAA/LP/93-24.*
- *T. Massam and Y. Zamora, "Some possible studies of many body states using the LPS", CERN/LAA/LP/93-31.*
- *P. Ford, "A constant tension spring device used for atmospheric force compensation", CERN/LAA/LP/94-08.*
- *T. Massam and Y. Zamora, "The leading proton spectrometer: high precision momentum calibration", CERN/LAA/LP/94-18.*
- *G. Anzivino et al., "Thermal properties of the leading proton multi-layer PCB prototype", CERN/LAA/LP/94-32.*
- *H. Larsen, "Leading proton spectrometer: fast fan-out", CERN/LAA/LP/95-01.*
- *H. Larsen ,"LPS: read-out controller", CERN/LAA/LP/95-02.*
- *H. Larsen, "LPS: serial read-out controller", CERN/LAA/LP/95-03.*
- *H. Larsen and C. Nemoz, "The ZEUS leading proton spectrometer: data acquisition system", CERN/LAA/LP/95-04.*
- *C. Nemoz, "The ZEUS leading proton spectrometer: hardware control system", CERN/LAA/LP/95-05.*
- *R. Ayad, "LPS stations alignment control", CERN/LAA/LP/95-06.*
- *H. Larsen "LPS: hybrid front-end power supply", CERN/LAA/LP/95-07.*
- *H. Larsen, "Leading proton spectrometer: seven channel silicon strip detector bias supply", CERN/LAA/LP/95-08.*
- *H. Larsen, "Leading proton spectrometer: FIC 823X to transputer link interface in VME format", CERN/LAA/LP/95-09.*
- *H. Larsen and A. Zampieri, "Leading proton spectrometer: beam position instability monitor", CERN/LAA/LP/95-10.*
- *H. Larsen, "LPS - Level 1 trigger pipeline, LPS pipe", CERN/LAA/LP/95-12*
- *H. Larsen, "LPS - Level 1 trigger interface - LPS_GFI", CERN/LAA/LP/95-13*

6 — *DATA ACQUISITION AND ANALYSIS (17)*

- *R.K. Bock et al, "How can we get intelligent systems close to experiments?", presented at the Snowmass Summer Study, July 5-7 1998.*
- *R.K. Bock et al, "Benchmarking computer architectures for HEP real time algorithms", CERN/LAA/RT/88-08.*

- *R.K. Bock et al., "The LAA real-time benchmarks", presented at the Workshop on triggering and data acquisition at the SSC, Toronto, Jan. 1989.*
- *S. Lone et al., "Fine-grain parallel computer architectures in future triggers", Nucl. Inst. Meth. A288 (1990) 507.*
- *M. Letheren et al., "Simulation of a macro-pipelined multi-CPU event processor for use in Fastbus", IEEE Trans. Nucl. Science 36,5 (1989) 1597.*
- *J. Christiansen and C. Ljuslin, "A programmable high-performance integrated fastbus slave interface chip", CERN, FASIC user manual, 1991.*
- *R.K. Bock et al., "Feature extraction in future detectors", Nucl. Inst. Meth. A289 (1990) 534.*
- *R.K. Bock, "Trigger and data acquisition questions in the LAA-3 detector", Proc. 9th Workshop on "Perspectives for New Detectors in Future Supercolliders", Erice, Italy, 17-24 October 1989, eds. L. Cifarelli, R. Wigmans, T. Ypsilantis, World Scientific Co., Singapore.*
- *R.K. Bock and J. Pfennig, "A transition radiation detector as triggering device", Proc. Symposium on Detector R&D for the SSC, Fort Worth, October 1990, World Scientific 1991, p. 196.*
- *W. Krischer, "Image processing in LHC detectors", ECFA Large Hadron Collider Workshop, Aachen, Germany, 4-9 October 1990, CERN 90-10, vol. III, p. 121.*
- *V. Buzuliou, "The "farm" approach to the second level triggering for future HEP detectors", CERN/LAA/RS/92-014.*
- *V. Buzuliou, "Real-time recovery of the amplitude and shift of a pulse from its samples", CERN/LAA/RT/92-015.*
- *V. Buzuliou, "An integrated high-performance fastbus slave interface circuit", CERN/LAA/RT/92-16.*
- *J. Badier et al., "Evaluating parallel architectures for two real-time applications with 100 kH_Z repetition rate", IEEE Trans. Nucl. Sc., 40/1 (1993) 45.*
- *L.J. Levinson et al., "A simple HiPPI destination interface for second level trigger prototypes", Proc. Comp. in High Energy Physics, Annecy, France; CERN Report 92-07 (1992) 904.*
- *B. Green et al., "A test system for second-level trigger and data acquisition architectures", Proc. Comp. in High Energy Physics, Annecy, France; CERN Report 92-07 (1992) 701.*
- *A. Gheorghe, "Highly parallel signal processing architectures for second-level trigger applications", Proc. Comp. in High Energy Physics, Annecy, France; CERN Report 92-07 (1992) 239.*

7 — *MONTECARLO SIMULATIONS AND PHENOMENOLOGY FOR PHYSICS AT EXTREME ENERGIES (65)*

- *J.M. Alberty et al., "Monte Carlo Simulation for the LAA Detector", Proc. 9th INFN ELOISATRON Project Workshop "Perspectives of New Detectors for Future Supercolliders", Erice, 17-24 October 1989, eds. L. Cifarelli, R. Wigmans, T. Ypsilantis, World Scientific Co., Singapore, p. 160.*
- *J.M. Alberty et al., "Monte Carlo Simulation in the Nineties", Proc. 9th INFN ELOISATRON Project Workshop "Perspectives of New Detectors for Future Supercolliders", Erice, 17-24 October 1989, ed. L. Cifarelli, R. Wigmans, T. Ypsilantis, World Scientific Co., Singapore, p. 170.*
- *J.M. Alberty et al., "FMC: Full Monte Carlo Chain", User Guide, version 1.2, February 1990. Preprint CERN/LAA-MSL/90-01.*
- *J.M. Alberty et al., "FMC: Full Monte Carlo Chain", User Guide, version 1.3, July 1990. Preprint CERN/LAA-MSL/90-06.*
- *J.M. Alberty et al., "An Integration of Physics Packages Data Modelling and Human Interface", Proc. First International Workshop on Software Engineering, Artificial Intelligence and Expert Systems in High Energy and Nuclear Physics: New Computing Techniques in Physics Research, Lyon Villeurbanne (France), March 19-24, 1990, eds. D. Perret-Gallix, W. Wojcik, Editions du Centre National de la Recherche Scientifique, Paris 1990, p. 117.*
- *J.M. Alberty and A. Bialas, "Intermittency and the Central Limit Theorem", Zeit. für Phys. C50 (1991) 315.*
- *J.M. Alberty et al., "A Super Monte Carlo for Multi-TeV Physics Simulation", Proc. 12th INFN ELOISATRON Project Workshop on "New Technologies for Supercolliders", Erice September 1990, ed. L. Cifarelli and T. Ypsilantis, Plenum Pub. Co., New York 1991, p. 307.*
- *F. Anselmo et al., "Event Generators for LHC", ECFA Large Hadron Collider Workshop, Aachen, Germany, 4-9 October 1990, CERN 90-10, vol. II, p. 130.*
- *K. Bos et al., "$H^o \rightarrow \tau^+\tau^-$ Detection at the LHC", ECFA Large Hadron Collider Workshop, Aachen, Germany, 4-9 October 1990, CERN 90-10, vol. II, p. 538.*
- *F. Carminati et al., "Standard Interfaces between Modules of Event Generators using dynamical Common Structures", ECFA Large Hadron Collider Workshop, Aachen, Germany, 4-9 October 1990, CERN 90-10, vol. III, p. 52.*
- *J.R. Cudell et al., "A Comparison of Bottom Production in different Event Generators", ECFA Large Hadron Collider Workshop, Aachen, Germany, 4-9 October 1990, CERN 90-10, vol. III, p. 164.*

- *J.M. Alberty et al., "Super Monte Carlo Simulations at 16, 40, 200 TeV", CCAST ELOISATRON Workshop, Beijing, China, 18-19 October 1990. Preprint CERN/LAA-MSL/90-12.*
- *J.M. Alberty, "Note on Random Multiplicative Processes and the Statistics of Rare Events", preprint CERN/LAA-MSL/90-14.*
- *F. Anselmo et al., "COSMOS: A Comprehensive Super Monte Carlo System", Proc. 14th INFN ELOISATRON Project Workshop on "Data Structures for Particle Physics Experiments", Erice, 11-18 November 1990, eds. R. Brun, P. Kunz, P. Palazzi, World Scientific Co., Singapore 1991, p. 70.*
- *F. Carminati et al., "Information Modelling for Monte Carlo Event Generators", Proc. 14th INFN ELOISATRON Project Workshop on "Data Structures for Particle Physics Experiments", Erice, 11-18 November 1990, eds. R. Brun, P. Kunz, P. Palazzi, World Scientific Co., Singapore 1991, p. 83.*
- *G. La Commare et al., "The CAB database", Proc. 14th INFN ELOISATRON Project Workshop on "Data Structures for Particle Physics Experiments", Erice, 11-18 November 1990, eds. R. Brun, P. Kunz, P. Palazzi, World Scientific Co., Singapore 1991, p. 77.*
- *J.M. Alberty et al., "Anomalous Dimensions in Multiparticle Collisions and the Empty Bin Effect", Zeit. für Phys. C52 (1991) 297.*
- *C. Maidantchik et al., "SIM: A Software Information Manager", Proc. "Computing in High Energy Physics", March 91, Tsukuba, Japan, eds. Y. Watase and F. Abe, Universal Academy Press, Tokyo 1991, p. 335.*
- *G. Xexéo et al., "CAB: The COSMOS Application Builder", Proc. "Computing in High Energy Physics", March 91, Tsukuba, Japan, eds. Y. Watase and F. Abe, Universal Academy Press, Tokyo 1991, p. 359.*
- *F. Anselmo et al., "MEGA: Monte Carlo Event Generator Adaptor", Particle World, vol. 2, No. 4 (1991) 108.*
- *F. Anselmo et al., "The Effective Experimental Constraints on M_{SUSY} and M_{GUT}", Nuovo Cimento 104A (1991) 1817.*
- *G. La Commare et al., "Particle and Decay Data Base (PDKDB)", preprint CERN-ECP/91-19 and CERN/LAA-MSL/91-07.*
- *A. Bassi et al., "SUD: an Advanced Set-up Descriptor for Complex Multi-TeV Experiments", Nuovo Cimento 105A (1992) 1207.*
- *F. Anselmo et al., "The Evolution of Gaugino Masses and the SUSY Threshold", Nuovo Cimento 105A (1992) 581.*
- *L. Cifarelli et al., "An Entity Relationship Approach to the Event and Detector Monte Carlo Simulation in High-Energy Physics", Nuovo Cimento 105A (1992) 597.*
- *F. Anselmo et al., "The Convergence of the Gauge Couplings at E_{GUT} and Above: Consequences for $a_3(M_Z)$ and SUSY Breaking", Nuovo Cimento 105A (1992) 1025.*

- *F. Anselmo et al., "The Simultaneous Evolution of Masses and Couplings: Consequences on Supersymmetry Spectra and Thresholds", Nuovo Cimento 105A (1992) 1179.*
- *F. Anselmo et al., "Analytic Study of the Supersymmetry-Breaking Scale at Two Loops", Nuovo Cimento 105A (1992) 1201.*
- *A. Bassi et al., "SUD: an Advanced Set-up Descriptor for Complex Multi-TeV Experiments", Nuovo Cimento 105A (1992) 1207.*
- *F. Anselmo et al., "A Study of the Various Approaches to* M_{GUT} *and* α_{GUT}*", Nuovo Cimento 105A (1992) 1335.*
- *J.L. Lopez et al., "Troubles with the Minimal SU(5) Supergravity Model", Phys. Lett. B291 (1992) 255.*
- *F. Anselmo et al., "A* χ^2*-Test to Study the* α_1, α_2, α_3, *Convergence for High-Precision LEP Data, Having in Mind the SUSY Threshold", Nuovo Cimento 105A (1992) 1357.*
- *F. Anselmo et al., "Predictions for Secondary Particle Production at Existing and Future Hadron-Hadron Colliders", Nuovo Cimento 105A (1992) 1371.*
- *A. Peterman and A. Zichichi, "On a Class of Finite Sigma-models and String Vacua: a Supersymmetric Extension", Nuovo Cimento, 106A (1993), 719.*
- *J.L. Lopez et al., "Proposed Tests for Minimal SU(5) Supergravity at Fermilab, Gran Sasso, Superkamiokande, and LEP", Phys. Lett. B299 (1993) 262.*
- *F. Anselmo et al., "Acoplanar Di-leptons and Mixed Events on the Basis of Two Supergravity Model predictions", Nuovo Cimento 106A (1993) 1389.*
- *J.L. Lopez et al., "Supersymmetry Tests at Fermilab: A Proposal", Phys. Rev. D48 (1993) 2062.*
- *J.L. Lopez et al., "Improved LEP Lower Bound on the Lightest SUSY Higgs Mass from Radiative Electroweak Breaking and its Experimental Consequences", Phys. Lett. B306 (1993) 73.*
- *J.L. Lopez et al., "SUSY Signals at HERA in the No-scale Flipped SU(5) Supergravity Model", Phys. Rev. D48 (1993) 4029.*
- *J.L. Lopez et al., "Towards a Unified String Supergravity Model", Phys. Lett. B319 (1993) 451.*
- *J.L. Lopez et al., "The Superworlds of SU(5) and SU(5)×U(1): A Critical Assessment and Overview", Proc. Internat. School of Subnuclear Physics: 30th course: "From Superstrings to the Real Superworld", Erice, Italy, 14-22 July 1992, World Scientific, p. 311.*
- *J.L. Lopez et al., "SU(5)×U(1): A String Paradigm of a TOE and its Experimental Consequences", preprint CERN-TH.6926/93 -REV.*
- *J.L. Lopez et al., "First Constraints on SU(5)×U(1) Supergravity from Trilepton Searches at the Tevatron", preprint CERN-TH.7107/93.*

- *F. Anselmo et al., "Neural Networks for Higgs Search", Nuovo Cimento 107A (1994) 129.*
- *A. Peterman and A. Zichichi, "A Search for Exact Superstring Vacua", Nuovo Cimento, Note Brevi, vol. 107A, N. 2 (Feb. 1994) 333.*
- *J.L. Lopez et al., "The Simplest, String Derivable, Supergravity Model and its Experimental Consequences", Phys. Rev. D49 (1994) 343.*
- *F. Anselmo et al., "Heavy Higgs Search with Hadron Supercolliders up to $\sqrt{s}$=200 TeV", Nuovo Cimento 107A (1994) 783.*
- *J.L. Lopez et al., "The Strongest Experimental Constraints on SU(5)×U(1) Supergravity Models", Phys. Rev. D49 (1994) 355.*
- *A. Peterman and A. Zichichi, "Proof of the Equivalence between Double Scaling Limit and Finite-size Scaling Hypothesis", Nuovo Cimento, 107A, (1994) 507.*
- *F. Anselmo et al., "Track Reconstruction with a Central Two-shell Scintillating Fibre Tracker (SFT)", Nucl. Instr. Meth. A349 (1994) 398.*
- *J.L. Lopez et al., "Scrutinizing Supergravity Models Through Neutrino Telescopes", Phys. Rev. D49 (1994) 3691.*
- *A. Peterman and A. Zichichi, "The Full Two-loop Approach to the Problem of the Light Supersymmetric Threshold", Nuovo Cimento 108A (1995) 105.*
- *J.L. Lopez et al., "A Layman's Guide to SUSY GUTs", Riv. Nuovo Cimento 17, N° 2 (1994) 1.*
- *D. Hatzifotiadou et al., "A Large Area Rich for Long Baseline Neutrinos", preprint LPC/94-29 and CERN-LAA/94-17.*
- *J.L. Lopez et al., "New Precision Electroweak Tests of SU(5)×U(1) Supergravity", Phys. Rev. D49 (1994) 4835.*
- *J.L. Lopez et al., "Experimental Aspects of SU(5)×U(1) Supergravity", Phys. Rev. D50 (1994) 2164.*
- *J.L. Lopez et al., "The Top Quark Mass in SU(5)×U(1) Supergravity", Phys. Lett. B327 (1994) 279.*
- *J.L. Lopez et al., "Status of the Superworld from Theory to Experiment", Prog. Part. Nucl. Phys. 33 (1994) 303.*
- *A. Peterman and A. Zichichi, "Explicit Superstring Vacua in a Background of Gravitational Waves and Dilaton", Nuovo Cimento 108A (1995) 97.*
- *J.L. Lopez et al., "New Constraints on Supergravity Models from $b \rightarrow s\gamma$", Phys. Rev. D51 (1995) 147.*
- *J.L. Lopez et al., "Experimental Consequences of One-parameter No-scale Supergravity Models", Int. J. Mod. Phys. A10 (1995) 4241.*
- *F. Anselmo et al., "Supersymmetry in Underground Labs and ν-Telescopes", presented at the 6th Internat. Symp. on Neutrino Telescopes,*

Venice, Italy, 22-24 Feb. 1994, ed. M. Baldo Ceolin, p. 69.

- *S. Kelley et al., "New Phenomena in the Standard No-scale Supergravity Model", preprint CERN-TH-7433/94.*
- *T. Taylor, H. Wenninger and A. Zichichi, "«Leading» Physics at LHC including Machine Studies plus Detector R&D (LAA)", Nuovo Cimento 108A, (1995) 1477.*
- *A. Peterman and A. Zichichi, "Renormalization Structure of the Transition to Chaos", Nuovo Cimento 109A (1996) 341.*

8 — *APPLIED SUPERCONDUCTIVITY, RADIATION HARDNESS AND OTHER MISCELLANEOUS STUDIES ON PARTICLE DETECTORS (46)*

- *P. Bruni et al., "Design study of a ∅19.5x36 m superconducting solenoid", Proc. 1990 Applied Superconductivity Conference (ASC 1990), Snowmass, Co, 24-26 September 1990, (IEEE Trans. Magnet 27, 1991), p. 1969.*
- *Q. Lin et al., "Development of a 13.5 Nb_3Sn dipole magnet for future particle accelerators", Proc. 25th Workshop of the INFN Eloisatron Project on Future Superdetectors and Coliders at the Frontier of High Energy and High Luminosity, Erice, Italy, 17-22 November 1992 (World Scientific, Singapore).*
- *Q. Lin et al., "Study of a low temperature electromagnetic valve", Cryogenics 1994, vol. 34, ICEC Supplement.*
- *A. Bonnano et al., "Effects of a partial orientation of CU^{++} complexes in $YBa_2 CU_3 O_{7-x}$ pellets", Applied Sciences, NATO Series E 176 (1990) 559.*
- *A. Bonnano et al., "Interpretation of electron paramagnetic resonance spectra of YBCO single phase pellets", Proc. 10th Ampere Summer School and Symposium of Magnetic Resonance and Relaxation, Portroz (Yugoslavia), 1988, p. 139.*
- *A. Bonnano et al., "Studio di Risonanza paramagnetica elettronica su pellet di YBCO", LXXIV Congresso Nazionale delle Società Italiana di Fisica, Urbino 1988, p. 16.*
- *G. Chiarello et al., "Unoccupied electronic states of CuO and Cu_2O studied by secondary electron emission", Journal of Electron Spectroscopy and Related Phenomena 70 (1994) 45.*
- *H. Schönbacher, "Radiation tests for the LAA Project, CERN/LAA/RA/88-3.*
- *H. Schönbacher, "Radiation damage studies for detector materials", ECFA Study Week on Instrumentation Technology for High Luminosity Hadron Colliders, Barcelona, 1989, eds. E. Fernandez and G. Jarlskog (CERN 89-10, Geneva 1989), vol. I.*

- *H. Schönbacher, "Radiation damage studies for detector materials within the LAA Project", CERN/TIS-CFM/90-06.*
- *H. Schönbacher, "Radiation hardness studies for LHC detector materials", ECFA Large Hadron Collider Workshop, Aachen, Germany, 4-9 October 1990, CERN 90-10.*
- *H. Schönbacher, "Radiation damage studies, irradiations and high dose dosimetry for LHC detectors", CERN/DRDC/90-06/DRDC/P24.*
- *E. Leon Florian and M. Tavlet, "Dose and neutron flux measurements in the PS-ACOL irradiation facility in 1990", CERN/LAA/RA/91-009.*
- *H. Schönbacher, "The PS-ACOL irradiation facility at CERN, Proc. Europ. Conf. on radiation and their effects on devices and systems, Montpellier, Sept. 1991.*
- *H. Schönbacher, "Review of radiation damage studies on scintillating materials used in high energy physics experiments", CERN/TIS-RP/202.*
- *E. Leon Florian and H. Schönbacher, "Facilities & dosimetry for detector material irradiation", presented at INFN Eloisatron Project 20th Workshop on GaAs Detectors and Electronics for High Energy Physics, Erice, Italy, 12-18 Jan. 1992.*
- *H. Schönbacher, "Materials in the radiation environment of high-luminosity colliders", presented at INFN Eloisatron Project 25th Workshop on Future Superdetectors and Colliders at the Frontier of High Energy and High Luminosity, Erice, Italy, 17-22 November 1992.*
- *F. Wulf et al., "Electronic components in a high luminosity environment", presented at INFN Eloisatron Project 25th Workshop on Future Superdetectors and Colliders at the Frontier of High Energy and High Luminosity, Erice, Italy, 17-22 November 1992.*
- *H. Larsen et al., "Radiation tests on service electronics for future Multi TeV detectors", CERN Report 93-04.*
- *F. Anghinolfi et al., "1 MHz sampling rate 12-bit low-power analog-to-digital converter for data processing in particle detectors", 2nd International Conference on advanced technology and particle physics, 11-15 June 1990, Como, Italy.*
- *M. Campbell et al., "A 10 MHz micro power CMOS front-end for direct readout of pixel detectors", 1989 IEEE nuclear science symposium, San Francisco, USA.*
- *F. Anghinolfi et al., "Monolithic CMOS front-end electronics with analog pipelining", presented at the IEEE Nuclear Science Symposium, Alington, USA, 23-26 October 1990.*
- *T. Ekelöf, "Megarad-hard readout electronics in SOS", ECFA Study Week on Instrumentation Technology for High-Luminosity Hadron Colliders, 14-21 September 1989, Barcelona, Spain.*
- *T. Ekelöf, "Radiation-hard microstrip read-out electronics", Proc. 9th Workshop on Perspectives for New Detectors in Future Supercolliders,*

Erice, Italy, 17-24 October 1989, eds. L. Cifarelli, R. Wigmans, T. Ypsilantis, World Scientific Co., Singapore.

- *N. Bingefors, "Radiation-hard SOS-VLSI for detector electronics", ECFA Large Hadron Collider Workshop, Aachen, Germany, 4-9 October 1990, CERN 90-10.*
- *J. Seguinot and T. Ypsilantis, "Particle Identification", CERN-LAA/89-1, p. 326.*
- *T. Ypsilantis and J. Seguinot, "Particle Identification at Hadron Colliders", CERN-EP/89-150.*
- *J. Seguinot and T. Ypsilantis, "Particle Identification at Hadron Colliders", Barcelona Workshop on LHC, CERN/89-10, p. 661.*
- *J. Seguinot et. al., "Reflective UV Photocathodes", Proc. 9th Eloisatron Workshop p.89; Nucl. Instr. & Meth. A297 (1990) 133.*
- *M. Hempstead et. al., "A Measurement of the Refractive Index of Sodium Fluoride in the far UV", Nucl. Instr. Meth. A306 (1991) 207.*
- *J. Seguinot and T. Ypsilantis, "Ring Imaging Cherenkov Counters", CERN-LAA/PI/91-004.*
- *J. Seguinot et. al., "AZEX: A Third Generation Solar Neutrino Experiment", 6/7/91, CERN-LAA/91-010.*
- *V. Peskov and A. Zichichi, "A new scintillator: $LiBaF_3$ preliminary tests", Proc. 12th INFN Eloisatron Project Workshop: "New Technologies for Supercolliders", Erice, Italy, 15-20 Sept. 1990, eds. L. Cifarelli and T. Ypsilantis, Plenum, New York, 1991.*
- *R. Arnold et. al. "A Fast Cathode Pad Photon Detector for Cherenkov Ring Imaging", Nucl. Instr. Meth. A314 (1992) 465.*
- *J. Seguinot et. al., "A High Rate Solar Neutrino Detector with Energy Determination", Proc. 4h Neutrino Telescopes, Venice, March 1992.*
- *G. Laurenti et. al., "A Study of a High Rate Solar Neutrino Detector with Neutrino Energy Determination", Proc. 2nd NESTOR Workshop, Pylos, Greece, Oct. 1992.*
- *M. French et. al. "A Fast Integrated Readout System for a Cathode Pad Photon Detector", Nucl. Instr. Meth. A324 (1993) 511.*
- *G. Laurenti et. al., "HELLAZ: A Solar Neutrino Detector with Neutrino Energy Resolution", Proc. 5th Neutrino Telescopes, Venice, March 1993.*
- *J.L. Guyonnet et. al. "Operation of a Fast-RICH Prototype with VLSI Readout Electronics", Nucl. Instr. Meth. A343 (1994) 178.*
- *M. French et. al. "A Fast Integrated Readout System for a Cathode Pad Photon Detector II", Nucl. Instr. Meth. A343 (1994) 222.*
- *J. Seguinot and T. Ypsilantis, "A Historical Survey of Ring Imaging Cherenkov Counters", Nucl. Instr. Meth. A343 (1994) 1.*
- *T. Ypsilantis and J. Seguinot, "Theory of Ring Imaging Cherenkov Counters", Nucl. Instr. Meth. A343 (1994) 30.*

- *F. Arzarello et. al., "HELLAZ: A High Rate Solar Neutrino Detector with Neutrino Energy Determination", Proc. Particles and the Universe, Budapest 30/7/93, World Scientific.*
- *J. Seguinot et. al., "Beam Tests of a Fast-RICH Prototype with VLSI Readout Electronics", Nucl. Instr. Meth. A350 (1994) 430.*
- *D. Hatzifotiadou et. al., "A Large Area RICH for Long Baseline Neutrinos", CERN-LAA/PI/94-17; College de France LPC/94-29.*
- *F. Arzarello et. al., "HELLAZ: A High Rate Solar Neutrino Detector with Neutrino Energy Determination", CERN/LAA/94-19, College de France LPC/94-28; Dark Matter and Cosmology School, Valencia, Oct. 1993, World Scientific.*
- *T. Ypsilantis and J. Seguinot, "Particle Identification for a Forward Collider B Experiment", Nucl. Instr. Meth. A351 (1994) 188.*
- *T. Ypsilantis and J. Seguinot, "Development of Ring Imaging Cherenkov Counters for Particle Identification", College de France LPC/94-48.*
- *J. Seguinot et al., "Flavor and flux determination of the pp solar neutrinos via ν-e elastic scattering", CERN/LAA/95-11, College de France LPC 95/08.*
- *T. Ypsilantis et al., "A long base line RICH with a 27 kton water target and radiator for detection of neutrino oscillations", Nucl. Instr. Meth. A371 (1996) 330.*
- *J. Seguinot et al.,"Tests of a TPC for solar neutrino detection (Hellaz-0)", CERN/LAA/96-05, College de France 96/31.*
- *E. Chesi et al., "Performance of a 256 pad hybrid photon detector for ring imaging", Nucl. Instr. Meth. A387 (1997) 122.*
- *T. Ypsilantis et al., "A long base line RICH with a 27 kton water target and radiator for detection of neutrino oscillations", presented at XXIII SLAC Summer Institute on Particle Physics, SLAC-R-194, p. 225.*
- *R. De Leo et al., "Electronic detection of focused Cherenkov rings from aerogel", Nucl. Instr. Meth. A401 (1997) 187.*
- *P. Antonioli et al., "AQUA-RICH, an atmospheric and long baseline neutrino experiment at Gran Sasso", CERN/LAA-AR/98-01.*

TOTAL NUMBER OF PUBLICATIONS
346

Section 4

NEW TECHNOLOGIES FOR PHYSICS AT LHC AND BEYOND

The results obtained with the LAA project have been of great value for the detectors to be used at the LHC. A review illustrating the impact of the LAA results on the LHC detectors is reported in this section, together with a review of the role played by Italy with the ELN project in the framework of multi-TeV colliders in Europe.

Horst Wenninger
CERN, Geneva, Switzerland

THE IMPACT OF LAA ON THE LHC DETECTORS

THE IMPACT OF LAA ON THE LHC DETECTORS

Horst Wenninger
CERN, Geneva, Switzerland

In this chapter we discuss the impact of the LAA results on the LHC detectors and the role played by Italy in the ELN project within the framework of multi-TeV colliders in Europe.

Comparing the onion-like structures — schematically shown in the layout of Fig. 1 — of the two large LHC experiments ATLAS and CMS, presently under construction and scheduled to start up for physics in the year 2005, with the components, goals and main achievements of the LAA Project, as described by A. Zichichi in the previous chapter of this book, one immediately recognises that the logic of the LAA Project in defining its R&D components perfectly matches the requirements of the LHC detectors. The impact of the various LAA R&D components, defined 10 years before technical choices were approved for LHC detectors and 20 years before LHC detector operation for physics is scheduled, clearly depends on many factors and differs for each LAA R&D component and the corresponding LHC detector component.

The direct impact of specific technical solutions, explored by LAA teams, on the LHC detectors clearly depends on (i) the final technical solutions adopted by the collaboration, which are in turn strongly influenced by the specific know-how and competencies existing in the institutes which take responsibility to build and finance the corresponding components of the detectors, (ii) the number of potentially equivalent technical solutions and particle detection methods available, e.g. electromagnetic calorimeters: liquid argon (ATLAS) vs. lead tungstate crystals (CMS) vs. solid scintillator avalanche counter or spaghetti calorimeter (LAA) and (iii) on the result and the evolution of the R&D work, which started with LAA around 1986 and extended into the CERN Detector R&D initiative from 1990 to 1996.

An indirect impact of LAA on the LHC detectors remains, even if complementary or completely new technical solutions are adopted for the construction of several of the LHC detector components. As we will describe below, many ideas and much of the technology and know-how, acquired through LAA via generic developments and prototyping and then followed up during the subsequent R&D phase at CERN, now benefit the LHC detector construction efforts. LAA staff, mostly working today for LHC detector construction support at CERN, contribute with know-how acquired during the LAA R&D work.

The strongest impact of LAA on LHC detectors results from the LAA work on high precision tracking and particle detection with related electronic circuit development. The LAA R&D work on high precision tracking included (i) the construction of multi-drift chambers, (ii) a vast study in the field of scintillating fibre trackers and novel read-out devices, (iii) tests of GaAs microstrip detectors capable of withstanding very high neutron radiation rates, and last but not least (iv) advanced designs of read-out electronic circuits.

The multi-drift modules showed high rate capabilities, excellent track resolutions, and high radiation resistance. The prototyping on scintillating fibre bundles led to large improvements in hit densities for fibre-bundles, giving track resolutions of 30 μm. For the read-out of the fibre-bundles, novel delay-tubes and ISPA (Imaging Silicon Pixel Array)–tubes were developed in collaboration with industry. A comprehensive review on the scintillating fibre technology resulting from the LAA work can be found in Nucl. Instr. and Meth. A364 (1995) 422-448. In the field of tracking detector read-out electronics, stringent requirements on rate capability, radiation resistance, large production volumes and low power dissipation were addressed. Demonstrator CMOS chips using pipelined architecture were built. Radiation-hard silicon-on-sapphire technology was tested.

The direct impact on final technical choices for LHC particle tracking detector devices (e.g. silicon-pixel) and on ASIC/VLSI chip detector read-out chip designs is easily explained by the fact that techniques and know-how in this field had to be built up nearly from scratch in particle physics laboratories and university institutes around the mid-80's when the LAA Project was implemented. From the beginning it was therefore realised that an important effort on silicon tracking detectors and micro-electronics should be part of the detector development project LAA.

A number of CERN expert engineers, working today on silicon pixel tracking detectors and electronics design and data acquisition systems in support of LHC experiments, were indeed recruited on LAA R&D posts at CERN and helped to acquire the techniques and implement the know-how. The LAA Project allowed the micro-electronics engineering groups in CERN EF Division to purchase the design tools and the electronic circuit simulation software required. LAA promoted collaborations between micro-electronics design teams in CERN Member State laboratories and industries, which are now collaborating on the final detailed designs for LHC detectors. CERN's R&D program 1990 to 1996 included in fact several proposals as an immediate follow-up of the LAA work (e.g.: ref. RD 8: "The GaAs Collaboration", RD 9: "Demonstrator analog signal processing circuit in a radiation hard SOI-CMOS technology, CERN Microelectronics group", RD 19: "Development of hybrid and monolithic silicon micropattern detectors").

A remarkable electronics project, developed with resources coming mostly from the LAA project was the "FASIC" chip, an integrated Fastbus Slave Controller, which involved commercial high performance gate array technology. This FASIC chip was the most complex digital integrated circuit design in the HEP community in the early nineties. This design has been created using state of the art tools acquired through the LAA project. In particular, the first modern CAE stations and design tools for simulation (Verilog) have been introduced to CERN thanks to LAA. The importance of these modern design and simulation tools has been appreciated by all designers involved in the development of advanced electronics for future detectors in HEP. Today these tools are becoming standard among the electronic designers in several HEP laboratories throughout the world. The FASIC chip design was released to Motorola company for fabrication and was successfully tested thereafter. This chip design project demonstrated best the great impact of the LAA project that provided CERN with the unique possibility to acquire the know-how of a most complex technology and VLSI methods, which requires a long training time for designers, which are now able to provide the necessary electronic designs for LHC experiments.

The development of scintillating fibres for tracking detectors — part of LAA R&D precision tracking component as outlined above — needed precise measurement of light transmission to characterise fibre qualities. Fibre quality measurements related to optoelectronic analog signal transfer for LHC detector studies (RD 23), specifications of scintillating fibres embedded in many calorimeter designs (RD 25, RD 40) and also optical quality control for scintillating crystals for calorimetry (RD 18 and now CMS electromagnetic calorimeter project) benefited, thanks to the LAA Project, from Laboratory facilities set up at CERN EF Division for this purpose.

A direct spin-off from the fibre tracking project of LAA is also the new development of highly sensitive photon counting detectors and read-out chips, part of the CERN/INFN collaboration on the ISPA tube, an imaging single photon detector, aiming at medical applications for X-ray imaging. A similar spin-off for medical application results from the intensive R&D studies on gaseous detectors, originating from the LAA collaborations with Charpak's team at CERN.

Another strong, though more indirect, impact of LAA on LHC detectors results from LAA component "Radiation Hardness". This LAA activity brought together CERN radiation protection experts with experts in various laboratories and industries and helped setting up test facilities for radiation damage studies on materials and electronic components for LHC detectors. Although radiation damage component tests and the development of radiation-hard read-out electronics by specialised industries for LHC detectors largely exceed today the original scope and modest resources available at the time, LAA nevertheless initiated an important collaborative effort now helping the LHC detector collaborations.

Even if the technical solutions adopted by LHC collaborations in the design of their respective calorimeters and the choices of sensitive media for their calorimeter detectors differ from original LAA prototype solutions (e.g.: ATLAS liquid argon calorimeters and plastic scintillator plates embedded in iron absorbers, read out by wave-length-shifting fibres, or CMS copper plates / plastic scintillator calorimeters, read out by wave-length-shifting fibres, as compared to the LAA high resolution spaghetti calorimeter), much of the experience gained on cost-effective large size mechanical construction contributes to the acquired know-how. The same argument holds for large size muon chamber prototypes, for LAA alignment studies, for leading particle detector developments, for particle identification solutions and for generic studies of gaseous detectors.

Part of the LAA calorimeter development component was the construction of a 13 ton lead-fibre calorimeter for electromagnetic and hadronic energy measurements (SPACAL). It showed 13% / SQRT(E) energy resolution of electrons and 30% / SQRT(E) for hadrons. The goal to achieve compensation was nearly obtained with an e/pi signal ratio of 1.028. The calorimeter was used for many detailed studies on lateral shower development, longitudinal shower leakage, signal timing, particle identification, pi-zero content of hadronic showers, trigger algorithm studies and radiation resistance. Hybrid photo diode detectors, capable of measuring calorimeter signals over eight orders of magnitude in a two Tesla magnetic field were developed in collaboration with industry. Work was also carried out in the field of high precision electromagnetic calorimetry with the study of BaF2 and KMgF3 solid scintillators, read out through CsI photocathodes and gaseous detectors. Detailed work on the solid photocathodes led to improved performances and 3% / SQRT(E) calorimeter energy resolution were reported. Totally active liquid Krypton and Xenon calorimeters were studied and their capability of measuring electromagnetic energy with 1.4% / SQRT(E) energy resolution was demonstrated.

Considering the contribution of LAA in the field of large area devices for muon detection, we have to mention the Blade chambers and gaseous pixel chambers but in particular the studies on single-gap and multi-gap Resistive Plate Chambers (RPC) with high rate capabilities (4kHz/cm2) and excellent time resolution well suited for the first level trigger requirements at LHC. Also a study of a muon spectrometer with a toroidal field of 3.6 Tesla was undertaken. It included field calculations, engineering design, alignment aspects, detector resolution and trigger.

Finally, for leading particle detection, far from the interaction point near the beam pipe, radiofrequency shielding and moving pots withstanding high radiation levels were built. The technique of cutting silicon detectors at appropriate elliptical shapes was mastered. Studies of a liquid Xenon detector for leading proton detection and a quartz calorimeter for leading neutron detection was carried out. A complete leading particle spectrometer was built and tested in a real experiment, the ZEUS experiment at HERA in DESY.

The LAA project also tackled the question of data acquisition and data analysis with special techniques using local processors and fast algorithms. Fastbus and VME cards, which integrate processors in the experimental read-out pipeline, were designed. For the simulation of high-multiplicity multi-TeV hadron events, a Super Monte Carlo System (SMCP) was set up. This Monte Carlo system provides a flexible modular framework connecting all necessary elements, such as parton generation and decay, detector geometry, particle propagation, hit generation, tracking, analysis and display using ADAMO relational data structure systems.

Considering the LHC experiments data acquisition main requirements (see Figs. 2, 3) and the fast evolution of data acquisition technology/architectures (see Fig. 4) driven by informatics/computing industry over the last decade including today's software revolution, moving away from FORTRAN to Object Oriented programming technologies with C++ or Java languages, it is obvious that the LAA R&D has the least impact of what will be used in LHC experiments and which even today is not completely defined. Nevertheless, even in this domain, original work on pattern recognition algorithms and extensions to the design of the SPARC chip from SUN Microsystems, started within the framework of LAA, pointed to the right future direction.

In conclusion, research and development carried out in the above mentioned fields can be seen as important contributions to the advancement of collider physics with great relevance and measurable impact to the LHC experiments. The know-how gained and the techniques learned and developed during the LAA project under the leadership of Professor A. Zichichi are being applied at this very moment in the largest and most sophisticated particle detectors ever conceived.

Onion-like Structure of HEP Experiments

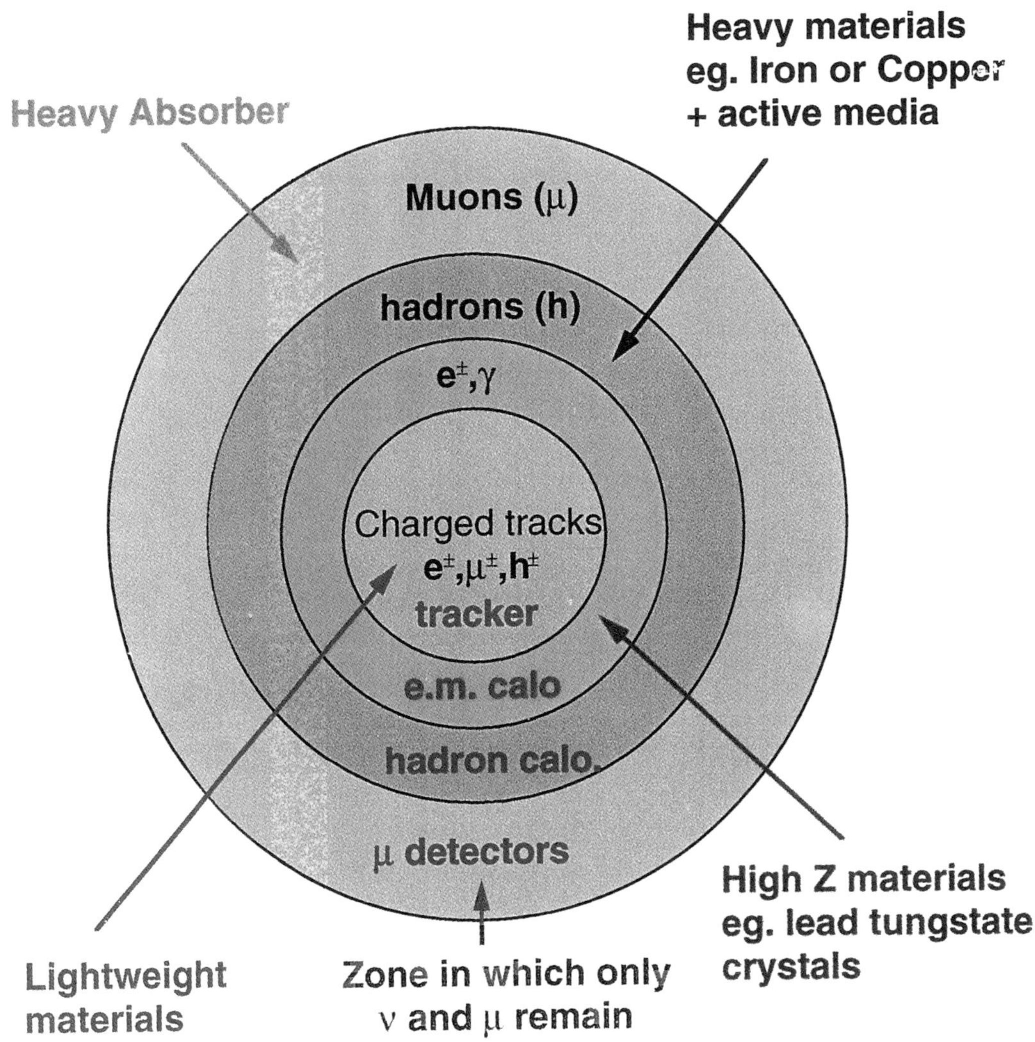

Each layer identifies and measures (or remeasures) the energy of particles unmeasured by the previous layer

Figure 1

http://cmsdoc.cern.ch/pictures/publicity/overview.html

LHC experiments DAQ main requirements

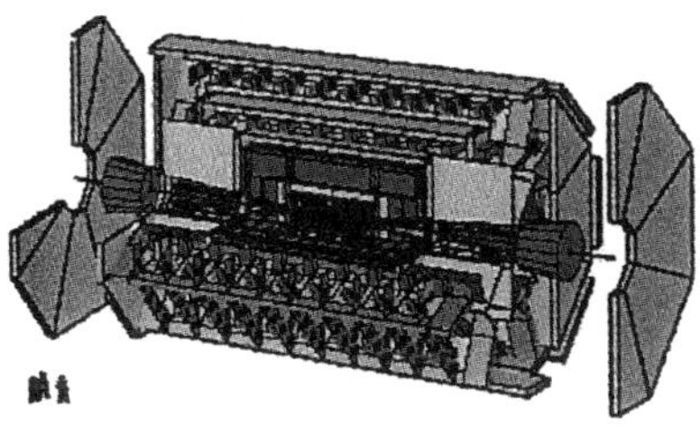

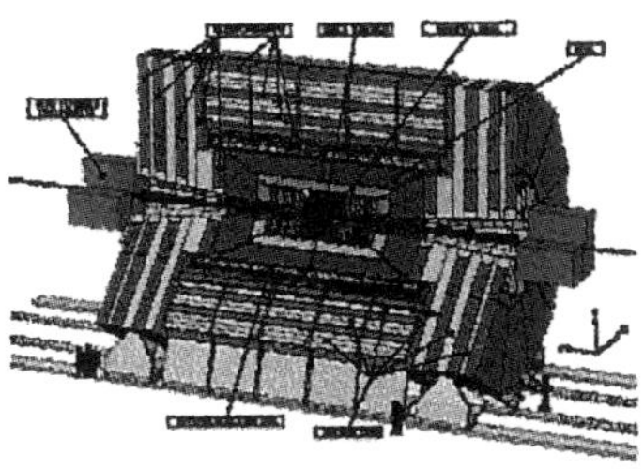

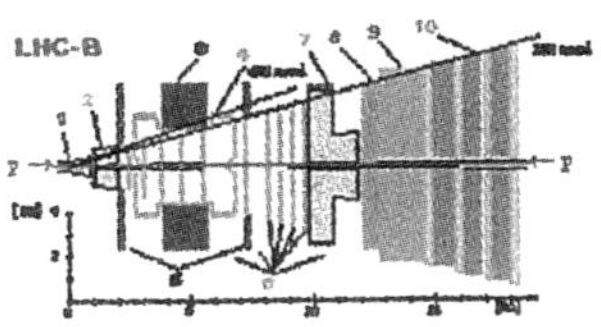

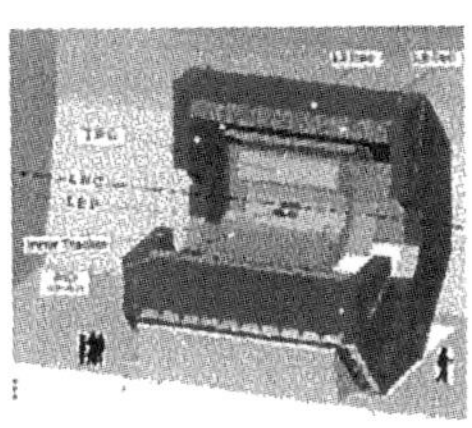

	Level-1 kHz	Event MByte	Storage MByte/s
ATLAS	100	1	100
CMS	100	1	100
LHCB	400	0.1	20
ALICE	1	25	1500

Figure 2

CMS trigger and data acquisition

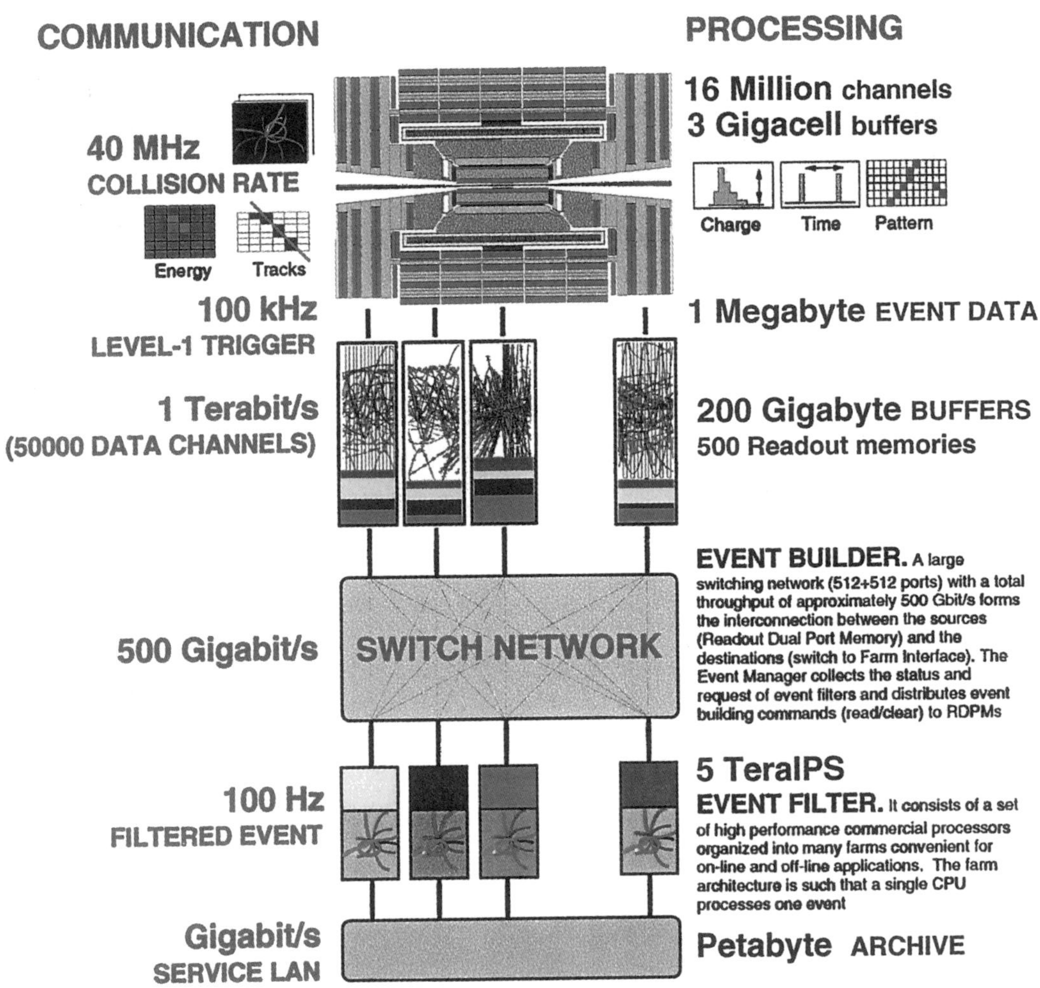

Tera : 10^{12}; Peta 10^{15}; IPS : Instruction Per Second; LAN : Local Area Network

Figure 3

The Compact Muon Solenoid. SC April 1998 108

Evolution of DAQ technology/architectures

1970-80 MiniComputers
- CAMAC first standardization
- **kByte/s**

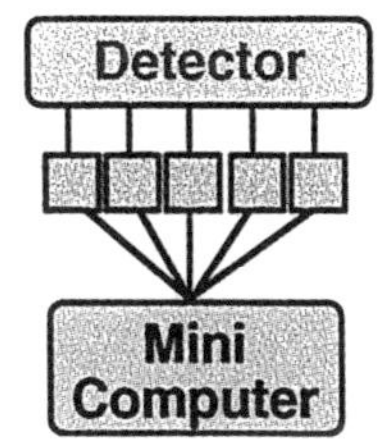

1980-90 MicroProcessors
- Parallel systems
- Distributed intelligence
- **MByte/s**

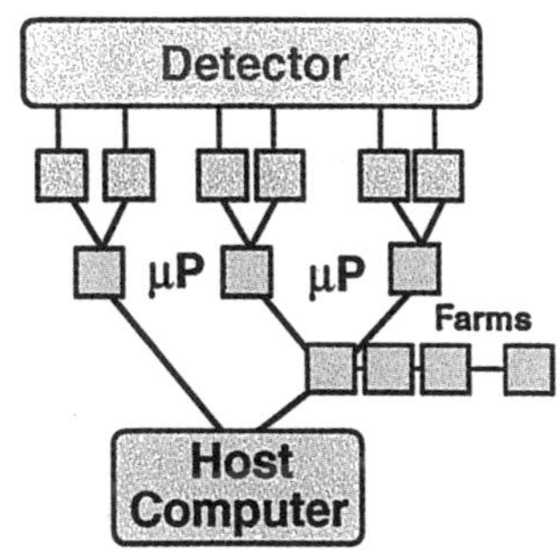

2000 Communication
- Embedded processing
- Data and control networks
- **GByte/s**

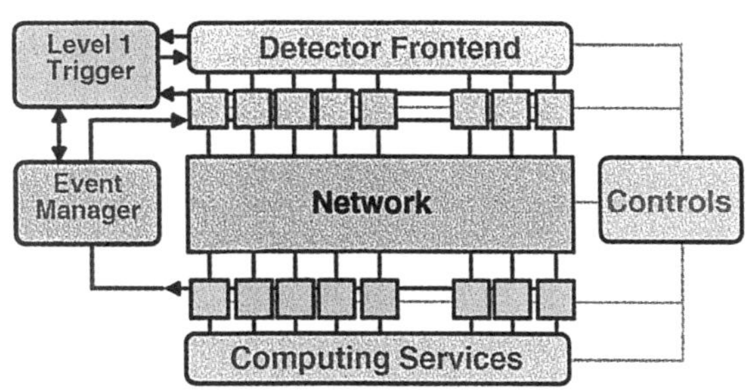

Figure 4

Horst Wenninger
CERN, Geneva, Switzerland

THE ROLE PLAYED BY ITALY IN THE ELN PROJECT WITHIN THE FRAMEWORK OF MULTI-TeV COLLIDERS IN EUROPE

THE ROLE PLAYED BY ITALY IN THE ELN PROJECT WITHIN THE FRAMEWORK OF MULTI-TeV COLLIDERS IN EUROPE

Horst Wenninger
CERN, Geneva, Switzerland

Around mid-1980, most particle physicists in university institutes and the engineering staff of DESY and of CERN were fully engaged in constructing detectors for approved experiments and in preparing accelerator components together with industry for the electron-proton collider (HERA at DESY) and for the electron-positron collider (LEP at CERN). These projects left no room for new technological developments.

Neither HERA nor LEP could have been built without prior technological developments, e.g. in superconducting magnet technology for the proton ring of HERA and in accelerating structures and power klystron technology for LEP. It is evident that timely investment in R&D for future projects assures the continuous success of CERN and DESY and can in general not await the completion of ongoing project work.

The Eloisatron project, launched in Italy by Prof. A. Zichichi in 1979, aims at continuously promoting advances in accelerator technology for proton machines, in particular the development of superconducting magnets for proton collider rings (e.g. HERA, LHC and beyond). It appeared therefore consequent to include in the Eloisatron concept a major technological initiative, the LAA project, focused on instrumentation and advanced detector ideas, on electronic designs, data read-out and recording systems, and on advanced computing.

Three major R&D initiatives, launched by Italy together with DESY and with CERN, allowed preparation for the future and paved the way for Europe's leading role in particle physics research, culminating in the construction of LHC with a major participation from other regions of the world, in particular the US, Russia, Japan. These R&D initiatives were: (i) an intensive R&D on superconducting accelerating cavity technology, which started at CERN in early 1980; (ii) joint efforts in Italy, CERN and DESY on R&D for superconducting magnets allowing to build HERA at DESY and move towards LHC at CERN; (iii) to the benefit of all European experimental subnuclear physics, the LAA project allowed the development of experimental techniques required for the proton accelerator program, thus preparing R&D for experiments ten years ahead of the approval of the LHC experimental programme. Both the work on superconducting cavity development for LEP 200 and the LAA detector R&D projects were implemented at CERN in the Experimental Facilities Division.

Once LEP construction was completed, the LAA project was complemented by a Europe-wide call for detector technology and R&D. Similar efforts were going on in the US to develop detector systems, electronics and computing for their ongoing SSC project; without the LAA project, Europe would have been left behind. Many of the early CERN DRDC (Detector Research & Development Committee) proposals of the years 1991-1994 were based on ideas and results obtained by the LAA activity.

The LHC experiments have now selected their technical solutions and start to freeze the technical specifications. The R&D on superconducting cavities was industrialized via a successful technology transfer programme and the LEP energy upgrade project is completed. The situation is today similar to that in mid-1980. Available resources are focused on the construction of LHC machine and experiments at CERN and in the HEP institutes.

In preparation of the future and in parallel to ongoing work in US and Japan, R&D on accelerator techniques has started at DESY in collaboration with Italy and France. CERN's engagement in CLIC (Compact Linear Collider) is gaining momentum. Some R&D work continues on electronics and computing for experiments. In times when a public debate on technology spin-offs from basic research work has started, HEP and CERN need a revival of the LAA initiative to prepare its future beyond LHC.

The continuation of the Eloisatron concept, initiated by Prof. A. Zichichi in Italy, the study on options for future colliders at CERN, the ongoing R&D work on linear colliders and in particular the collaboration on TESLA (TeV Superconducting Linear Accelerator) between DESY and INFN, signed by Prof. Björn Wiik and Prof. L. Maiani, should go hand in hand with a project looking beyond the LHC experiment techniques and into the future of 10-100 TeV proton machines, i.e. the ELN Project.

Albrecht Wagner
DESY and University of Hamburg

OUTLOOK

OUTLOOK

Albrecht Wagner
DESY and University of Hamburg

The progress in the understanding of the basic building blocks of matter and of the forces acting between them has been the result of a highly productive interaction between theorists, experimentalists and accelerator builders. In this way theoretical ideas could be confronted with experimental data, taken with high technology detectors at powerful accelerators. All this work has led to a surprisingly powerful and complete theory of the microscopic world, the Standard Model, which during recent years has been proved to describe the experimental data with unprecedented accuracy. We know however that this theory cannot be the final word. Physicists are therefore preparing for the next generation of experiments at accelerators of the highest energies, to prove theories which lead beyond the Standard Model.

At CERN the Large Hadron Collider (LHC) is presently under construction, to be completed in 2005. In the LHC, proton beams of 7 TeV each are made to collide. The large energy gives access to many of the open questions of particle physics such as: What is the origin of electroweak symmetry breaking? Is there a theory uniting particles and forces such as supersymmetry?

The large energy and high luminosity of the LHC poses outstanding challenges for the detectors, which not too many years ago seemed impossible to meet. The rate of interactions per collision, the resulting flux of particles and the repetition rate of the accelerator will be so high, that conventional technologies to measure the energy and tracks of particles would fail.

Considerable research and development work was needed to overcome these problems.

It was A. Zichichi who very early realised the need for better detectors and in 1986 through the LAA project paved the way for a systematic development program, stimulating new ideas and technologies. Of special importance for the subsequent applications were the activities on high precision tracking, calorimetry and radiation hardness. This program was well matched to the challenges of LHC and eventually led to a European detector research and development program, co-ordinated and reviewed by CERN through the DRDC. The LHC detectors, as presently under construction, are based to a large extent on the know-how and technologies developed through these programs.

At DESY, an international collaboration is preparing the technology for the electron-positron linear collider, TESLA, which will complement the scientific program of the LHC. While a lot of work has gone already into the development of the accelerator, in particular into the highly successful research program for superconducting RF cavities of very high gradient, the necessary effort for the development of new detector technologies is only starting. Considering the successful and stimulating activities started by A. Zichichi for the LHC, DESY has decided recently to initiate a similar program for detectors for linear colliders.

Major progress in particle physics has been made whenever new technologies for accelerators and detectors have become available, providing higher energies and better resolution. Developing new technologies is therefore one key element for the future progress in science. It is important to continue this development.